Megatrends in Food and Agriculture

Megatrends in Food and Agriculture

Technology, Water Use and Nutrition

Helmut Traitler
California, United States

Michel Dubois
France

Keith Heikes
Wisconsin, United States

Vincent Pétiard
France

David Zilberman
California, United States

This edition first published 2018

Registered Offices
John Wiley & Sons, Inc., 111 River Street, Hoboken, NJ 07030, USA
John Wiley & Sons Ltd, The Atrium, Southern Gate, Chichester, West Sussex, PO19 8SQ, UK

Editorial Office
9600 Garsington Road, Oxford, OX4 2DQ, UK

For details of our global editorial offices, customer services, and more information about Wiley products visit us at www.wiley.com.

Wiley also publishes its books in a variety of electronic formats and by print-on-demand. Some content that appears in standard print versions of this book may not be available in other formats.

Library of Congress Cataloging-in-Publication data applied for

ISBN: 9781119391142

Cover Design: Wiley
Cover Images: Photos by Helmut Traitler and Therese Meyer Traitler

Set in 10/12pt Warnock by SPi Global, Pondicherry, India
Printed and bound in Malaysia by Vivar Printing Sdn Bhd

10 9 8 7 6 5 4 3 2 1

Contents

Foreword *xiii*
Acknowledgments *xv*

Part 1 Agriculture and the Food Industry *1*

1 The Role of Agriculture in Today's Food Industry *3*
1.1 Introduction *3*
1.1.1 The Four Building Blocks *4*
1.1.2 Some History of Agriculture *5*
1.1.3 Eat More and Increase the Likelihood for Survival *6*
1.1.4 Food Can Be Grown and Plants Can Be Bred: What's Next? *7*
1.1.5 From Very Old to Rather Recent Food-Preservation Techniques *9*
1.2 Agriculture: The Main Supplier to the Food Industry *10*
1.2.1 Artificial Ingredients *10*
1.2.2 The Main Raw Material Sources *11*
1.2.3 Milk's the Star *12*
1.2.4 Milk…What Else? *13*
1.2.5 Other Excursions from Food *14*
1.2.6 Noncompeting Alternatives *15*
1.3 Agriculture's New Role in Light of Food and Health *16*
1.3.1 Decades of Food Safety Rules and Regulations *18*
1.3.2 More Rules: What Do We Do? *20*
1.3.3 Raw Materials and Processes Become More Sophisticated *21*
1.4 Most Likely Drivers for Change in the Agriculture Industry *23*
1.5 Summary and Major Learning *25*
References *27*

2 **Water Management in Modern Agriculture: The Role of Water and Water Management in Agriculture and Industry** *29*
2.1 Introduction *29*
2.2 Multiple Dimensions of Water *30*
2.3 On the Evolution of Water Institutions and Policies *33*
2.4 Reforming Water-Resource Management at the Micro-Level (Farm and Field) *35*
2.5 Reforming Regional Water-Allocation Regimes *38*
2.6 Improved Water Project Design *42*
2.7 Improved Water Quality *43*
2.8 Climate Change *44*
2.9 Summary and Major Learning *46*
References *47*

3 **Innovation in Plant Breeding for a Sustainable Supply of High-Quality Plant Raw Materials for the Food Industry** *53*
3.1 Introduction *53*
3.2 Challenges for Future Agricultures and Food Industries *54*
3.2.1 Strongly Growing Food Needs *54*
3.2.2 Energy Issues *56*
3.3 Genetic-Based Techniques for Plant Breeding in the Context of Agricultural Production *59*
3.3.1 Genetic Innovation and Agronomic Practices *59*
3.3.2 The Process of Plant Breeding and Its Main Limitations *61*
3.3.3 Preliminary Conclusions *67*
3.4 Trends: Shift in Allocation of Resources to Global Needs? *68*
3.4.1 Methodology *68*
3.4.2 Analysis of Investment in Seed Research and Development *69*
3.4.3 Analysis of Deviations and Distortions of R&D Investments and Production Volumes *71*
3.5 A First Set of Conclusions and Recommendations *77*
3.6 Summary and Major Learning *80*
3.7 Appendix Tables *81*
References *85*

4 **The Agriculture of Animals: Animal Proteins of the Future as Valuable and Sustainable Sources for the Food Industry** *87*
4.1 Livestock and Animal Husbandry *87*
4.1.1 How We Got to Now *88*
4.2 Animals: A Source of High-Quality Proteins *89*
4.3 Animal Protein Demand in Emerging Markets *90*
4.4 Optimal Animal Welfare: Sustainable, Humane, and Healthy *93*

4.4.1 Animal Production Increase *95*
4.5 Animal-Breeding Programs *97*
4.5.1 Genomic Breeding of Animals *98*
4.6 The Use of Big Data for Management and Genetic Evaluations *102*
4.7 Summary and Major Learning *106*
References *108*

Part 2 The Future of the Food Industry *109*

5 The Food Trends—The New Food—Enough Food? *111*
5.1 Historical Food Trends: From Then to Now *111*
5.1.1 Food and Beverages during the Period of Classical Greece *111*
5.1.2 Food and Beverages in the Roman Empire *113*
5.1.3 Food in Medieval Times in Central Europe *115*
5.1.4 From European Renaissance and Enlightenment to the First Industrial Revolution *118*
5.1.5 Food in the 20th Century: The Real Food Revolution *121*
5.2 Present-Day Food Fashions and Trends: A Never-Ending Story *124*
5.2.1 Food and Nutrition Trends: A Story of Perception, Deception, and Beliefs *125*
5.3 New Food Sources: New Protein Sources *128*
5.3.1 Insects: A New Food Source? *129*
5.3.2 Increased Food Security through Exploiting New Protein Sources *130*
5.3.3 A "Crazy" Idea for Other Food Sources: Beyond Proteins *131*
5.4 Vegetarian Food and Its Potential Societal and Economic Impact *132*
5.5 Urban Gardening and Urban Agriculture *134*
5.5.1 The Urban Bee-Highways *136*
5.6 Summary and Major Learning *137*
References *138*

6 The New Food Industry Business Model: From B2C to B2B, from Product Manufacture to Selling Know-How, and from Now to Then *141*
6.1 The Old: Develop, Manufacture, and Sell ("Demase") *141*
6.1.1 The Fall of the Righteous *142*
6.2 The New: The Customer Is King, the Consumer Is an Enabler, and from B2C to B2B *144*
6.2.1 Slotting Allowance *145*
6.2.2 Retailers Become the Most Important Partners for Food and Beverage Companies *146*
6.2.3 How This Could Work: A Possible Path and Examples *147*

6.3 From Selling Products to Selling Know-How *150*
6.3.1 The Knowledge-Centric Company *152*
6.3.2 Engaging, Interacting, and Selling: The New Etiquette *154*
6.4 The Community of Consumers: It's What They Want that Counts! *155*
6.4.1 The Consumers Become Involved *158*
6.5 Food-Related Trends and Hypes in Today's Societies: An Outlook to the Future *161*
6.6 Summary and Major Learning *163*
References *165*

7 The Internet of Just about Everything: Impact on Agriculture and Food Industry *167*
7.1 Modern Cooking: Forward to the Past *167*
7.1.1 The Role of Robotics and Connectivity *169*
7.2 Everything Is Online and Everyone Is Online—All the Time *171*
7.3 Food and Agriculture: The New Hardware and Software *174*
7.3.1 Big Data Are Here to Stay *176*
7.3.2 Agriculture and Space Science: The New Connection *176*
7.3.3 Impact on the Food Industry and the Consumer in the Middle *178*
7.4 An Attempt at Peaking Ahead: Will There Still Be an Agriculture or Food Industry? *179*
7.4.1 Bigger Is Not Always Better *180*
7.4.2 Elements that Will Stay and Others that Might Disappear *182*
7.5 Summary and Major Learning *186*
References *189*

8 Nutrition: The Old Mantra … the New Un-Word *191*
8.1 Nutrition: What's All the Fuss about? *191*
8.1.1 The Hottest New Food Trends *192*
8.1.2 The Debate Continues: What's Good and What's Not Good for You? *195*
8.1.3 And Here We Go Again: Fasting Can Do You an Awful Lot of Good *196*
8.1.4 A Few Simple Tips When It Comes to Healthy and Happy Eating *197*
8.2 A Bit of Nutrition History *198*
8.2.1 Low and Reduced, Lower and "Reduced-Er": Low or Reduced Fat *202*
8.2.2 Low or Reduced Salt *203*
8.2.3 Low or Reduced Sugar and No Sugar *204*
8.2.4 Low Saturated Fats, Good Monounsaturated Fats, More Polyunsaturated Fats, and Lots of Ω3 Fats *206*

8.3 Typical Nutrition Controversies *208*
8.3.1 So Many Recommendations…Too Many? *208*
8.3.2 More Controversies *210*
8.4 Food and Claims, Food and Benefits *211*
8.5 Summary and Major Learning *214*
References *217*

Part 3 The New Food World *219*

9 A Food Company Transforms Itself *221*
9.1 The Not-So-New Realities *221*
9.1.1 Automation Is Here…For Quite Some Time Actually *223*
9.1.2 The Novel Directions in Food and Agriculture are Governed by Regulatory Involvement *225*
9.1.3 All-Natural Industrial Food Products: The Way Forward? *226*
9.2 From Product to Know-How Seller: An Encore *227*
9.2.1 Some Assumptions as to How This May Function *228*
9.2.2 What are Possible Consequences for Food Ingredient Suppliers? *230*
9.3 Anticipating the Inevitable: Possible Scenarios *234*
9.3.1 Possible Future Models and Scenarios *238*
9.3.2 The Return Of Medical Food? *239*
9.4 Reality or Fiction? Reality and Fiction! *241*
9.4.1 A New Manufacturing Reality *242*
9.5 Summary and Major Learning *244*
References *246*

10 Food for the Future: A Future for Food *247*
10.1 Proactive Agriculture *247*
10.1.1 What If Agriculture Anticipated Real Food Requirements and Trends? *249*
10.2 Democratized Agriculture *252*
10.2.1 Agrihood *253*
10.2.2 Permaculture *254*
10.2.3 From Large to Small *259*
10.2.4 The Growing Role of Urban Agriculture: Self-Centeredness or Community Driven? *261*
10.3 Agriculture and Food Manufacture in Exotic Places *264*
10.3.1 An Ice Cream Factory in Greenland? *265*
10.3.2 A Chocolate Factory in Ghana? *266*
10.4 A Future for Food *269*
10.4.1 What about the Role of Restaurants? *269*

10.4.2 Pet Food Is Food, Too *270*
10.4.3 Will We Eat Food in Pill Format? *271*
10.5 Summary and Major Learning *273*
References *274*

11 Summary and Outlook *277*
11.1 Introduction *277*
11.1.1 The Role of Agriculture in Today's Food Industry *278*
11.1.2 Food-Preservation Techniques *279*
11.1.3 Agriculture Is the Main Raw Material Supplier to Be Transformed to Food *280*
11.1.4 Nonfood Uses of Agricultural Raw Materials *280*
11.1.5 Agriculture in a World of Rules and Regulations *280*
11.1.6 Food Raw Materials and Process Became More Sophisticated and Complex *281*
11.2 Water Management in Modern Agriculture *281*
11.2.1 The "Water Reform" *282*
11.2.2 Water Productivity *282*
11.2.3 Water-Related Government Policies *283*
11.2.4 Getting It Right: Policies and Price *283*
11.2.5 Controlling Water Quality *284*
11.3 Innovation in Plant Breeding: High-Quality Plant Raw Materials for the Food Industry *285*
11.3.1 Agricultural Plant Output: The Essential Raw Material Source for the Food Industry *285*
11.3.2 Demand Forecast Based on Food Requirements *285*
11.3.3 Genetic Improvement of Cultivated Crops *285*
11.3.4 The Major Crops versus "Orphan Crops" *286*
11.4 The Agriculture of Animals: Valuable and Sustainable Sources for the Food Industry *286*
11.4.1 Growing Population: Growing Amount of Livestock *287*
11.4.2 Animal Health and Intensive Farming *287*
11.4.3 Animal Breeding *287*
11.4.4 Good Farm Management: Good Data Management *288*
11.5 The Food Trends—the New Food—Enough Food? *288*
11.5.1 Food and Beverage Fashions and Trends of the Past *289*
11.5.2 The "Real" Food Revolution of the 20th Century *289*
11.5.3 Present-Day Food and Nutrition Trends *290*
11.5.4 New Food Sources: New Protein Sources *290*
11.5.5 Vegetarian Food and Its Impact on Society *291*
11.5.6 The Role of Urban Agriculture and Bees *291*
11.6 New Business Models for the Food Industry *292*

11.6.1 From "Consumer Is King" to "Customer Is King": Retailers Become Real Partners *292*
11.6.2 Good-Bye to Selling Products and Hello to Selling Know-How *293*
11.6.3 Consumers Become Involved *294*
11.7 The Internet of Just about Everything and What This Means for Agriculture and Food *295*
11.7.1 Modern Cooking: A Brief Look to the Past *295*
11.7.2 Robotics and Connectivity *295*
11.7.3 Food and Agriculture: Big Data *296*
11.7.4 Will There Still Be Agriculture and Food Industries? *296*
11.7.5 What Will Remain, and What Will Disappear? *297*
11.8 Nutrition: What Else? *298*
11.8.1 Healthy and Happy Eating *299*
11.8.2 A Short History of Nutrition *300*
11.8.3 Nutrition Controversies *300*
11.8.4 Claims and Benefits *301*
11.9 The Company Transforms Itself *301*
11.9.1 The Role of Automation: Threat or Blessing? *302*
11.9.2 Regulatory Involvement in the Industry *303*
11.9.3 The New Business Model 2.1 *303*
11.9.4 Scenarios of Relevance for Food and Agriculture *303*
11.9.5 Medical Food: A Future? *304*
11.9.6 Reality or Fiction? *304*
11.10 Agriculture Listens, Finally? *305*
11.10.1 Agriculture and Farming at the Fingertips of Everyone *306*
11.10.2 Small Is Beautiful *306*
11.10.3 Is Urban Agriculture a Sign of Self-Centeredness or Is It Community Driven? *307*
11.10.4 Manufacturing Food Where It Makes Sense *307*
11.10.5 What Role Do Restaurants Play? *308*
11.10.6 The Role of Pet Food in the Food Industry *308*
11.10.7 Food in the Format of Pills? Will Consuming Food Pills Be Part of Megatrends? *308*

Index *311*

Foreword

The right to food is a human right. A free exchange of ideas and knowledge that support culturally appropriate food selections produced with the sustainable needs of the grower, consumer, and land "top of mind" are global issues and inherent duties that resonate across all sectors of today's agriculture and food industry. This is especially relevant in Hawai'i's agriculture and food industry, where actions to ensure future generations the gift of a communal table that is abundant, vibrant, tasteful, and nutritious are challenged by insecurity. Stimulating local, national, and global food production levels based on sustainable, environmental, social, and economic virtues will continually play a significant role in reducing worldwide food security fears.

In Hawai'i, which in small ways mirrors much of the wider world of agriculture and food, collaboration is occurring across our island state that seeks to rebuild cultural linkages among agriculture, food, and community values. Discussions are increasingly focused on new thought patterns that would encourage a return to agricultural practices that support sustainable food systems and water-management practices that place nature's goodness and personal wellness in the forefront as local residents and island visitors make informed decisions toward virtuous food choices.

That the State of Hawai'i currently imports 85% of its food is an ongoing dilemma. The cost of living in Hawai'i, at 167% the US national norm, is the highest in the nation. The annual mean wage is well below the national average, driving many residents to work multiple jobs to simply afford the high costs of food. Agriculture and food production have been identified as the number-one-targeted industries required to strategically diversify and strengthen the economy by providing jobs with a living wage, increase GDP to lower local food costs, and increase the long-term sustainability of Hawai'i as a geographically isolated island in the event of a natural disaster.

Aggressive measures to reverse imported food reliance as well as increase capacity for local food production and food product development is necessary for Hawai'i's agricultural future, and not surprisingly, reflects a much more global theme. Value-added options for local agricultural crops have been identified as a key need. The ability to manufacture new and value-added food products is especially critical on neighbor islands such as Maui, where agriculture plays a significant role in the economy and markets for fresh produce are limited. In other states and nations, Food Hubs & Innovation Centers have created millions of dollars in revenue by helping food producers achieve higher value from culls and excess produce by extending shelf-life through quality processing, offering high pressure and reduced oxygen packaging systems, opening new export markets for sales, and providing training programs to farm and food entrepreneurs to meet or exceed today's stringent food safety regulations.

Maui County, in particular, is at a critical economic juncture with the closure of Hawai'i's last remaining sugar plantation, the Hawai'i Commercial & Sugar Company. The closure brings the loss of hundreds of jobs along with the opportunity for small farm, diversified agriculture, and food-related business growth on 36,000 acres of now vacant land. The County of Maui is at the brink of economic distress and at the same time unlimited opportunity to develop small businesses in the food and agriculture industries to foster job creation, promote private investments, and diversify the financial options and stability of its residents.

Hawai'i has a deep responsibility to provide high-quality affordable food to advance our people, our communities, and our islands. Such a large mission requires a celebration for diversity in production, respect for land and water and care for those committed to wholesome food and agricultural practices.

Helmut Traitler's and his co-authors' *Megatrends in Food and Agriculture* serves as a "best in field" resource to help all sectors of Hawai'i's food industry—from large mono-crop producers, major political decision makers, and young food businesses—identify their strengths, weaknesses, opportunities and threats, then address those issues successfully to ensure that the right to food is continually valued.

Beyond the shores of Hawai'i, *Megatrends in Food and Agriculture* offers a borderless canvas to organically seed-stimulating study, conversation, and action concerning the future development of sustainable agriculture systems, new emerging food industries, and innovative technological services that rightfully place the human condition first in all thoughtful agriculture practices and food pursuits.

Chris Speere
Site Coordinator & Food Specialist
Maui Food Innovation Center, December 2016

Acknowledgments

Once again, this was not an easy book to write. Although I consider myself a fairly cognizant expert in most matters food, agriculture, yet so close to food, always seemed so detached and far away from all these ever-important considerations and thoughts regarding food. So this is my first foray into the world of agriculture as it links to food, and it was clear to me from the onset of this project that I had to invite top-class experts in the relevant fields of plant and animal agriculture, and at the same time cover the vastly important field of water.

I was lucky enough to find the right experts who have contributed in important ways to the successful creation of this book. My thanks go to David, who with help from Ben Gordon, Research Assistant at University of California–Berkeley, has so skillfully crafted Chapter 2 and thoroughly highlighted all critical topics around water and water management of importance to the well functioning of any farming activity and beyond. My good friend Vincent and his colleague Michel, both top-class experts in their field, and with the help of Fatma Fourati-Jamoussi, Associate Professor at UniLaSalle, INTERACT Research Unit, Beauvais, France, have painted an important picture in Chapter 3 regarding modern plant agriculture, especially the themes of advancements in modern plant breeding outside the still highly contested area of genetically modified plants. Let me also express my special gratitude to Keith Heikes, who despite a busy calendar filled with many professional obligations, found the time to write Chapter 4, on the agriculture of animals.

In pursuing this project, I have again learned so much and gotten a better understanding of the intimate connectedness of agriculture and food. I am convinced, more than ever, that we have to increase our vigilance and level of caring for an often fragile agriculture industry, and especially the productive part of farming in all its facets. For this recognition I am particularly grateful.

This is my fourth book on a topic related to the food industry in a rather short period of time, and I am always surprised that after having written the first page I actually make it to the end. A very special thank goes to my wife

Thérèse, who like in the previous books, was the one who brainstormed with me on chapter outlines and contents, and all this from an unsuspected and untainted, just pragmatic and rather innocent position. My son Nik Traitler, like in my other book projects, helped me designing most of the figures.

I would also like to thank my dear colleague Chris Speere, who took it to write a fitting foreword to this book. Being from Hawai'i meant that Chris brought an important insular and well-defined view to this story.

Finally I would like to express my sincere gratitude to our publisher Wiley Blackwell and the entire team behind this project for their continued trust in the ability of me and my co-authors to not run out of ideas to share with our readers. For being here at this point of reading, I want to send you, the readers, my special thanks!

Part 1

Agriculture and the Food Industry

1

The Role of Agriculture in Today's Food Industry

> *The ultimate goal of farming is not the growing of crops, but the cultivation and perfection of human beings.*
>
> —Masanobu Fukuoka

1.1 Introduction

In every form, agriculture has always been and most likely will always remain the twin sibling of food and especially the food industry. We all seem to be rather clear on what "agriculture" stands for but let me provide a short definition, so that we are on the same page.

> Agriculture means to use natural resources to produce commodities which maintain life, including food, fiber, forest products, horticultural crops, and their related services. This definition includes arable farming or agronomy, and horticulture, all terms for the growing of plants, animal husbandry and forestry. ("Agriculture," n.d.)

Merriam-Webster gives a slightly shorter, yet rather similar definition:

> The science, art, or practice of cultivating the soil, producing crops, and raising livestock and in varying degrees the preparation and marketing of the resulting products. ("Agriculture," n.d.)

There are some important keywords to be found in both of these definitions, such as: *arable farming, cultivating the soil, crops, horticultural crops, fibers, forest products, growing plants, livestock, animal husbandry, preparation and marketing of resulting products*, and *food.*

Megatrends in Food and Agriculture: Technology, Water Use and Nutrition, First Edition.
Helmut Traitler, Michel Dubois, Keith Heikes, Vincent Pétiard and David Zilberman.

The most important message though is using natural resources to produce commodities (mostly food but also fibers for cloths and wood for shelter and fuel, at least in part), which maintain life. Maintaining life and all that goes with it is really the major driver here; agriculture in all its forms and shapes and all that derives from it is the basis of life. Without it we could not really exist, let alone survive.

Although all this is pretty obvious, it is definitely worthwhile to remind us of these various elements and roles and to put them in a right perspective. So let's briefly list and define the four major building blocks that are necessary for life and agriculture to exist in the first place.

1.1.1 The Four Building Blocks

These supporting building blocks are necessary not only for life at large but also agriculture in any form.

The *first* building block is water and hasn't been mentioned yet. Water plays a crucial role; it is one of the cornerstones of life and growth on this planet. Interestingly enough, whenever my colleagues at JPL/NASA launch a rover to Mars, one of the great questions they want to find answers to is: "Is there or was there water on that planet?" Seems that there was and actually still is—at least traces—and that could mean that, although quite some time ago (maybe hundreds of millions or even billions of years back), there was enough water on Mars to sustain growth of agricultural matter of one kind or another.

This just shows how critical the presence of water is for any organic growth—water, together of course, with other compounds such as oxygen, hydrogen, carbon, nitrogen, sulfur, and phosphor, to name just the most important ones. These atoms, in their various molecular permutations form the backbone of every living matter as we know it and, together with water could be called the "dirty half dozen" of forming and sustaining life, including the growth of plants and animals. This combination of a multitude of such molecules forms the *second* building block.

Let me add one more critical element, light, that is, mainly visible light. It is the *third* building block. Without light, the two first building blocks would not be able to work together and support the growth of life, organic matter, and animals and plants, let alone humans. I do realize that there are life forms on our planet earth that grow and thrive in the absence of light, but when it comes to traditional agriculture for both plants and animals, light is a crucial element.

The *fourth* building block is temperature. Life as we know it does not really thrive in extreme temperatures, such as below 273 K (−0.15 °C) or above 325 K (~52 °C). Yes, there are bacteria or even specially adapted types of frogs that are known to grow or survive below 273 K (−0.15 °C) or even close to temperatures at which water is boiling in volcanic environments. The latter goes for

- Water → H2O
- The “appropriate” basic molecules (just the major ones) → O_2, H_2, N_2, C, S, P
- Light → visible wavelengths from approx. 400 nm to 700 nm (possibly 380 nm to 750 nm)
- Temperature → most comfortable range from >273 K to <325 K / 220 K to 370 K is possible

Figure 1.1 The four building blocks of growth.

bacteria, not the frog! Animals in arctic or Antarctic environments can survive at temperatures as low as 220 K (−53 °C) because they have developed survival strategies, both genetic as well as behavioral. So the range in which life as we know it can exist is probably larger than described here and could range from 220 K (−53 °C) to 370 K (97 °C).

Enough of this excursion into these basics; I just want to make sure that we have the same basic understanding of the topic ahead and accept these definitions of the major required building blocks of water, molecular composition, light, and temperature as baseline for this book. I know that this is rather short and only scratches on the surface, but believe that this is setting the tone of this book appropriately.

Figure 1.1 depicts in a simplified form the four main building blocks for growth and sustenance of life.

1.1.2 Some History of Agriculture

In case you know it all already or are typically not interested in history of any kind, including the one that describes agriculture at large, then please skip this section and move on to the next section. Although I do hope that I can convince you to read on through this section.

The drive for survival, which includes searching and finding food and water as well as breathing air, is the most important basic physiological need according to Maslow (1943): “Air, water, and food are metabolic requirements for survival in all animals, including humans. Clothing and shelter provide necessary protection from the elements.”

It can safely be said that since the dawn of time, or since the beginning of animal life in any form, this drive for survival has always existed and has preoccupied all life forms in important ways. Even plants, solidly rooted in the ground, strive for growth and existence by feeding on nutrients in the ground “breathing” oxygen, assimilating carbon dioxide and reaching out for light. Animals, irrespective of whether they are prey or predator, are in constant search of food and water, while breathing air and, for many different reasons, they also need light.

Humans, by most definitions, are predators, although they might be hunted as prey by large carnivores in some parts of the world. Their first and foremost task during all periods of evolution, and to this very day, is to find enough “fuel” or enough food and water to grow, to survive, to exist, and be able to expand their activities to whatever they have chosen them to be. However, satisfaction of basic physiological needs, or in simpler words, making the body function properly, is still the main driver. Unless we all become robots, this is not likely to change any time soon.

Although debated by some, humans, at least in more recent times, are omnivores (i.e., eating just about anything, from plants to animals and most recently even cheap and crappy food). With some exceptions, every type of food is welcome into the stomachs of human beings. Not surprisingly, exceptions are rather numerous and include people who don't eat certain foods for ethical or religious reasons (e.g., vegetarians, vegans, halal, kosher) or health reasons (e.g., free of lactose, gluten, sugar, salt). I shall discuss this and other related topics in much detail in Part 2, especially in Chapters 5, 8, and 10.

The story goes that human beings started out as predators with opportunistic strategies to obtain food and find water. Hunting was part of this strategy and being able to read and understand surrounding nature intimately was of the utmost importance. Moreover, there was no guaranteed and continued supply for food, especially enough food for all. Community was, and is, important for physiological needs and safety needs escribed by Maslow (1943): "Once a person's physical safety needs are relatively satisfied, their safety needs take precedence and dominate behavior."

1.1.3 Eat More and Increase the Likelihood for Survival

There is one more, often overlooked element to this story of the development of mankind, namely those who were able to eat more food and drink more water than others in the community or village—whenever food and water became available—had a better chance to for longevity and longer-term survival than those who were second in line. In actual terms, this simply means that the more you could eat and keep, the fuller your stomach became, the better your chances were that you were still around say in 2 weeks or so, when the next successful hunt was brought in or when you or your friends came across enough berries and other edible plants in your environment.

I used the *village* loosely because those were also times when groups of people, hunters and other members of the tribe, roamed along after game, edible plants, and sources of water in opportunistic ways. Once everything that was within reach was hunted and all the berries were foraged, one had to move on to the next location where game and plants were again abundant. In addition, seasonal variations made life even more difficult and the uncertainty of regular and sustained food and water availability ultimately drove such hunters and gatherers to become more sedentary and start the "business of agriculture."...

Well, maybe not so fast. I wouldn't call it business yet because domestication of a variety of plants, mostly grains as well as animals, happened for reasons of personal survival first. After the last Ice Age some 20,000 years ago, depending on climate and region of the world, animals such as bison, goats, later sheep, and cows were domesticated, first for personal use and probably only much later as a business—goods for trade—for some in ways similar to those that we know today.

Fairly early on in the process of domestication we find dogs, evolving from wolves. It is interesting to note that wolves and humans hunted (and for wolves this is true to this day) in groups or packs, led by males, and were all members of the same family who were friendly to each other but suspicious of outsiders and competed for the same prey. It is likely that out of this common pattern of hunting, domestication of wolves toward dogs happened at a rather early stage, and respective bone finds date back some 12,000 years supporting this ("History of the domestication of animals," n.d.).

So, what do we have until now? Agriculture, as previously defined, is not a really old activity pursued by humans. However, agriculture in the larger sense of existence of natural resources, plants, and animals, is rather old—arguably as old as plants and animals, and through predator–prey relationships, always made use of each other, although not in organized ways.

1.1.4 Food Can Be Grown and Plants Can Be Bred: What's Next?

Once humans knew how to grow seeds and other plants and once they had mastered domestication of "useful" animals through more or less controlled breeding, it quickly became apparent that food might be prepared in better ways than through hunting and gathering. Apart from preparing food of vegetable or animal origins through any kind of cooking, preservation of food became an increasingly important element. Preserving food by either appropriately storing it or by applying simple preservation methods such as cooking, drying, salting, fuming, or other types of preservations such as cold storage in dug-out cellars became the name of the game.

These methods were created rather early on during the gradual development of humankind and can be linked to the "discovery" of fire, most likely a few hundred thousand years ago, and much later, to the mining of salt from either rock or seawater, and their appropriate application for food preservation. Although in the early days of our history cooking, fuming, drying, and salting dominated food-preservation techniques, many more such preservation methods have been used in more recent centuries as a result of supporting technological developments, scientific discoveries, and the recognition of specific nutritional needs.

Methods such as industrial type of cooling, sugaring, pickling, alkalizing, canning, jellying, curing through fermentation, pasteurization, UHT, sugaring, vacuum packing, canning, bottling, adding of food additives (artificial or natural), ionizing radiation, pulsed electric or magnetic fields, modified atmosphere, ultra-high pressure, and bio-preservation using specific biota are the most prominent ones in this long list of food-preservation techniques. Additionally, food preservation can be achieved through appropriate so-called hurdle techniques, which are mainly, although not exclusively, composed of smart modulation of pH, water activity, and temperature.

Table 1.1 depicts a typical list of food- and beverage-preservation techniques (not necessarily complete).

Table 1.1 The major food preservation techniques.

Drying	Oldest documented technology; sun drying as early as 12,000 BC
Cooling	Slowing down microbial growth and reproduction
Freezing	For preservation, processing, and distribution
Boiling	Kills microbes
Heating	Below boiling yet just hot enough to kill microbes
Salting	Draws moisture, reduces water activity
Sugaring	Preserving mainly fruit in antimicrobial syrup
Smoking	Aids drying and reducing water activity
Pickling	Edible antimicrobial liquid either chemical or through fermentation
Alkalinization	Food becomes too alkaline for microbial growth
Canning	Sealing in of foods
Jellying	Solidification, improves water activity
Curing	In general: dehydration to reduce water activity
Fermentation	Replace bad microbes by good ones; creates new products
Pasteurization	*See* heating
Vacuum packing	Air tight environment: exclusion of oxygen
Food additives	Antimicrobials, antioxidants
Ionizing radiation	Kills microbes (sometimes affects food product)
Pulsed fields: electric/ magnetic	Can destroy bacterial cells
Modified atmosphere	Exclusion/reduction of oxygen
Ultra-high pressure	For preservation of freshness and appearance of food
Bio-preservation	*See* fermentation
Hurdle technology	Avoid microbial growth through pH and water activity management

1.1.5 From Very Old to Rather Recent Food-Preservation Techniques

I do realize that I have jumped ahead quite a bit, but I really want to keep this section on history of food preservation as short as possible. I needed to build the framework for the coexistence, almost cohabitation, of agriculture from its beginnings to this day and the series of important steps toward industrial food and the food industry. Most of the old fire-related preservation techniques, especially cooking and curing through techniques such as smoking, are almost lost in time when trying to put a date on them; others such as industrial cooling can more easily be related to more recent technological achievements such as the invention of the mechanical/electrical fridge, which dates back to the early 20th century. On the other hand, use of ice for cold storage and prolonged preservation dates back to prehistoric times. It was only through mechanization of the creation of cold temperatures for either processing or storing or both, that the food industry could create products such as ice cream, frozen entrees, or freeze-dried soluble coffee on an industrial level.

Anyhow, let me describe a few old food and beverage industries that can be dated back through written records and that have been operational since their creation. The oldest records of a beverage company and that is still manufacturing and selling products to this day is the "Weihenstephan Monastery Beer Brewery" ("Klosterbrauerei Weihenstephan, n.d.) in Bavaria, Germany. Although founded probably even hundreds of years earlier by monks, the first written and still-preserved records date back to the year 1040. This seems to be the record-holder when it comes to carrying the banner of oldest, and still operational food or beverage company, with almost 1000 years of age! Some of the "middle aged" giant Sequoia trees in the Californian Sierra Nevada were just adolescents during those years.

There are others, such as the "Salumeria Giusti" in Modena, Italy, which dates back to 1605. The mineral water company "Aqua Panna" was founded in Tuscany, Italy, in 1564 and still produces bottled mineral water. I mention these to demonstrate that food is history and history is food; it's a longstanding relationship and dates back hundreds and even thousands of years, such as to the first traces of wine-making in giant, air dried amphora-like clay vessels in Georgia, South Caucasus, as far back as 8000 years ago.

All this leads to one more surprising, or maybe not so-surprising, observation: beverages, and especially the alcoholic ones, seem to have a far longer history compared to the much-younger (mainly) food companies such as the Nestlé Company, which incidentally commemorated its 150 years of existence in 2016. Most other large or small food companies are younger than Nestlé, with the exception of "Salumeria Giusti."

So let me focus at the task ahead, namely the role of agriculture in today's food industry and proceed to the next section that discusses and analyzes the "supplier role" of the agriculture industry for the food industry in more detail.

1.2 Agriculture: The Main Supplier to the Food Industry

The section head and its underlying meaning could almost be characterized as a simple and obvious truism; however, it's not as clear-cut as this. The relationship between agriculture and the food industry has not always been an easygoing and trusted one. Let me expand on this. Although it is clear that without agricultural raw materials, both plants and animals, no food or beverage product could ever be made and this is true for both the individual food provider at home and the entire food industry. So, things should be simple and straightforward, yet they are not always so smooth.

In my most recent book *Food Industry Research and Development: A New Approach* (Traitler, Coleman & Burbridge, 2016), I mentioned the story of so-called single-cell proteins, proteins that were derived from oil—oil that comes from the ground that is. This story begins in the late 1950s and ends the mid-1970s. It is one of the rare examples when the raw material for edible food did not come from the field or an animal off the field but straight out from the deep ground. This mostly happened because the large oil companies felt that they had a surplus of oil that could serve goals other than being fractionated and burned or made into chemicals and could actually feed a growing world population. On the other side of the coin, it was feared that traditional agriculture could not keep up with this growing world population and other, artificial sources would have to be tapped into.

Making sugar from wood, not a typical food source, is possibly another example in which traditional agriculture, farming, and animal husbandry have not worked hand in hand in a supplier–receiver and transformer relationship. Although most of sugar obtained from this source —glucose in a yield of more than 20%—is used to be fermented to ethanol, and in theory, this sugar would, after refining, be good for human consumption if need arises. Needs had arisen in past wars and nobody tells us that it might not happen again. On the other hand, refining such substrates and rendering them safe for human consumption would be so expensive that thus far such sugars from wood, either through chemical or biochemical pathways, were and are exclusively used to generate ethanol.

1.2.1 Artificial Ingredients

I do agree that these examples are rather outlandish and rare, and in almost all instances, agriculture really is the main supplier of raw materials to the food industry. I say main, because in the past, food manufacturers sourced quite a number of minor food ingredients from the chemical industry, especially food colors, many food flavors, and to some degree even emulsifiers and stabilizers were of "non-natural" origin. Over the years, the trend moved to the so-called "nature identical" and then most recently to "all natural" food ingredients. It has been quite a lengthy ride and was almost exclusively initiated and driven by the consumers and consumer organizations. It was, and still is, not an easy feat to find all natural equivalents for all product-critical minor ingredients, such as vibrant and shelf-stable natural food colors for products such as Smarties® or M&Ms®.

Parent organizations and members of the medical and scientific communities feared and suspected that artificial food colors were responsible for some health disorders that may have befallen our kids after consumption of such colored products—although there are no entirely proven and well-established relationships that would point to a clear-cut cause and effect. As always in the world of science, more clinical and similar studies are required to make a clear case. There also appears to be a genetic disposition for such sensitivity to artificial food colors. There is still much debate in the scientific community as to what degree certain artificial food dyes are linked to children's disorders such as hyperactivity or attention deficits.

Older studies, such as "Artificial Food Colors and Childhood Behavior Disorders" by Silbergeld and Anderson (1982), have found that exclusion of artificial colors from the diet did not demonstrate an important enough beneficial effect. And the article suggests that a genetic basis might be responsible for certain neural responses that should be considered in future studies.

Another more recent scholarly article by Harrington (2015) concluded that certain color additives to food products contribute to hyperactive behavior in children. Further, Kanarek (2011) describes attention deficit disorder (ADHD) as one of the most prevalent behavioral disorders in children. Risk factors are genetic as well as of environmental nature.

I discuss this here not to stir the pot and take a position on the topic but to show that the food industry's approach to these matters is not always driven by reason but rather by margin considerations, and that the agricultural industry, properly involved and connected early on in these matters, could have helped and solved this issue much earlier. Often, voluntary anticipatory actions are probably the best ways forward to constructively and positively tackle any subject and help everyone concerned to build and maintain trust and mutual respect.

These excursions into topics such as food ingredients and health are just a few examples of the importance of this relationship and will be discussed in the next section and more in depth in Chapters 8 and 10. It is probably the most controversial topics of all: how can and does food influence our health, both in the immediate and in the medium to long-term future.

1.2.2 The Main Raw Material Sources

Thus far, I have attempted to describe a few examples—which are the outliers—of the exceptions in which agriculture is not or far detached of being the raw material supplier to the food industry. These represent a minority, a really small, yet influential share; influential here in the sense that they negatively impact the standing and reputation of the food industry, at least for the chemically derived minor food ingredients of the artificial type. The vast majority of food ingredients, however, stems from agriculture of any kind, plants, animals, and animal derivatives such as dairy products and eggs. Focus on breeding, sustainability, and the economics of plant and animal agriculture will be discussed in Chapters 3 and 4, respectively. Given that water and its availability are key ingredients in *any* form of agriculture, Chapter 2 will discuss this topic in depth.

However, let me introduce these various elements here so that you, the reader, are being prepared to these topics of what is ahead. The simple facts are the following: the major crops such as wheat, sugar beet, maize, corn cob, potatoes, rice, and soy probably represent the vast majority of the total plant agricultural raw material input. The remainder—always on the vegetal side—such as fruits and the minor crops (also called "orphan" crops) make up for the rest. A large proportion of orphan crops, especially, such as coffee and cacao, go towards the food industry, whereas the split of the larger crops between industrial and personal use is more equilibrated. Large food companies such as Nestlé or Mars absorb large percentages of these two orphan crops of up to and around 15% of the total world production each; almost the opposite is true for the use of milk and dairy products.

The more recent numbers from Eurostat show the following picture for the European Union (Eurostat, 2015):

- Cereals (wheat, spelt, barley, grain maize, corn cob mix): 324 million tonnes
- Sugar beet: 128 million tonnes
- Oilseeds (mainly rape seed, turnip rape, sunflower): 24 million tonnes
- Tomatoes: 17 million tonnes
- Carrots: 5.5 million tonnes
- Onions: 6.4 million tonnes
- Fruits: 14 million tonnes
- Grapes: 23 million tonnes
- Olives: 8 million tonnes

1.2.3 Milk's the Star

Although all dairy products such as cheese and processed milk and milk powders of any kind are clearly the children of the food industry, the global shares of large food companies in the dairy segment are much smaller, often below 2% or less.

Not wanting to bore you too much, however, here are a few numbers concerning the milk industry from a global perspective (all numbers based on milk from mainly cows and buffalo; FAO, n.d.).

- It is estimated that worldwide milk production rose by an annual 13% between 2002 and 2007.
- It is widely assumed that too much milk may be produced, especially in developed countries.
- Total annual milk production accounted for approximately 700 million metric tons in 2007.
- The largest milk producers in 2007 were: South Asia (mainly India and Pakistan) with 160 billion liters (23% of global production); the European Union (mainly Germany and France) with 150 billion liters (21%); the United States with 85 billion liters (12%); Russia and Ukraine with 70 billion liters (10%); Latin America (mainly Argentina, Brazil, Colombia, and Mexico) with 70 billion liters (10%); China and Japan with 56 billion liters (8%); and New Zealand and Australia with 28 billion liters (4%).

Milk is an important and special ingredient in the overall ingredient mix in the small and large food industry. Typical products derived from milk are butter, cream, cheese, whey proteins at large, any type of milk powder such as nonfat dry milk powder (NDM) or the almost identical skim milk powder (SMP), and a few more applications such as ingredients in ice creams, dairy creamers, and infant formula. Milk and its subproducts are also found in beverages (smoothies, milk shakes, cocoa beverages) as well as in milk chocolate and many culinary recipes for sauces, binders, and simply as taste and texture improvers.

1.2.4 Milk…What Else?

There are, by the way, uses of milk and dairy products that are to be found outside the food and beverage range of products. Milk derivatives can be found in the manufacture of plastics, textile fibers, glues, ethanol, and methanol. Casein-based glues were already know in ancient Egypt but caseinate-based polymers are still used in protective coatings, paper coatings, foams, adhesives, and injection molded items. Caseinates can also be found as surface-active agent in soaps.

Whey and whey proteins can be used as important precursors to be fermented to methane, ethanol, butanol, acetone, organic acids, amino acid, vitamins, a number of polysaccharides, and oils. Much more detailed information on this topic can be found in an excellent article by Audic and colleagues (2003).

In the light of apparent overproduction of milk in many parts of the world and because of much resistance by farmers—not to mention transforming industry and the trade—it would be advisable to look into profitable and useful alternative usages of milk and its derivatives that can help to improve sustainability and gradually reduce dependence on fossil fuels. I do realize that it might be most desirable to reduce milk production, thereby reducing the number of animals and ultimately reducing the carbon dioxide as well as methane emissions, thereby reducing overall greenhouse gas emissions. However, as long as this is not happening and before much of the surplus of dairy products goes to waste, they may be used for nonfood, alternative applications.

1.2.5 Other Excursions from Food

On a similar subject, but in the context of other agricultural raw materials such as corn, sugar beet, or sugar cane, much debate has arisen around the topic of using food raw materials for nonfood applications such as to create ethanol and to some degree methane (mostly from the "less edible" neutral biomass) as well as plastics materials such as poly lactic acid (PLA), all through fermentation. This approach has aroused much public debate, especially questions about having arable land competing for either food or other nonfood applications.

It was speculated that perfectly edible and useful agricultural raw materials such as corn and sugar cane were and are taken out of the food raw material stream, thereby decreasing the amount of food available for human consumption and in turn increasing raw material costs. Although there is some value to this argument, the more prevalent reason to reject this approach is the reduction of land available to grow crops for food. The fact is that countries like Brazil cover a large portion of their mobility needs by substituting gasoline for ethanol, obtained from fermenting cane sugar.

> Brazil is the world's second largest producer of ethanol fuel. Brazil and the United States led the industrial production of ethanol fuel in 2014, together accounting for 83.4 percent of the world's production. In 2014 Brazil produced 23.4 billion liters (6.19 billion U.S. liquid gallons), representing 25.2 percent of the world's total ethanol used as fuel" ("Ethanol fuel in Brazil," n.d.)

Currently there are more than a dozen automakers that offer so-called flexible fuel vehicles in Brazil, which can run on any mixture of E25 (25% ethanol) or E100 (100% ethanol). It is estimated that ethanol amounts to approximately 50% in the fuel mix for vehicles in Brazil.

There would, however be alternative pathways to escape this debate over arable land between food and fuel or plastics. The first one is to use halophytic plants, which are plants that are saltwater resistant and grow on land that is otherwise not fit for growth of regular agricultural material. Thus creating an entirely new stock of available biomass for transformation first to methane and, depending on the plants that are either already available or would need to be developed by breeding, to ethanol in a later stage. I developed a set of ideas as early as 2010 about this and have tried to find investors for this idea. I was unsuccessful.

1.2.6 Noncompeting Alternatives

A few years ago, I attempted to get a business going in the area of saltwater-resistant plants, so-called *halophytes*. These could be bred and grown in agriculturally solid and sustainable ways, leading to a noncompeting, side-by-side coexistence of growing food and feed on the one hand and biomass for fuel on the other. I have suggested the following:

> Halophytes—A New Lifeline
> Today's biofuels and bioplastics are mainly derived from plant materials that are either directly competing with the food and/or feedchain such as corn and to some degree sugar cane or grow on lands that could be used for food crops such as switch grass.
>
> It is a fact that large biofuel transformers today exclusively use corn, sugar cane and switch grass as raw materials.
>
> Algae are another alternative and today companies develop special algae, however mainly for the production of specialty lipids or other high added value ingredients.
>
> We can potentially escape this dilemma by using [saltwater-resistant] plants, so called halophytes, as the starting bio mass for further transformation to biofuels.
>
> Underlying are two basic ideas:
> 1) The development, optimization and usage of [saltwater-resistant] plants that serve as biomass to be transformed to either methane or ethanol or both.
> 2) The usage of land with high degrees of salinity, which cannot be used to grow traditional food crops anymore and which can give livelihood and personal wealth to farmers and their families who otherwise would not have a basis for farming and income anymore. In Bangladesh alone, according to a 2010 report from IRIN news (Integrated Regional Information Networks), there are 20 Mio people at risk losing their livelihood due to high soil salinity.
>
> Within the next 50 years, over 20 million people could be displaced and become "climate change refugees," if sea and salinity levels rise in Bangladesh, according to the government's 2009 Bangladesh Climate Change Strategy and Action Plan."
>
> The challenges are complex, yet surmountable and can be summarized as follows:
>
> - Development of appropriate halophytic plants that have optimal agronomic traits such as yield per acre, speed of growth, harvestability, resistance against disease and infestation, overall handling and storage of biomass and fairly short distances to transport such biomass to biofuel converters.

> - Especially the agronomic development, ideally through selection and fast propagation is the heart of this development and business proposal, resulting in optimized plantlets and seeds that can be brought to farmers in areas with critical farm land conditions such as for example Bangladesh, Bengal and many other borderline lands unsuitable for growing crops around the world.
> - The ideal plant should contain both, cellulosic material for fermentation to methane as well as starch/sugar containing fruits that can be used as a basis for transformation to ethanol.
> - Initially, the seeds can be made available to farmers through the help of micro credits, NGOs [nongovernmental organizations] and other appropriate organizations until such time that the farmers can develop independence by becoming the important first step in the value chain of biofuel production.
>
> This will, however, just be the starting point for an entire area of other applications such as: applying the developed IP (traits of plants that are responsible for [saltwater] resistance and to some degree drought resistance) to cash crops such as corn, wheat or others but also to plants that are the starting raw materials for the flavor and fragrance industry. (IRIN, 2010; Ministry of Environment and Forests, 2009; H. Traitler, personal communication, August 2010)

In the preceding paragraphs, I mentioned the second alternative; it is not based on traditional agricultural practices but uses the oceans to grow algae in controlled ways. However, to the best of my knowledge, to this day, marine algae are mainly used as raw materials to extract specialty oils of the Ω3 polyunsaturated families such as eicosapentanoic acid (20:5Ω3) or docosahexanoic acid (22:6Ω3) for nutrition- and health-related applications.

As far as the creation of noncompeting biomass for fermentation to any kind of fuel is concerned, the halophyte pathway appears to be the much more sustainable one, with the additional advantage of giving millions of farmers in adversely affected areas (e.g., flooding, oversalting through irrigation with high salt content water) a future and livelihood for their families. So, this could be an elegant, although neither easy nor fast way, out of this dilemma of competition for arable land.

1.3 Agriculture's New Role in Light of Food and Health

Health has become one of the dominating concerns of large parts of the population. I realize that this sounds rather bombastic, but I believe that for many populations in developed countries, this may even be an understatement.

Health and wellness have become an obsessional pastime for many, if not all. There is nothing profoundly wrong with this trend, however, there is probably still much debate what can improve health and how can our good health be sustainably supported both, in the medium and, especially, the long term.

For many, physical activity is the key. For others, it's more the mental activity that counts. For most, however, outspoken or not, it's the combination of both, as the Roman poet Decimus Iunius Iuvenalis so famously and appropriately said, *mens sana in corpore sano* (MSICS; "a healthy mind in a healthy body"). Most likely every one of us would agree to this; however, we may have different approaches as to how to best achieve such a balanced status. It is my firm belief that food and drinks do not yet play the role that they not only deserve but also actually need to play in getting to the so desirable MSICS status! But why is that so? Well, simply because we still have the mentality of "everyday low prices" and "greed is cool" when it comes to shopping for our food.

It has been mentioned that there is a negative correlation between the low-price performance of a food discounter and the size and luxury of cars parked in the parking lot of such discounters: the lower the food prices, the more expensive the cars parked in front. I admit that it is a simplistic view and does not happen everywhere, but I have observed this many times. It is also clear that people need to be able to afford food for their own and their families' daily needs and often typical salaries are just not sufficient to purchase valuable food—valuable in a nutritional sense that is. There is still too much disparity between nutritious and inherently healthy costly food versus so called "cheap calories", affordable for those who have little or need to use their government food stamps for their food purchases.

It's not easy to break out of this vicious circle, and I discuss this topic here because agriculture has its fair share in this and has an important role to play. Let us get back to the desire and need of achieving and maintaining good health through means of physical activity and mental training paired with the right food—the right nutrition for everyone. Everyone means two things here: first, the largest possible number, ideally the totality of the population to be fed properly, and secondly, the recognition that not everyone is equal to everyone else and a good degree of nutritional personalization would be the ideal end state. Part 2 of this book will discuss this apparent dilemma and great challenge in much more detail.

Suffice it to say that with today's means finding solutions to this riddle will be difficult. That does not hinder us to at least try and think it through and propose possible solutions to this problem of healthy versus unhealthy, more valuable food versus cheap calories, high costs of food versus affordability, implication of food in the health cycle and especially in the larger healthcare business environment, and last but not least, how to convince everyone involved in this field to agree to commonly acceptable solutions. Again, agriculture has a major role to play in this.

1.3.1 Decades of Food Safety Rules and Regulations

It's a simple fact that food, and in turn, ingredients and raw materials that go into the manufacture and production of food and beverage products have been and still are increasingly regulated and scrutinized from all angles, not least of all food safety and health-related aspects. This is a good thing and should not be reversed by any means. It is maybe interesting to observe that the number of food and food ingredients rules and regulations have increased pretty much at the same pace as analytical sciences have developed to being able to detect smaller and smaller amounts of just about anything, relevant or not.

Organic analytical techniques have made quantum leaps of improvement, even more so since the full integration of IT qualification and quantification techniques for the better part of 30 or so years. Signal-to-noise detection has vastly improved, and all analytical techniques have participated in this improvement. This has gone hand in hand with being able to discover more and more food-related hazards since the 1960s.

Let me quote from an excellent report published by the European Commission (2007):

> The first EU food hygiene rules, which were adopted in 1964, were limited to requirements for fresh meat. Over the decades, however, further hygiene legislation was developed and implemented for other food groups, including eggs, milk products, poultry meat, fishery products and game meat. (p. 16)

Today, food regulations cover all food groups, all food and beverage ingredients, regardless of origin. Again, it is a good thing for the safety of consumers to have an appropriate number of rules and regulations in place, yet one negative result of this situation is the abuse of these rules for political or financial gain. Scare tactics are often used and consumers are always in the middle of any such debate. I do not want to go into much further detail, but I can say with conviction that I have seen that abuse exists and that it does not necessarily help the cause of protecting the consumer in the best possible ways, and at the same time, leaving room for nutrition and health-related improvements in the food and beverage industry.

The story of the "evolution" of the mentioned food hygiene rules and food health and safety issues could have been written by Agatha Christie or Stephen King but can rather be found in the report of the European Commission (2007). Let me go through the decades as well as topics.

The 1960s saw the first important push into the food safety issues of that decade such as:

- Salmonella in eggs
- *Escherichia coli* (*E. coli*) in undercooked meat

- *Clostridium botulinum* in improperly canned foods
- *Listeria monocytogenes* in unpasteurized milk (boiling up importantly in the mid- to late 1980s)
- *Staphylococcus aureus* in ham
- *Trichinella spiralis* in undercooked pork
- Hepatitis A virus in shellfish

Much of this was simply that more modern food processing, preserving, packaging, storing, and transporting food technologies were especially developed and improved during these years.

The 1970s saw an important push toward mastering safety issues and alerting the food and government communities to better protect consumers. It was also during those years that the public begun to link unhealthy food products and unhealthy eating with disease and premature death. The important issues of those years were (and still are, by the way and this is valid for the list of the 1960s too!):

- Pesticides and their residual levels
- Food additives
- Flavorings
- Contaminants and their maximum levels

The main food safety issues of the 1980s were the following (with a continued close eye on all of the preceding, too):

- Salmonella outbreaks
- Cases of botulism
- Bovine spongiform encephalopathy (BSE)

The biggest food safety–related topic during the 1990s was probably the large BSE outbreak in the United Kingdom at the end of that decade.

- BSE
- Dioxins

Lastly, in this context, the topic of genetically modified organisms (GMOs) was, and still is, the most prominent and most hotly debated topic of that decade. The major push, unfortunately came from the agricultural industry, which wanted to push apparent solutions to the consumers, without really explaining what it was all about and what benefit the consumers would take from GMOs.

Yet another big issue discussed and dissected in many was the pushback from regulatory authorities to the inflation of health-related claims for food and some of its ingredients. This caught many in the food industry, who had spent millions to support such soft health claims, really by surprise.

Maybe the most important topic of debate during that decade was the push toward proper information, traceability, and education and training for all members of the food and beverage industry. This holds true to this day and is probably the best way forward toward innovative, safe and healthy foods and beverages (European Commission, 2007).

1.3.2 More Rules: What Do We Do?

One can confidently say that the world of food, food ingredients, and processes became more complicated in the last 50 years, probably even more so in the last quarter of a century. Over the years I could observe different answers to this increased complication of the world of food.

I don't want to point any fingers at any food company in particular, but the typical patterns of response to crisis were also to be found here. I am referring to Elizabeth Kübler Ross's (1969) response pattern to grief: denial, anger, bargaining, depression, and acceptance.

The individual descriptions of these five stages, "translated" to the tightening regulatory situation in the food industry could be the following.

1) **Denial:** The first reaction is often denial by refusing the premise. In this stage companies, and especially middle management, might believe that the newly proposed rule is somehow mistaken, going far too far, or is simply wrong, and they might cling to a false, preferable reality.
2) **Anger:** When the company and its management, middle and higher, recognize that denial cannot continue, they become frustrated, especially at "proximate individuals"—both inside and outside the company—especially at representatives from the regulatory authority. Typical company responses could be: "Why our industry, why especially our company? It's not fair! How could this happen to us? Who is to blame? Why could this happen and why were we not better prepared? Our people really sleep and didn't earlier pick up, what was coming our way?"
3) **Bargaining:** The third stage involves the hope that the company might just be able to avoid hardship. Usually, the negotiation for more lenient judgment is made in exchange for a promise of developing better and healthier products for all consumers.
4) **Depression:** Here we are probably deviating most from the original grief management concept and language. In typical food companies, such state of depression might simply be expressed by announcing yet another cost-cutting exercise, stopping projects, or worst, letting people go.
5) **Acceptance:** "It's going to be okay; we can't fight it, we cannot get away by not complying, so we may as well prepare for it and even do better than our competitors. We are going to be the really good guys!"

I agree that maybe not all five steps can be seen in all difficult, crisis-like situations, but I personally have seen and heard many of these elements and arguments in numerous cases over my many years in the food industry.

The best and most efficient answer to the increasingly difficult regulatory environment in the food and beverage industry is simply to be ahead of the game. Sounds easy but it's not. The consequences of not complying with regulatory requirements are pretty dire and can lead to really dramatic situations for the company, even leading to its demise. Therefore, and despite the humanly understandable reactions as described in the five stages, the straightforward reaction should always be compliance.

However, being ahead of the game is far better. How can my company—any company—be ahead of the game? Well there are two possible approaches, which do not exclude each other and are probably most efficiently applied in complementary fashion.

Here's the first approach: become an active and constructively contributing member of important food regulation associations, thereby helping to drive the agenda together with the other representatives, such as governments and consumer organizations. Thereby, your company is not only well aware what is being discussed and soon to be cast in a new rule but you can also hear and experience "what is cooking" and how this might affect your company in the future. The mantra is constructive input and output and collaboration; it must never be controversy, although debates may be heated.

My second approach to make life simpler and more enjoyable for you and your company: simplify. What do I mean by this? Well, simple (no pun intended): Demystify and simplify the list of ingredients that are needed to make a specific product; keep it proverbially stupidly simple and more easy to understand for the consumer and at the same time reduce the number of possible regulatory touch points. I do realize that this is not an easy feat and especially not easy for well-established products in your portfolio. But it can definitely be applied to any new product development in the future and I shall discuss this topic in later chapters.

Reality, however is a slightly different one: the world of food, for many reasons, has become more complex and more complicated, and agriculture and the production of agricultural raw materials for the food industry has an increasingly important partnership role to play.

1.3.3 Raw Materials and Processes Become More Sophisticated

For those readers who do not like the word *sophisticated,* I could offer alternatives such as fancy, complex, functional, traceable, sustainable, healthy, nutritional, affordable, tasty, safe, or simply a combination of all of these. I believe you see, where I am going with this. And all this, in my eyes, applies to the development, growth, harvesting, storage, and distribution of agricultural raw materials but also to the various unit operations happening in the food and beverage industry. Typical such operations are procurement, inbound supply chain, manufacturing, packaging, storage, and outbound supply chain. It would be a fair comment to say that this is slightly oversimplified as the entire value chain for both partners—agriculture and the food industry—is far more complex and is comprised of many more elements and activities happening in both areas. So let me focus on the agricultural raw materials first and discuss and analyze some of the most important aspects.

This discussion has much to do with today's expectations that many or most consumers have when it comes to food, *their* food that is! Consumers have become sophisticated, too. They have sophisticated expectations with regard to food and beverages.

When looking into the public discourse regarding food in more detail, the consumer, or rather the majority of consumers with the economic possibilities to afford food, has a certain set of expectations and requirements. I do not speak about developed countries compared with less developed ones because the boundaries are not between countries any more when it comes to being able to buy appropriate food for oneself and one's family but they have fault lines between the various segments within one country.

Let me briefly discuss the United States as an example because it might be the most striking one, given the fact that overall, on the grounds of most economic factors, the United States is a rather rich country. That said, many people do not have enough financial means to buy food for them and their families.

Let me "bore" you with a few numbers. When looking back over a period of 40 years, the number of people had received food stamps to purchase their daily food amounted to approximately 21 million in 1975 and rose to an all-time high of 47.6 million in 2013. It has since then slightly decreased and in July of 2016 stood at 43.6 million. This means that slightly more than 14% of the U.S. population, or every seventh person, receives food stamps and, at least in part, lives on support from the government.

The average monthly value of food stamps per person was around $17 versus $127 in 2015 or about seven times higher (SNAP, 2017; Trivisonno, 2017). Part of the increase in value is simply explained by monetary inflation over a period of 40 years and part of it might be based on authorities' recognition and acceptance of the fact that proteins simply cost more than bulk carbohydrates. When looking into the consumer index average annual numbers, we see an increase of a factor of 4 during these same 40 years (Consumer Price Index, 2017). The larger increase can hopefully really be attributed to a better nutrient mix in food that can be purchased with such food stamps.

These numbers are not only striking but they also prove that, as far as access to and availability of nutritious and healthy and affordable food is concerned, there are no country distinctions, such as developed or developing, and it is more important to look into societies and even communities within such societies in much more detail.

Let me briefly come back to the already mentioned requirements for modern food and how these take roots, literally, in the preceding agricultural practices, directions, and strategies and how they can and must feed into the food industry but especially into the consumers' mouths and bodies.

Modern food, if there is such a thing, is supposed to be healthy and nutritious. Modern agricultural output in terms of crops and other semi-finished products demands efficiency and stability and preservation during transport and distribution to the various manufacturing sites. Although the wine industry has more or less voluntarily reduced the output of grapes per vine or per square feet or meter, to enhance quality of the end product, all other branches of agriculture have, as far as I am aware, gone in the exact opposite direction.

Chapter 3 will expansively discuss, analyze, and scrutinize modern plant agriculture, including topics such as modern breeding methods and the economic and societal dimensions and importance of the major food crops as well as so-called orphan crops. Chapter 4 has at its core the agriculture of animals, especially the theme of animal proteins of the future and the sustainability of such proteins in an increasingly complex humanity. Chapters 5 and 10 will discuss and analyze topics such as food trends, new (not novel) foods and the availability of healthy and nutritious food of today and discuss possible future scenarios.

1.4 Most Likely Drivers for Change in the Agriculture Industry

Change is palpable, yet change is feared as previously described in Kübler-Ross's "angst" scenarios and subsequent reactions. However, even without notice, change does and will happen in all domains of life, including the world of agriculture. At the risk of repeating what has been said elsewhere and even before in this book, let me list the main drivers for change—subtle and gradual or more revolutionary.

As we know by now and for quite some time, the world population is not only at a constant but also fast paced growth. Although the estimated total world population accounted for "just" 1 billion persons, it took until 1927 to reach double that amount. However, it only took less than half a century to double that number again to 4 billion in 1974, and it is taking even less time to almost reach 8 billion, which by all estimates should happen around 2020. These are just numbers, but behind them are people of all genders, age, and race, and they all need food and water. However, our planet has not become bigger, at least not that anyone noticed. Yes, some have said that by submitting every single person in this world to the same living space conditions as people in Hong Kong, the entire world population might fit in the state of Texas. Yet their needs as far as food and water go still need to be fulfilled through modern agricultural techniques and usage of arable land for a growing number of people in our world.

I do not want to get in this debate whether the world is "overpopulated" or just "badly distributed"; after all we are talking about human beings and their rights to live a good life and be able to feed themselves and their families in the best possible and most healthful and nutritious ways. I do insist on "most healthful and nutritious" because I truly believe that this a basic human right and is not negotiable. I do realize that this may sound utopian, or even naïve to some, but I truly believe in this mantra, which was one of the major reasons I worked almost my entire professional life in the area of food.

Let's not get into politics, but just state a few simple facts: world population stands at close to 7.5 billion in 2016; arable land, useable to grow crops or graze cattle, has not really increased, and yet agricultural practices have been devised over the years that have improved crop output per acre, have in some instances been able to switch to more drought-resistant plants, have, in part thanks to a warming climate in many parts of our world, made it possible that in moderate zones, two harvests of rape seed oil can be grown now in central Europe. The latter is of course also made possible because of the abundance of water in these areas.

Other, much warmer and much dryer areas, such as the Mediterranean can often not even get to one full crop. Although temperatures and abundance of light maybe just perfect, water is missing in the equation. Some of my best friends are avocado and citrus growers in Santa Barbara and San Luis Obispo counties of California. Much has been written and said about the ongoing severe drought in most parts of California, and although we as private citizens do our fair and honest share in the mandatory water-conservation efforts, even saving 20% in areas outside agricultural use of water, is just a drop in the entire equation. It is estimated that agriculture in California uses up to 90% of all water resources in the state, and so 20% of the remaining 10% only account for a 2% overall reduction of water usage.

To come back to my avocado- and citrus-grower friends, I know for a fact that they are really low on water, so much so that they had to stump a good portion of their avocado trees and, on top of this are just beginning to see the onslaught of the "Huanglongbing" or greening disease caused by the Asian citrus psyllid, which is just arriving at their doorstep. So, the conditions are not the best for these farmers, and there are many, many like them, with no or very little water and bugs that may potentially wipe out entire industries, which is something that has already partly happened in Florida. The topic has become so hot that an article in the Bloomberg News (Perez, 2015) mentioned that oranges in Florida are possibly doomed unless ways can be found to stop the citrus psyllid. Both scientists and governments need to collaborate on this important issue.

I do not want to end this chapter with a flood of lamentations but rather lay the foundations for what is to come in the following chapters and sketch in a broad stroke the topics that are paramount to better understanding the topic of megatrends in food and agriculture by highlighting themes such as technology, water use, and nutrition. The ultimate goal is always the same: to optimize and modernize these elements in such a way that the people in this world can have better lives through an improved access to affordable, safe, healthy, and nutritious food, and yes, it should taste good, too.

1.5 Summary and Major Learning

Food and agriculture are truly important topics to discuss and especially in a book like this one that attempts to combine these two areas in ways that help to better understand not only the megatrends but also the interaction and interdependence of these.

The following topics were discussed and analyzed in this chapter.

- When trying to define what agriculture means and stands for, a recurring pattern can be observed with keywords such as: *using natural resources, produce commodities to maintain life, arable farming, agronomy, horticulture, crops, fibers, forest products, livestock, animal husbandry,* and last but not least, the link to the *food industry, preparation and marketing of resulting products* and *food.*
- Nature's four building blocks to support life and also agriculture in any form are: water; molecular permutations of oxygen, hydrogen, nitrogen, carbon, sulfur, and phosphorus; visible light of 400 to 700 nm, possibly 380 to 750 nm; and appropriate temperature range from more than 273 K to less than 325 K, possibly from 220 to 370 K.
- Some history of agriculture was discussed, and from Maslow's theory of human motivation, it was quoted that air, water, and food are metabolic requirements for survival in all animals, including humans. This drive for survival is probably as old as animals and humans on this planet.
- When humans were hunter-and-gatherers, opportunistic strategies toward agriculture were the rule and humans ate what they could come across—animals, seeds, berries, and roots. Those who could eat more whenever food was available had a better chance of survival, given that the next successful hunt or forage was probably several days or even longer out.
- Later on, the opportunistic approach to agriculture gave way to a more planned approach, both in the area of domestication of animals and in organized farming of crops.
- Once there was some know-how and technology available to breed domesticated animals for food and to grow and harvest crops, approaches had to be devised to the render agricultural output stable, at least for short periods of time. Some of the oldest such preservation technologies are drying, salting, and fermenting.

- Fermentation led to one of the oldest industrial food products and related companies, the oldest of which was founded in the year 1040 and still produces beer to this day under the same name "Weihenstephan." Many of the longest existing—and still operating—companies today are food and beverage companies, although the majority much younger being only 50 to 150 years old.
- The role of agriculture as the main raw material supplier to the food industry was introduced and discussed, as was the attempt to create food (especially proteins) from other than agricultural sources, such as oil.
- Other examples of food and beverage raw materials were discussed, such as artificial colors, flavors, and texturizing agents. Some of them are still used to this day.
- Main agricultural raw material sources for the food industry were introduced and discussed, and they account for the vast majority, especially when it comes to volumes.
- Although the food industry acquires large portions of some specialty crops, such as coffee and cacao, the use of milk is more widespread, and the largest food companies today purchase 2% or less of the total global milk production.
- It was also discussed that milk was, and is, used to manufacture products other than food, such as plastics, textile fibers, glues, ethanol, and methanol.
- Other deviations of food raw materials towards nonfood applications are sugar and other fermentable carbohydrates, which are used to manufacture ethanol for transportation and polylactic acid for packaging purposes.
- It was briefly discussed that side-streaming edible carbohydrates toward nonfood application can have a true or "perceived" price impact because competition for arable farmland. A solution of which would be the further development and smart applications of halophytic crops that would grow in agronomical worthwhile speed and yield on overly salty or seawater-flooded low-elevation, otherwise unusable, farmlands.
- The important topic of agriculture's new or newly defined role in the light of the relationship of food and health was introduced and discussed. The old Roman concept of a healthy mind in a healthy body was mentioned, and in today's everyday low prices for food state-of-mind it is difficult to appropriately adhere to this mantra.
- The all-important topic of food safety and food regulations was discussed as was the food industry's reaction, which includes the agriculture industry as well. This reaction has not always been an enthusiastic one and historically has a shaky path toward fully embracing necessary and appropriate rules and regulations.
- Food and especially food processing has become more sophisticated, much more complex and complicated. This has, in part, led to the distinction into "cheap, bad calories" and "good, valuable calories," with a rather clear distinction when it comes to paying the price for these.

- Today's consumer societies' boundaries are not any longer country boundaries and do not follow the distinction between developed and developing nations, but the fault lines are within one and the same country, irrespective of its position in the affluence scale. It was discussed that in the United States every seventh person lives on food stamps.
- Lastly, this introductory and tone-setting chapter honed in on the question of most likely drivers for change in the agriculture industry. Population growth, paralleled by the fact that arable land has not really increased lately, more sustainable innovation has to come from this industry. Increasing efficiency of water usage in any shape and form and being able to fight and eradicate pandemic plant as well as animal diseases are key topics; however, they are not the only ones.

References

Audic, J-L., Chaufer, B., Daufin, G. (2003). Non-food applications of milk components and dairy co-products: A review. *Lait*. 83, 417–38.

"Agriculture." (n.d.). Available from: https://en.wikipedia.org/wiki/Agriculture [Accessed July 23, 2016].

"Agriculture." (n.d.). Available from: http://www.merriam-webster.com/dictionary/agriculture [Accessed May 27, 2017].

Consumer Price Index. (2017). US Inflation Calculator. Available from: http://www.usinflationcalculator.com/inflation/consumer-price-index-and-annual-percent-changes-from-1913-to-2008/ [Accessed May 27, 2017].

"Ethanol fuel in Brazil." (n.d.). Available from: https://en.wikipedia.org/wiki/Ethanol_fuel_in_Brazil [Accessed May 27, 2017].

European Commission. (2007). 50 Years of Food Safety in the European Union. Available from: http://eeas.europa.eu/archives/delegations/china/documents/eu_china/food_safety/50years_foodsafety_en.pdf [Accessed May 27, 2017].

Eurostat, (2015). Agricultural production—crops. Available from: http://ec.europa.eu/eurostat/statistics-explained/index.php/Agricultural_production_-_crops [Accessed May 27, 2017].

FAO. (n.d.). Available from: http://www.fao.org/docrep/012/i1522e/i1522e02.pdf [Accessed June 4, 2017].

Harrington, R. (2015). Does Artificial Food Coloring Contribute to ADHD in Children? *Scientific American*. Available from: https://www.scientificamerican.com/article/does-artificial-food-coloring-contribute-to-adhd-in-children/ [Accessed May 25, 2017].

"History of the domestication of animals." (n.d.). Available from: http://www.historyworld.net/wrldhis/PlainTextHistories.asp?historyid=ab57 [Accessed May 27, 2017].

IRIN. (2010). Government against climate aid via World Bank. Available from: http://www.irinnews.org/news/2010/02/16 [Accessed June 5, 2017].

Kanarek, R. B. (2011). Artificial food dyes and attention deficit hyperactivity disorder. *Nutrition Reviews.* 69 (7), 385–91. doi: 10.1111/j.1753-4887.2011.00385.x.

"Klosterbrauerei Weihenstephan," (n.d.). Available from: https://www.weihenstephaner.de/en/our-brewery/history/ [Accessed May 27, 2017].

Kübler-Ross, Elizabeth. (1969). "Kübler-Ross Model." Available from: https://en.wikipedia.org/wiki/Kübler-Ross_model [Accessed May 27, 2017].

Maslow, A. H. (1943). "A theory of human motivation." *Psychological Review.* 50 (4), 370–96.

Ministry of Environment and Forests. (2009). Bangladesh Climate Change Strategy and Action Plan 2009. Available from: http://www.climatechangecell.org.bd/Documents/climate_change_strategy2009.pdf [Accessed May 27, 2017].

Perez, Marvin G. Florida's Orange Industry Is in Its Worst Slump in 100 Years." Available from: http://www.bloomberg.com/news/articles/2015-11-24/in-florida-the-oj-crop-is-getting-wiped-out-by-an-asian-invader

Silbergeld, E. K., & Anderson S. M. (1982). "Artificial Food Colors and Childhood Behavior Disorders." *Bulletin of New York Academy of Medicine.* 58 (3), 275–95.

SNAP (2017). Available from: http://www.fns.usda.gov/sites/default/files/pd/SNAPsummary.pdf [Accessed May 27, 2017].

Traitler, H., Coleman, B., & Burbridge, A. (2016). *Food Industry Research and Development: A New Approach.* Chichester, UK: Wiley Blackwell.

Trivisonno, Matt. (2017). Food Stamps Charts. Available from: http://www.trivisonno.com/food-stamps-charts [Accessed May 27, 2017].

2

Water Management in Modern Agriculture: The Role of Water and Water Management in Agriculture and Industry

Whiskey is for drinking, water is for fighting over.

—Mark Twain

Man is a complex being; he makes deserts bloom, and lakes die.

—Gil Stern

2.1 Introduction

Water is essential for survival and economic activities. As the world population grows and standards of living increase, the demand for water increases as well. But water resources are not changing, and thus a big challenge is improved access and management of water. Water management institutions and techniques have evolved throughout human history (Sedlak, 2014) and continue to evolve today. This chapter analyzes some of the challenges of water management in agriculture and provides some policy prescriptions for sustainable, equitable, and efficient water systems. Although there is a perception of an imminent water crisis, where water resources are no longer adequate for human needs, we will argue that we have water-management problem rather than water-resource crisis, and with improved policy and institutional design, we can address water challenges facing humanity. Our analysis will emphasize the importance of multiple disciplinary perspectives that incorporate understanding water technology and political economic situations and recognizing the importance of changes and adaptation over time. Finally, we will emphasize the heterogeneity and multiple dimensions of water systems and the need to develop a portfolio of solutions that address specific situations.

The first section of the chapter provides an overview of the multiple dimensions of water and patterns of changes in water-resource use in agriculture over time. An overview of the factors that affect the emergence of water institutions to explain institutional changes in water-resource management follows. This

Megatrends in Food and Agriculture: Technology, Water Use and Nutrition, First Edition.
Helmut Traitler, Michel Dubois, Keith Heikes, Vincent Pétiard and David Zilberman.

will lead to a discussion of economic-based approaches to water-management reform that will have four elements: micro-level and technological choices, water-pricing allocation and conveyance, water-project design, and water-quality regulation.

2.2 Multiple Dimensions of Water

Water resources are rich and diverse. It is important in analyzing water-resource management to identify sources of heterogeneity to develop focused strategies that address specific problems and integration that span multiple dimensions. The first distinction of water resources is by source, which can be distinguished by precipitation (rain-fed), surface water, and groundwater. Precipitation is a flow, whereas groundwater are stocks. Surface water can be divided to oceans and seas (salt), freshwater seas and lakes, all of which are stocks, and then rivers, which are flows. This overview suggests that water-resource management has to recognize dimensions of space and time as well as randomness.

Thus, the literature on management of ground- or surface water aquifers (see survey Schoengold & Zilberman, 2007) suggests that the use, pricing, and withdrawal or contribution to aquifers must recognize that

- the basic dynamics of the stock of water in an aquifer where change in stocks are equal to inflows (precipitation, runoff) minus outflows (withdrawals, leakages); and
- the net economic benefits of water use that take into account the impact both on markets and the environment, discounted over time. In analyzing water projects, it is quite important to emphasize spatial considerations. For example, Chakravorty and colleagues (1995) showed that design of water projects that have a strong element of conveyance must determine both allocation and pricing of water over space and time. Pricing may increase further away from the source and decrease or increase over time depending on changing economic conditions. These examples suggest that water-resource management is manifested by two types of outcomes: physical actions (e.g., irrigation at a given location and associated output), and monetary valuation associated with them, which are dependent on institutions, choices by agents, and technologies.

These outcomes vary by space and time, and moreover it is worthwhile to distinguish between variation within season (years) and between seasons. Therefore, water allocation at a given location is changing throughout the year and water pricing is changing within the year and over time. Furthermore, water pricing changes across location; it is not meaningful to speak about price of water in, say, California but rather speak about price at a given location at a given time. In the case of groundwater, a major component of the cost to farmers depends to a large extent on the depths of aquifer and the cost of pumping. The cost of pumping from a deep aquifer or lifting water to a high plateau (say lifting more than 1000 feet) may be ten times or greater than the cost of pumping from a shallow well (below 100 feet). Similarly, the cost of water for irrigation in the late winter or spring may be much lower than late in the season. It will be much higher during droughts than during wet years. Thus, within agriculture, water prices vary by order of magnitude of 10 and across locations and over time (Sunding 2000).

Water resource management can be divided into three categories: (1) supply management that may include investments in water storage, withdrawal (including desalination and recycling), and conveyance, (2) demand management that includes establishing allocation rules, through for example queuing and pricing, and investing in water consuming or saving activities, and (3) water-quality management, including both investment in physical activities (e.g., filters, treatment) and regulations. The water economic literature (Schoengold & Zilberman 2007) develops criteria to develop strategies and they will be discussed herein.

It is also useful to distinguish between different water-use activities. The first distinction is between consumptive and nonconsumptive uses. Consumptive uses include agricultural, industrial, and municipal applications, and nonconsumptive include environmental, hydrological, and recreational. Different uses have different valuation strategies, and the pricing varies across space and time. Water-strategy management must recognize the multiple uses of water for system design and valuation.

To understand water policies, we need to incorporate political dimensions to heterogeneity of water-resource availability. Some countries, like Canada, are water-rich, and others, like Jordan, are water-poor. This results in different pricing schemes, innovations, and institutions (Dinar & Pochat, 2015). For example, Israel and Cyprus, who are relatively water scarce, have strong monitoring mechanisms, groundwater management, and high water pricing to all sectors, including agriculture, that may result in modern irrigation technologies and emphasis on high-value crops. Heterogeneity may occur within a country. Although Canada has overall abundant water, it has regional deserts that require reliance on irrigation to maintain agricultural systems. The spatial differences in water situations may result in different priorities; some regions may emphasize investment in flood control and others in systems to adapt to droughts.

Heterogeneity and diversity may lead to conflicts. For example, there are continuous conflicts over management of the Nile, as well as the Mekong and Himalaya regions (Warner, Zeitoun & Mirumachi, 2014). But regional differences may also lead to transfers that serve as engines of growth, but also lead to conflicts and controversies. California has one of the most advanced agricultural sectors in the world, but most of the farming is done in desert regions using water exported from wet regions (Israel &Lund, 1995). Farming in the desert allows California to produce high-value crops with less reliance on chemical pest control than farmers in other regions. At the same time, there have been concerns about the environmental side effects of water transfer projects and the damage caused to the source regions (Reisner, 1993). Thus, optimal design of water projects that balance economic well-being with monetary and social costs is a topic discussed in this chapter.

Although water has multiple uses, the majority of diverted fresh water ("blue" water) is used for irrigation at roughly 90% (Siebert & Döll, 2010). Irrigation has played a crucial role in enhancing supply of food to a growing global population. While global population has tripled since 1950, food production per capita has increased faster (Hazell & Wood, 2008). At the same time, productivity has increased faster than agricultural land. Much of the increase in productivity is through increased use of fertilizer and improved varieties and other practices. Furthermore, land available for irrigation increased from 63 million hectares (Mha) in 1900 to 306 Mha in 2005, with 46% in arid regions, 26% rice, and remaining 28% (Siebert et al., 2015). In addition, because irrigated land can achieve more than one harvest per year, harvested irrigated area is actually higher, at 421 Mha (Rosegrant, Ringler & Zhu, 2009).

During the same time, global population shifted with arid regions growing from 19% to 26% of people, and decreasing in wet regions from 46% to 35%. Globally, irrigated agriculture represents approximately 19% of land in use, while growing 43% of cereal production (Siebert & Döll, 2010; Bruinsma, 2009). Siebert and Döll (2010) also computed that absent irrigation would reduce production by 47%, which is 20% of total global production, with much higher losses in North Africa (66%) and Southeast Asia (45%).

Bruinsma (2009) shows that yield increases are expected to play a significant role (77%) in meeting growing food demand, with crop intensification and arable land expansion to a lesser extent. Further, while actual crop yield is close to 60% in North America and Western and Central Europe, it is roughly 30% in sub-Saharan Africa as well as Central America, Eastern Europe and the Caribbean (FAO, 2012). Further, roughly 46% of global arable land is not suitable for rain-fed agriculture (Valipour, 2015). Therefore, modern irrigation systems play a key role in achieving yield enhancement.

Furthermore, irrigated agriculture often produces higher value crops, including fruits and vegetables, with the value of output per hectare of irrigated land at least 6 times greater than rain-fed (Schoengold & Zilberman, 2007; Makombe, Kelemework & Aredo, 2007). Although agricultural land may contribute to pollution through runoff and soil erosion, the higher productivity has slowed expansion of agriculture and prevented deforestation and other negative side effects (Ruttan, 2002). In addition, with 40% of Africa and Asia and 63% of Southeast Asia in the agricultural labor force, higher value production provides important income-generating activities (Valipour, 2015).

The expansion of irrigation and water projects has been costly. Water projects have been costly and suffer from a high degree of cost overruns from design flaws and changes, misallocation of resources, and delays (Inocencio, 2007). They have also resulted in human and environmental costs, for example, the human cost of water projects in China like Three Gorges Dam (Jackson & Sleigh, 2000; Scudder, 2012). In many regions of the world, irrigation has been excessive. Irrigation has contributed to increased salinity and waterlogging (Wicheins, 1999). Altogether, there is a perception of water supply and availability crisis. But with the combination of policies, institutions, and technologies, water can be managed in a sustainable way and meet the needs of a growing world. Before we provide specific solutions, we need to understand the basic factors that contribute to the emergence of existing water-management strategies.

2.3 On the Evolution of Water Institutions and Policies

Economists have recognized the importance of institutions in shaping resource allocation and the important role that political economy considerations play in development of institutions, policies, and incentives (Acemoglu, Johnson & Robinson, 2005). Institutions emerge in response to economic, technological, and political conditions, and changes in conditions require institutional changes. However, institutional transitions are difficult to implement and they are the major challenge of water-policy reform. Three major factors are likely to affect water institutions and organizations; they are:

- relative scarcity of water,
- government ability to tax and finance public projects, and
- policy objectives of government and society. The history of U.S. water policies can illustrate these points.

In the early stages of evolution of U.S. agriculture (1800s), the government was financially weak, water and land were relatively abundant, and there was desire for expansion and growth. In this period, the government introduced a system of homesteading that endowed settlers with ownership of land. It also established a similar system of water rights in the West (the prior appropriation system) that functions as a queuing system, where seniority is based on first-in-time, first-in-right, coupled with "use it or lose it" (Cochrane, 1979). Under this system, private individuals developed and invested in water projects that diverted bodies of water to irrigate relatively arid zones. In California, early water projects were developed for gold mining and later on, farmers established water districts for irrigating crops (Mercer & Morgan, 1991). As the government became richer, was able to accumulate resources through taxes, and while water resources were abundant and the desire for growth continued, the government invested in water projects and subsidized expansion of irrigated agriculture. This is what happened in the United States during the early part of the 20th century, where the Bureau of Reclamation, Army Corps of Engineering, and the State of California built water infrastructure and provided the water to farmers at subsidized prices. The same processes have occurred globally where water projects have been established throughout the world.

Water projects serve multiple purposes. Some of them are important for flood control, others serve to provide electric power, and many provide water for drinking and irrigation. Although the economic returns to many water projects have been questionable, and there has been a consensus that water projects have been excessive, there is no doubt that without them, cities like Los Angeles and the aerospace industry in Seattle wouldn't have been feasible (Reisner, 1993). However, changes in economic conditions are leading to changes in new institutions and arrangements. For example, increased water scarcity may lead to development and adoption of improved water-use efficiency technologies and practices. It may also lead to the emergence of increased water trading. As government's capacity to raise taxes and resources is reduced, it may lead to increased privatization of water resources and infrastructure. There is growing tendency of countries to engage private sector to invest in water-supply expansion, even though pricing may be regulated by the public sector (Dinar & Subramanian, 1997). Concern about depletion of groundwater resources may lead to government monitoring and regulation of groundwater pumping. Environmental consideration of pollution may lead to water-quality regulations, and concern of endangered species and wildlife may lead to regulation of water extraction and conveyance. Similarly, concern about equity and welfare may lead to regulated prices, for example tiered pricing (Schoengold & Zilberman, 2014).

There are many barriers to transitions in water institutions. Although there has been growing awareness, for example, of the benefits of transition from water rights to water markets in California and elsewhere, the actual transition hasn't occurred in many places (Dinar & Pochat, 2015). One barrier is the transaction cost associated with design and implementation of new institutions. For example, markets may require more concise measurement of water flow and investment in improved conveyance (Shah, Zilberman & Chakravorty, 1993). Another is political economy constraints, and especially when assuming higher degrees of loss aversion by interested parties, losers from a transition will put much more effort per dollar to block the change (Rausser & Zussman, 1991). There is also concern of uncertain, irreversible outcomes that may delay introduction of change (Dixit & Pindyck, 1994). Nevertheless, transitions have occurred, and quite frequently during times of crisis. Droughts led to water reform, including increased reliance on trading and water markets, including Australia, Israel, United States, Spain and Canada (Dinar & Pochat, 2015).

Although thus far we have emphasized large institutional and policy reforms, much of the transitions in agriculture occur and start at the micro-level: consumers, firms, and farms. These micro-level decisions are driven by technology and policy changes. We will emphasize changes in water use at the farm because agricultural water is the largest segment of freshwater consumption.

2.4 Reforming Water-Resource Management at the Micro-Level (Farm and Field)

We will start with analyzing changes in agricultural practices and technologies in irrigated agriculture. Irrigation is one of several agricultural inputs and can be used to extend the growing season, supplement rainfall, and make farming possible in arid regions (Boggess, Lacewell & Zilberman, 1993). For several thousand years, irrigation took the form of gravitational methods of flood and furrow dating back to early agriculture in Egypt and Mesopotamia. Irrigation technologies have changed significantly over the past 60 years since the introduction of sprinkler systems in the 1950s and drip in in the 1960s. But the adoption of these technologies has been relatively slow. For example, Taylor and Zilberman (2017) show that in California, while drip irrigation was available beginning in 1968, it took until the drought of 1986–88 to reach 15% of irrigated agricultural area, and today it is 40%. Globally, drip represents less than 10% of irrigated land, sprinkler is about 30%, and the remainder is furrow and flood. Further, the use of information technology and genetics are two important complements to modern irrigation systems.

There is a wide literature on the *economics of irrigation*. Irrigation allows for the expansion of agriculture to arid regions that may be productive with irrigation. Adoption of technologies, like pumps, has allowed for increased productivity on farms as well as increasing the area of irrigated agriculture. Agriculture is the output of multiple inputs, often complementing one another (e.g., soil, water, fertilizer, seeds). Thus, increased productivity of fertilizer may increase the use of water as long as prices are the same. The adoption of irrigation leads to indirect gain in output when it leads to the increased use of other inputs. In some cases, though, the yield effect of improved irrigation may lead to increased supply and reduced prices, and that may curtail overall agricultural land (Boggess, Lacewell & Zilberman, 1993).

At the same time, improvements of other inputs may lead to increased use of irrigation. For example, improved seed varieties may increase the use and benefits of irrigation, both directly and indirectly because of complementarity. Plant breeding through various techniques, from selective breeding to transgenic varieties and gene editing (CRISPR), enable the selection and modification of seed traits most desired for specific conditions as well as qualities (e.g., taste, flavor, texture, and transportability). The Green

Revolution witnessed the simultaneous introduction of modern seed varieties with the use of irrigation (Feder, Just & Zilberman, 1985). The introduction of transgenic varieties, in cases where they increase yield (Klümper & Qaim, 2014) can also lead to adoption of irrigation and further increase supply (Barrows, Sexton & Zilberman, 2014). However, improved seed varieties may also directly augment the impacts of irrigation through drought-tolerant traits and ability to deal with low water quality. For example, new varieties in sugarcane and maize can increase tolerance of plants to drought or higher levels of saline, thus reducing yield losses (Kumar et al., 2014; Kulanyi, 2016).

The literature on the economics of *irrigation technologies* analyzes the diffusion of the various forms of the technology to explain the impact of adoption of various irrigation technologies, the factors that affect adoption, and empirical patterns of adoption (see survey in Taylor & Zilberman 2017). Irrigation technologies impact agricultural productivity by directly affecting the productivity of water use, as well as the productivity of other inputs. Agricultural inputs interact with soil and weather to produce crops. However not all applied inputs are taken up by crops. For example, some applied water may end up as runoff or deep percolation. Thus, it is important to distinguish between applied inputs and effective inputs (i.e., used by the crop). Although the water-use efficiency (i.e., effective/applied) in California of gravitational methods is roughly 0.6, it may rise to 0.8 for sprinklers and 0.9 for drip. There is also significant heterogeneity of land quality *between* regions. The input use efficiency of traditional technologies is much smaller on sandy soil or steep terrain. The residue, in this case, may be drainage that causes waterlogging and reduce productivity (Caswell, Lichtenberg & Zilberman, 1990). Thus, modern irrigation technologies (e.g., drip, microsprinklers) increase water-holding capacity of soils. With drip irrigation, one can also apply fertilizer and other chemicals (e.g., fertigation and chemigation) and increase their input use efficiency. Another advantage of drip irrigation is that it allows more precise application of water and other inputs throughout the season. With traditional technologies, large quantities are applied a few times a year, but with drip irrigation, application is more precise and results in increases in yield above and beyond the water-holding capacity (Shani et al., 2009).

Adoption of modern irrigation technologies tend to increase yields and in many cases save water and reduce drainage and runoff. The yield effect is higher and input savings are lower in locations with low water-holding capacity. However, there is a trade-off between augmentation of yield, saving of water and less pollution, and cost of equipment. Modern irrigation technologies are more efficient but also more costly. Therefore, adoption is not optimal everywhere. It is more likely to occur at locations with low land quality, expensive water, and high-value crops. Furthermore, the savings of inputs, like fertilizers, can also lead to adoption of drip and other irrigation technologies. It is important to distinguish between adoption of irrigation that leads to switching technologies in production of a given crop (intensive margin effect) and adoption of technologies that change the use patterns of land (extensive margin effect). For example, the introduction of drip irrigation in California allowed expansion of avocado and grape production to steep hills where they hadn't been produced before. Furthermore, drip was first introduced in high-value crops, like avocados, strawberries, and vegetables for fresh markets, but over time, as the technology improved and cooperative extension modified cultural practices to accommodate drip and modern technologies, it was adopted on lower value crops, like processing tomatoes. Furthermore, adoption tends to occur during periods of high water scarcity (e.g., drought periods) and adoption is more likely in locations that are closer to markets, and locations with higher water prices, for example, locations that rely on groundwater from deeper wells (Taylor & Zilberman, 2017; Sunding, 2000; Baerenklau & Knapp, 2007).

Adoption of modern agricultural technologies is also associated with shifts in crop selection. Modern irrigation technologies are more likely to be adopted with high-value crops. Moreno and Sunding (2005) documented the importance of environmental conditions, like slope, that are crucial as factors in adoption of irrigation technologies. But the relative prices between and among crops and technologies affect choice of technologies, crop selection, and the combination of crop/technology choices. The adoption of irrigation technologies depends not only on average prices, but it may also increase when water prices and availability are uncertain. Furthermore, when the existing technology is older, replacement is more likely with a modern technology (Baerenklau & Knapp, 2007).

One of the major benefits of modern irrigation technologies is that they are part of, and enhance, the adoption of precision agricultural techniques. Precision agriculture seeks to increase net benefits of agriculture through the use of information technology, inputs, and management decisions over space and time (McBratney et al., 2005). There is a high degree of heterogeneity of land productivity even *within* fields, and precision agriculture, including drip irrigation, can modify input use within a given field. Modern irrigation is more often adopted in location with heterogeneous land quality within fields (Feinerman, Letey & Vaux, 1983). There is also significant heterogeneity within seasons and ability to optimize water application throughout the season can significantly increase productivity.

Development of weather stations that can provide precise information on evapotranspiration, and other weather indicators, over space and time can lead to increased productivity. However, the effectiveness of these technologies increases with the use of modern irrigation technologies. Thus, the California Irrigation Management Information System (CIMIS) was a key component in the diffusion of drip irrigation in California because it provided data on evapotranspiration that allowed farmers to adjust water scheduling and quantity to local weather conditions (Taylor & Zilberman, 2017). CIMIS has also played an important role in pest-management systems by providing data needed for timing of pesticide application (Parker et al., 1996) and the complementarity of pest control and water led indirectly to increased likelihood of adoption of drip.

Finally, water-use efficiency may also be enhanced by land-management practices, rather than irrigation technologies. For instance, techniques such as terracing and laser-leveling increase the efficiency of applied water, nutrients, and agro-chemicals (Jat et al., 2006) and have similar effects to modern irrigation technologies. So, these techniques may actually substitute modern irrigation technologies. For example, laser-leveling throughout the western United States increased productivity of irrigation (Boggess, Lacewell & Zilberman, 1993). But in areas with sandy soils, there is a higher tendency to adopt more advanced irrigation technologies, like center pivot or low-energy precision application (LEPA). Land-management practices like terraces can improve the productivity of rain-fed agriculture. Similarly, soil augmentation with substances like biochar may increase the water-holding capacity of soil, as well as nutrient availability to crops (Akhtar et al., 2014). Conceptually, biochar is a land quality augmenting technology for rain-fed agriculture, while drip enhances land quality in irrigated farming.

Although irrigation has spread throughout the developing world, adoption of modern irrigation technologies has lagged significantly compared to California, Israel, and Australia. One major reason is that in developed countries, the adoption of modern irrigation technologies has been associated with adaptation of crop systems to the specific features of the technology that became feasible because of engagement of cooperative extension and the higher education of farmers. Thus, adoption of modern irrigation technologies in developing countries will require strengthening of overall agricultural outreach and education systems.

2.5 Reforming Regional Water-Allocation Regimes

Efficiency of agricultural water use is dependent on the pricing of variable inputs as well as the availability and timing of water. These factors are the outcomes of regional water institutions and policies. For example, the literature suggests that a critical element that affects water-use efficiency is the ability to trade water, and the allocation of water among farms. This is especially true when it comes to surface water. As we mentioned previously, in many regions of the world, water is allocated by water-rights systems (e.g., the prior appropriation system). A strict interpretation of this system disallows water trading because once farmers sell their water to others and don't use it, they lose their rights to the water. Thus, under water-rights mechanisms, the gains from trading are lost. Transition from water rights to water trading would allow higher efficiency.

The gain from transition from queuing systems to water trading depends on the scarcity of water, which can be measured by the water-to-land ratio. In cases where water is abundant, farmers may not use all their agricultural land and apply water with gravitational systems. Generally, these systems are used to establish water rights. But as demand for agricultural products produced on the land, and the potential for water use increases, the gain from switching from queuing to trading is increasing. Higher water scarcity is likely to result in higher prices of water once markets are introduced and may lead to switching to high-value crops, as well as modern irrigation technology. Dinar and Letey (1991) have shown that introduction of water markets are likely to increase the adoption of modern irrigation technologies in California. The transition to water trading is likely to generate economic welfare gains, which is larger when water scarcity is higher. But this transition may require significant transition costs, in terms of establishing water rights, quantifying water movements, and reducing barriers to moving water over space when needed. Thus, if the surplus gain is greater than the transaction cost, it is desirable (Schoengold & Zilberman 2007).

The introduction of trading may encounter significant design challenges (Chong & Sunding, 2006; Schoengold & Zilberman, 2007). One problem is the "third-party effect." Because efficiency of traditional irrigation can be quite low, runoff from the system generates benefits to other farmers and environmental uses (e.g., wetlands). So, if the owner of water rights is allowed to sell water applied in the past, the third-party beneficiary downstream will lose. Therefore, some water reform that introduces trading allows for the transfer of only a certain percent, say 75–80%, to buyers. The rest of the water is transferred to a local authority to facilitate use by third parties. A second issue is who owns the water. In some proposed designs, the government that originated the water rights now have the right to sell the water through bidding, etc. However, such designs will encounter significant objection from the current water-rights owners, and thus it may be politically more feasible to move from water rights that disallow trading to transferable rights (even though some percentage of the water cannot be sold because of third-party considerations).

A third issue is whether to allow permanent or temporary sale of rights. Temporary transactions allow adjustment to crisis situations. For example, during a drought, growers of annual or low-value crops may sell their rights to cities or growers of high-value crops. This was the case in California's drought of 1991–93 (Zilberman et al., 2011). However, permanent transfer of rights enables the buyer to make long-term investments and result in higher economic benefits. A fourth issue is whether transfers will be restricted within a water basin or more broadly. Moving water away from a region may benefit the sellers of the water rights, but harm the rest of the economy. But moving water across regions may result in significant benefits socially. For example, the city of Los Angeles, agriculture in the Central Valley of California, and the wine industry in Chile wouldn't be possible without interregional transfer of water.

A key decision to be made once water trading is introduced is the price of water. From an economic perspective, the price of water should take into account the social cost associated with supply of water. Water-supply systems have several elements, including extraction, conveyance, and distribution across locations. In addition, the extraction of water may have environmental costs, and if the water is pumped from an aquifer, it may also have future scarcity costs. Therefore, in a socially optimal equilibrium, water price should be set to the sum of marginal extraction, conveyance, distribution, externalities, and future costs. In equilibrium, this price should be equal to the marginal benefit of the buyer (Tsur 2009). In many water systems, the externality costs are ignored, and water prices are set lower than optimal. In other cases, water prices are heavily subsidized and are much below the social cost. This underpricing results in excess use of water, underinvestment in conservation, modern irrigation technologies, and overproduction of low-value crops (Johansson et al., 2002).

Another challenge in designing water systems is the spatial distribution of water. Conveyance of water over space has resulted in significant losses through deep percolation or evaporation due to underinvestment in the conveyance system. Thus, it is important to design conveyance systems that will maximize expected net present value of social benefits from water use minus conveyance costs. Furthermore, water pricing may need to vary over space to reflect increasing transmission costs away from the source. Establishing optimal conveyance and pricing systems frequently requires an institutional set-up, namely a water user association that maximizes overall benefits (Chakravorty, Hochman & Zilberman, 1995). There is significant social cost for underinvestment or conveyance or for the use of uniform pricing for water over space. These suboptimal systems result in lower production levels, higher output price for consumers, greater conveyance loss, less adoption of modern irrigation technologies, and shorter canals. Uniform pricing results in underproduction close to the water source and overproduction further away (Chakravorty et al., 2009).

Appropriate water pricing is important in the case of groundwater. In many regions of the world groundwater aquifers are shared by many producers, thus their use patterns may suffer because of the "tragedy of the commons" problem. Namely, each individual user doesn't take into consideration how their action will affect the overall dynamics of the source in the future. Thus, early in the utilization of an aquifer, cost of pumping may be low, farmers may overuse water, use cheap and inefficient pumping equipment, and that will result in the slow depletion of aquifers. For example, levels of aquifers have declined significantly in India because of overpumping (Roy & Shah, 2002). One solution is that the government charge a price per unit of water pumped reflecting the social cost of pumping. An alternative policy is creation of tradable groundwater quotas that limit the amount of pumping in a given region per time period. To introduce and enforce such policies, governments need to monitor groundwater use. Indeed, in some countries, for example Israel and Australia, a key element of water reform has been monitoring and control of groundwater use. These policies have led to improved productivity of water (Dinar & Pochat, 2015).

Precipitation events are random and therefore the same region may experience water shortage in some periods and flood in others. One way to address this system is through conjunctive use of surface- and groundwater where water is stored during periods of excess rainfall and pumped during periods of shortages (Gemma & Tsur, 2007). Conjunctive use systems require management of water-containment facilities and varying water prices over time to take account of water availability and economic conditions. Thus, efficient water pricing needs to vary both over space and time.

Thus far, we emphasized principles of efficient water pricing that aim to maximize the net social benefit of water resource utilization. However, other key elements must enter into pricing considerations, namely equity and finance. There are differences in ability to pay for water among consumers, and the use of marginal water pricing may prevent many smallholders or poor consumers from access to water for production or even direct consumption. One solution is tiered pricing where several pricing levels are established. They include a "life-line" of low price that enables subsistence and higher levels including the marginal cost for larger users. The efficiency loss as a result of tiered pricing depends on the distribution of income among water users and the design of the tiered-pricing system. The resource allocation is likely to be more efficient if the life-line price is low, but even so the amount of water supplied this way is small (Schoengold & Zilberman, 2014). In some cases, tiered pricing may be accompanied by subsidies that allow the adoption of more efficient water pumping and use technologies. Because water users may be lacking the knowledge and capacity to monitor water situations and design optimal water-use systems, water agencies provide informational assistance (Dinar & Pochat, 2015).

Water agencies have significant fixed costs that reflect capital, fixed assets, and operational costs as well as costs associated with water extraction and delivery. Marginal cost pricing based on the cost of water may not cover the fixed costs. To cover these fixed costs, many water agencies are using formulas combining fixed charges plus volumetric charges. Given fluctuations in availability and cost of obtaining water, developing pricing formulas that balance efficiency, equity, and financial stability is a challenge, and therefore different agencies around the world have used different approaches with varying success. The operation and financial stability of water utilities tend to improve with better monitoring of water use, improved maintenance of conveyance efficiency, and assistance to users to improve their water-use efficiency (Dinar & Pochat, 2015).

2.6 Improved Water Project Design

Development of water resources requires significant investment in infrastructure. Traditionally the emphasis was on: (1) dams for flood protection, hydroelectric power, and irrigation, (2) storage facilities, (3) conveyance canals, (4) pumps, (5) distribution facilities, and (6) water-treatment facilities. In recent years, there has been growing emphasis on desalination plants and recycling and reuse facilities. Historically, decisions about water projects have been political and in many cases, reflect pork-barrel policies, with many water projects that were unsound economically (Reisner, 1993). The notion of benefit-cost analysis was introduced to develop more economically sound criteria for evaluation of water projects. The U.S. government agencies (e.g., Army Corps of Engineers, Bureau of Reclamation) develop guidelines for product evaluation relying on cost-benefit analysis to evaluate water projects (Armah et al., 2009).

The benefit-cost analysis for water-project evaluation aims to maximize expected net present value of economic and environmental benefits minus costs of proposed water projects. However, development of criteria is quite challenging because it requires monetizing nonmarket benefits and costs (Freeman, Herriges & Kling, 2014). Further, Dixit and Pindyck (1994) suggest that a key question in project evaluation is not only whether to produce a product but also the timing of it. They suggest, instead to use a real-option approach for project evaluation; this approach may lead to delay in execution of a project with positive expected net present value and even cancellation to take advantage of emerging information. The timing and magnitude of water projects can also be affected by considerations of economies of scale in project construction and the possibility of costly restoration (Zhao & Zilberman, 1999). For example, economies of scale may lead to oversizing a project that may not be used for an initial period and may even be eliminated if new technologies or preferences emerge and show the project to be not worthwhile. Traditionally engineers would come up with water project designs that emphasize structural solutions; today there is a growing recognition of the need for multidisciplinary teams, including economists and other social scientists that include nonstructural solutions. For example, one may reduce the need for large infrastructure investment if part of the design includes an efficient water pricing system (Gleick, 2000).

Although traditionally expansion of water supply was done through diversion of surface water as well as utilization of new aquifers, recently there are two new sources of supply expansion. First is desalination of seawater and other saline water bodies. These technologies are energy intensive, but through technological change have become much cheaper, and even economically viable in regions with high water scarcity for municipal, industrial, and high-value agricultural crops. The costs of desalination have declined in Israel and have become a major source of drinking water and agriculture, and even allowed for restoration of aquifers that suffer from seawater intrusion (Spiritos & Lipchin, 2013).

Desalination is a continuously improving technology and must meet challenges regarding its greenhouse gas emissions and other environmental impacts. There are also growing opportunities to use renewable energy in the process (Kalogirou, 2005).

Recycling of urban water is another source of water supply, and assessment of such projects must take into account both economic benefit as well as externalities (Hernández et al., 2006). One of the main challenges in designing desalination projects is identifying the optimal location to reduce conveyance and transport costs as well as identifying the institutional arrangements for recycled water distribution. Recycled water has become a main source of agricultural water supply in Israel and its use is growing in Spain and Australia (Dinar & Pochat, 2015). The combined use of desalinated and recycled water will allow both expansion of existing water supply and also enable reducing overpumping and overcome problems of groundwater desalination. In some places, it may allow replacement of surface water sources and using these resources instead for environmental amenities.

2.7 Improved Water Quality

Maintaining water quality is a major challenge globally. Water-borne diseases account for roughly 4% of all deaths in developing countries (World Health Organization, 2009). Further, 41% of people lack access to basic sanitation, and unsafe environmental conditions cause 25% of all children deaths. Water is contaminated by human, animal, and industrial by-products and design of policies to reduce environmental health risks has to control the risk-generation process that includes several processes: disposal, transfer, exposure, and vulnerability to waste. Each of these processes can be affected by policy. The disposal of toxic residues can be reduced both by reducing their use (e.g., by using more precise irrigation technologies, switching away from more toxic chemicals), which can be affected by regulations and incentives (e.g., pollution tax). The transfer of waste can be controlled by establishing effective sewerage and drainage systems, and either recycling the waste product or depositing them in areas where they cause minimal damage. The exposure to low-quality water can be reduced through filtering and control of water supplies. Finally, vulnerability to contaminated water can be affected by medical treatment (Lichtenberg, 2010). Optimal policy design has to balance cost and benefit and recognize heterogeneity among locations, randomness of damage, and the economics of water clean up and recycling. Sometimes it is optimal to maintain several water systems providing different levels of quality, for say drinking, industrial and agricultural use (Zivin & Zilberman, 2002).

Management of animal waste is, in particular, a major water quality challenge. Animal waste can be a source of acute risk as well as continued deterioration of groundwater by accumulation of nitrates. As contract farming becomes a major part of animal agriculture, developing joint liability for animal waste by both farmers and integrators, can be a major priority in overcoming animal waste challenges (Ogishi, Zilberman & Metcalfe, 2003). Management of animal waste needs to vary by location and treat different types of waste differently (Innes, 2000). It also should recognize that waste products can be used as benefits providing inputs in other activities, for example as fertilizer (Iho & Laukkanen, 2012). Research that can enhance the value of waste products is beneficial both because of the direct gain from the product as well as reduction in externalities. Such research may be underprovided if penalties on externalities are not introduced. Thus, in the same way that society gains from introduction of trading and appropriate pricing of water quantity, it benefits from introducing polluter-pays principles and social valuation of water quality.

Water-resource management has a significant impact on the environment. Diversion of water from the river can affect fish and run-off contaminating bodies of water can increase risk for public health and reduce supply of clean water. Although pollution taxation is desirable, it may not be feasible politically. One mechanism to enhance water quality is payment for environmental services where agents that prevent pollution or improve water quality receive payment. One reason for the need for this mechanism is that frequently polluters have property rights that allow them to pollute, and thus it compensates for not using their rights. These payments have been used to assure water qualities to cities and access to drinking water for wildlife, among many other activities (Bulte et al., 2008).

One of the major challenges in water-resource management is that much pollution is from nonpoint sources. Assigning responsibility for specific pollution is impossible, but policies that take advantage of other information, like input use and environmental conditions, can be used to estimate attributable pollution to a source. In some cases, it is useful to rely on communal responsibility for pollution control when central authorities have limited information on individual behavior (Segerson, 1988; Freeman, Herriges & Kling, 2014). The notion of nonpoint pollution is technology dependent. Technological process that enables better monitoring of individual action allows for policy design that leads to adoption of precision monitoring and improved attribution to polluters, and thus increases efficiency of water-quality control (Millock, Xabadia & Zilberman, 2012).

2.8 Climate Change

Agricultural water resources and the environment will be stressed by climate change. Climate change has four major impacts on agriculture. First, temperatures will increase and be more unstable; in a sense it can be viewed as migration of weather from the equator to the poles. For example, some parts of Oklahoma and Texas will become less fit for farming while parts of Canada become more fertile. The same will apply to parts of China and Russia. Extreme weather means more droughts and more floods. A second problem is melting ice, which may result in floods and lowered capacity of water storage (Barnett, Adams & Lettenmaier, 2005). Finally, rising water levels may eliminate agriculture in fertile, coastal regions, while also contaminating groundwater aquifers (Lipper et al., 2017). Each of these impacts will require different adaptation strategies, but these strategies may be related, as the new literature on climate change and water suggests (Zilberman, 2015). "Migrating weather" may require changes in crop, relocation of farms, adoption of modern technologies, and immigration. Much of the adaptation that affects water systems will be indirect; changes in crop may reduce or increase water productivity. In some areas, there will be needs to increase irrigation, while in others, irrigation systems will be scrapped. Thus, migrating weather will require significant reassessment of agricultural and irrigation infrastructure. In some areas, climate change will lead to significant decline in agriculture and water use, and in others the expansion of agricultural production will provide opportunities to modernize water systems.

The basic principles of water system design mentioned in this chapter should guide the introduction of these changes. In particular, the ability to trade water will reduce the necessity to migrate or significantly reduce economic welfare in regions with reduced water supply (Kahil, Dinar & Albiac, 2015). The ability of water-constrained agriculture to adapt to climate change will depend on building resilience through improved crop varieties more tolerant to extremes, as well as exposure to new pests. Thus, the tools of modern biology will be important in enabling agricultural regions to adapt to changing weather and water situations.

Rising water levels will lead to eliminate agricultural production in some regions, while in others in costly adaptation. This will include development of technologies to address soil salinity and modernizing roads and infrastructure (Dasgupta et al., 2015). A major challenge will be to obtain the resources to enable regions to invest in infrastructure associated with adaptation to climate change. Allocation of resources to affected areas should be compared to the cost of migration and resettlement of people living in these regions.

Both adaptation to rising sea level and melting snow requires significant investment in infrastructure. The storage capacity provided by snow will need to be replaced by physical storage and the increased likelihood of flooding will require investment in flood protection activities, as well as technologies to improve water-use efficiency (Bates et al., 2008). However, increased storage capacity and water-use efficiency are not necessarily substitutes. When productivity of water doesn't decline significantly with the amount of water use, increased water-use efficiency may lead to increased storage. Thus, with climate change and additional infrastructure for water storage, regions that have a potential to increase their production, for example by export, may actually expand their irrigated agriculture, while other regions that face limited demand will need to curtail activities (Xie & Zilberman, 2016).

One of the major challenges in adaptation to climate change, especially in the case of water-resource management and infrastructure, is uncertainty. Our ability to predict the extent and impact of climate change is limited; to design new infrastructure (e.g., storage facilities), it is important to have good spatial and temporal patterns of changes implied by climate change. This uncertainty may lead to reluctance to invest by risk averse decision makers, which may increase the cost of tackling climate change events once they occur. Therefore, there is a significant gain from research that leads to better understanding of the dynamics of climate change that will trigger more precise responses.

Given the high costs associated with climate change, there is significant social benefit to mitigation activities that aim to delay and reduce magnitude of its impacts. Mitigation can be enhanced by agricultural activities that sequester carbon as well as by substitution of fossil fuels with renewable sources. This may increase the demand of agricultural activities, and thus the demand for water and need to further enhance water use efficiency. Msangi and colleagues (2007) present an early assessment of biofuels on water resources. Assessing the impact of biofuels on climate change has to incorporate their indirect effect on natural resource use, which is subject to a growing literature, and when the direct greenhouse gas reduction as a result of biofuels is significant, as the case in sugarcane shows, the indirect effect cannot override it (Khanna & Crago, 2012). Emissions reduction through agricultural water use can be enhanced by adoption of improved seed varieties.

2.9 Summary and Major Learning

With population and income growth along with concern of climate change, there is a growing perception that water will be a binding factor on economic and agricultural growth. This chapter argues that water problems can be solved by better management and taking advantage of technological development. Reform of water-management practices can increase both economic and environmental benefits of water. The following was discussed and analyzed in some detail in this chapter.

- A key to water reform is to introduce policies that will enhance incentives to increase water productivity and decrease pollution. Irrigation is crucial for production of food, but it is mismanaged. Allowing trading that allows water allocation between regions and pricing, and other mechanisms, that will allow water users to pay for the consequences of their action, both in terms of cost of water delivery system and impact on the environment—and future resource availability is likely to lead to increased adoption of efficient water technologies and use of irrigation when it provides value.

- Water productivity can be enhanced by improved use of complementary inputs, including seed varieties, fertilizers, and other inputs. So investment in research and development and extension services, improved infrastructure, and regulatory systems that increase overall agricultural activity will improve productivity of water use.
- Increase in water productivity, in turn, may reduce water use over time. Improved information technologies should be taken advantage of in design of water, and agricultural technologies in general. Expanding the integration of global agriculture within the knowledge economy can lead to increases in precision and efficiency.
- Government policies and water projects will continue to play a major role in water systems. Water-project design should be a multidisciplinary exercise that incorporates structural as well as nonstructural measures.
- Projects should be evaluated using benefit-cost analysis that takes into account environmental considerations, emphasize adaptive learning, and is opportunistic with respect to timing of investment. Policy design should emphasize equity considerations and develop mechanisms, like tiered-pricing, as well as subsidies for adoption of technologies that allow lower income households and farmers to access and benefit from water.
- Measurement of agricultural systems is essential for effective policies and water policy should emphasize improving water quality and reducing the environmental side effect of water. Getting the price right is important for managing water quality as much as quantity. When possible, it is useful to adhere to the polluter-pays principle or to use mechanisms like payment for ecosystem services to reduce pollution.
- Control of water quality should recognize market structure considerations as well as information imperfections and develop mechanisms and technologies to overcome these constraints. Water policies must be modified periodically to take advantage of new technological capabilities that increase precision and attributional capability.

References

Acemoglu, D., Johnson, S., & Robinson, J. A. (2005). Institutions as a fundamental cause of long-run growth. *Handbook of Economic Growth.* 1, 385–472.

Akhtar, S. S., Li, G., Andersen, M. N., & Liu, F. (2014). Biochar enhances yield and quality of tomato under reduced irrigation. *Agricultural Water Management.* 138, 37–44.

Armah, J., Ayan, H., Bernard, C., Blumenthal, A., Fortmann, L., Garretson, L. R., Godwin, C., et al. (2009). Principles and guidelines for evaluating federal water projects: US army corps of engineers planning and the use of benefit cost analysis. A Report for the Congressional Research Service. Available from: https://fas.org/irp/agency/dhs/fema/evans.pdf [Accessed May 27, 2017].

Baerenklau, K. A., & Knapp, K. C. (2007). Dynamics of agricultural technology adoption: Age structure, reversibility, and uncertainty. *American Journal of Agricultural Economics.* 89 (1), 190–201.

Barnett, T. P., Adam, J. C., & and Lettenmaier, D. P. (2005). Potential impacts of a warming climate on water availability in snow-dominated regions. *Nature.* 438 (7066), 303–309.

Barrows, G., Sexton, S., & Zilberman, D. (2014). Agricultural biotechnology: The promise and prospects of genetically modified crops. *The Journal of Economic Perspectives.* 28, no. 1, 99–119.

Bates, B., Kundzewicz, Z. W., Wu, S., & Palutikof, J. (2008). *Climate change and water.* Technical Paper VI. Intergovernmental Panel on Climate Change (IPCC). Available from: http://www.ipcc.ch/pdf/technical-papers/climate-change-water-en.pdf [Accessed May 27, 2017].

Boggess, W., Lacewell, R., & Zilberman, D. (1993). Economics of water use in agriculture. In G. Carlson, D. Zilberman, & J. Miranowski (eds.), *Agricultural and Environmental Resource Economics* (pp. 319–91). New York: Oxford University Press.

Bruinsma, J. (2009). The resource outlook to 2050. Paper presented at the FAO Expert Meeting, 24–26 June 2009, Rome on "How to Feed the World in 2050." Available from: ftp://ftp.fao.org/agl/aglw/docs/ResourceOutlookto2050.pdf [Accessed May 27, 2017].

Bulte, E. H., Lipper, L., Stringer, R., & Zilberman, D. (2008). Payments for ecosystem services and poverty reduction: Concepts, issues, and empirical perspectives. *Environment and Development Economics* 13 (3), 245–54.

Caswell, M., Lichtenberg, E., & Zilberman, D. (1990). The effects of pricing policies on water conservation and drainage. *American Journal of Agricultural Economics.* 72 (4), 883–90.

Chakravorty, U. N., Hochman, E., & Zilberman, D. (1995). A spatial model of optimal water conveyance. *Journal of Environmental Economics and Management.* 29, 25–41.

Chakravorty, U., Hochman, E., Umetsu, C., & Zilberman, D. (2009). Water allocation under distribution losses: Comparing alternative institutions. *Journal of Economic Dynamics and Control.* 33 (2), 463–76.

Chong, H., & Sunding, D. (2006). Water markets and trading. *Annual Review of Environmental Resources.* 31, 239–64.

Cochrane, W. W. (1979). *The development of American agriculture: A historical analysis.* Minneapolis: University of Minnesota Press.

Dasgupta, S., Hossain, M. M., Huq, M., & Wheeler, D. (2015). Climate change, soil salinity and road maintenance costs in coastal Bangladesh. *Water Economics and Policy.* 1 (03), 1550017.

Dinar, A., & Subramanian, A (1997). *Water pricing experiences: An international perspective.* Washington, D.C.: The World Bank.

Dinar, A., & Letey, J. (1991). Agricultural water marketing, allocative efficiency, and drainage reduction. *Journal of Environmental Economics and Management.* 20 (3) 210–23.

Dinar, A., & Pochat, V. (2015). *Water Pricing Experiences and Innovations.* Edited by José Albiac-Murillo. New York: Springer International Publishing.

Dixit, A. K., & Pindyck, R.S. (1994). *Investment under Uncertainty.* Princeton, NJ: Princeton University Press.

FAO. (2012). "The state of food and agriculture." Available from: http://www.fao.org/docrep/017/i3028e/i3028e.pdf [Accessed May 27, 2017].

Feder, G., Just, R. E., & Zilberman, D. (1985). Adoption of agricultural innovations in developing countries: A survey. *Economic Development and Cultural Change* 33 (2), 255–98.

Feinerman, E., Letey, J., & Vaux, H. J. Jr. (1983). The economics of irrigation with nonuniform infiltration. *Water Resources Research.* 19 (6), 1410–14.

Freeman III, A. M., Herriges, J. A., & Kling, C. L. (2014). *The measurement of environmental and resource values: Theory and methods.* London: Routledge.

Gemma, M., & Tsur, Y. (2007). The stabilization value of groundwater and conjunctive water management under uncertainty. *Applied Economic Perspectives and Policy.* 29 (3), 540–48.

Gleick, P. H. (2000). A look at twenty-first century water resources development. *Water International.* 25 (1), 127–38.

Hazell, P., & Wood, S. (2008). Drivers of change in global agriculture. *Philosophical Transactions of the Royal Society B: Biological Sciences.* 363 (1491), 495–515.

Hernández, F., A., de las Fuentes, U. L., Bis, B., Chiru, E., Balazs, B., & Wintgens, T. (2006). Feasibility studies for water reuse projects: an economical approach. *Desalination,* 187 (1–3), 253–61.

Iho, A., & Laukkanen, M. (2012). Precision phosphorus management and agricultural phosphorus loading. *Ecological Economics.* 77, 91–102.

Innes, R. (2000). The economics of livestock waste and its regulation. *American Journal of Agricultural Economics.* 82 (1), 97–117.

Inocencio, A. B. *Costs and performance of irrigation projects: A comparison of sub-Saharan Africa and other developing regions.* Available from: https://core.ac.uk/download/pdf/7115731.pdf [Accessed May 27, 2017].

Israel, M., & Lund, J. (1995). Recent California Water Transfers: Implications for Water Management. *Natural Resources Journal.* 35 (1), 1–32.

Jackson, S., & Sleigh, A. (2000). Resettlement for China's Three Gorges Dam: socio-economic impact and institutional tensions. *Communist and Post-Communist Studies* 33 (2), 223–41.

Jat, M. L., Chandna, P., Gupta, R., Sharma, S. K., & Gill, M. A. (2006). Laser land leveling: A precursor technology for resource conservation. *Rice-Wheat Consortium Technical Bulletin Series.* Available from: http://www.knowledgebank.

irri.org/csisa/images/FactsheetsAndReferences/Techbulletins/lasertechbull.pdf [Accessed May 27, 2017].

Johansson, R. C., Yacov, T. L. Roe, R. D., & Dinar, A. (2002). Pricing irrigation water: A review of theory and practice. *Water Policy.* 4 (2), 173–99.

Kahil, M. T., Dinar, A., & Albiac, J. (2015). Modeling water scarcity and droughts for policy adaptation to climate change in arid and semiarid regions. *Journal of Hydrology.* 522, 95–109.

Kalogirou, S. A. (2005). Seawater desalination using renewable energy sources. *Progress in Energy and Combustion Science.* 31 (3), 242–81.

Khanna, M., & Crago, C. L. (2012). Measuring indirect land use change with biofuels: Implications for policy. *Annual Review of Resource Economics.* 4 (1), 161–84.

Klümper, W., & Qaim, M. (2014). A meta-analysis of the impacts of genetically modified crops. *PloS One.* 9 (11), e111629.

Kulanyi, S. Uganda Promotes New Drought Tolerant, Disease Resistant Maize Varieties. *Genetic Literacy Project.* Available from: https://geneticliteracyproject.org/2016/12/21/uganda-promotes-new-drought-tolerant-disease-resistant-maize-varieties/ [Accessed May 27, 2017].

Kumar, T., Khan, M. R., Abbas, Z., & Ali, G. M. (2014). Genetic improvement of sugarcane for drought and salinity stress tolerance using Arabidopsis vacuolar pyrophosphatase (AVP1) gene. *Molecular Biotechnology.* 56 (3), 199–209.

Lichtenberg, E. (2010). Economics of health risk assessment. *Annual Review of Resources Economics* 2 (1), 53–75.

Lipper, L., McCarthy, N., Zilberman, D., S. Asfaw, and G. Branca (Eds.). (2017). *Climate Smart Agriculture: Building Resilience to Climate Change.* Rome: FAO. Springer International Publishing.

McBratney, A., Whelan, B., Ancev, T., & Bouma, J. (2005). Future directions of precision agriculture. *Precision Agriculture.* 6 (1), 7–23.

Makombe, G., Kelemework, D., & Aredo, D. (2007). A comparative analysis of rainfed and irrigated agricultural production in Ethiopia. *Irrigation and Drainage Systems.* 21 (1), 35–44.

Mercer, L. J., & Morgan, W. D. (1991). Irrigation, drainage, and agricultural development in the San Joaquin Valley. In A. Dinar & D. Zilberman (eds.), *The Economics and Management of Water and Drainage in Agriculture* (pp. 9–27). New York: Springer.

Millock, K., Xabadia, A., & Zilberman, D. (2012). Policy for the adoption of new environmental monitoring technologies to manage stock externalities. *Journal of Environmental Economics and Management.* 64 (1), 102–16.

Moreno, G., & Sunding, D. L. (2005). Joint estimation of technology adoption and land allocation with implications for the design of conservation policy. *American Journal of Agricultural Economics.* 87 (4), 1009–19.

Msangi, S., Sulser, T., Rosegrant, M., & Valmonte-Santos, R. (2007). Global scenarios for biofuels: Impacts and implications for food security and water use. Paper presented at the Tenth Annual Conference on Global Economic

Analysis special session on "CGE Modeling of Climate, Land Use, and Water: Challenges and Applications" Purdue University, West Lafayette, Indiana 7–9 June 2007.
Ogishi, A., Zilberman, D., & Metcalfe, M. (2003). Integrated agribusinesses and liability for animal waste. *Environmental Science & Policy.* 6 (2), 181–88.
Parker, D., Zilberman, D., Cohen, D., & Osgood, D. (1996). *The economic costs and benefits associated with the California Irrigation Management Information System (CIMIS): Final report.* Sacramento, CA: California Department of Water Resources.
Rausser, G. C., & Zusman, P. (1991). Organizational failure and the political economy of water resources management. In A. Dinar & D. Zilberman (eds.), *The Economics and Management of Water and Drainage in Agriculture* (pp. 735–758). New York: Springer.
Reisner, M. (1993). *Cadillac desert: The American West and Its disappearing Water.* New York: Penguin.
Rosegrant, M. W., Ringler, C., & Zhu, T. (2009). Water for agriculture: Maintaining food security under growing scarcity. *Annual Review of Environment and Resources.* 34 (1), 205.
Roy, A. D., & Shah, T. (2002). Socio-ecology of groundwater irrigation in India. *Intensive use of groundwater challenges and opportunities.* Available from: http://publications.iwmi.org/pdf/H029653.pdf [Accessed May 27, 2017].
Ruttan, V. W. (2002). Productivity growth in world agriculture: Sources and constraints. *The Journal of Economic Perspectives.* 16 (4), 161–84.
Schoengold, K., & Zilberman, D. (2007). The economics of water, irrigation, and development. *Handbook of Agricultural Economics.* 3, 2933–77.
Schoengold, K., & Zilberman, D. (2014). The economics of tiered pricing and cost functions: Are equity, cost recovery, and economic efficiency compatible goals? *Water Resources and Economics.* 7, 1–18.
Scudder, T. (2012). *The Future of Large Dams: Dealing with Social, Environmental, Institutional and Political Costs.* London: Earthscan.
Sedlak, D. (2014). *Water 4.0: The past, present, and future of the world's most vital resource.* New Haven, CT: Yale University Press.
Segerson, K. (1988). Uncertainty and incentives for nonpoint pollution control. *Journal of Environmental Economics and Management.* 15 (1), 87–98.
Shah, F., Zilberman, D., & Chakravorty, U. (1993). Water rights doctrines and technology adoption. In K. Hoff, A. Braverman, & J. Stiglitz (eds.), *The Economics of Rural Organization: Theory, Practice, and Policy* (pp. 478–99). New York: Oxford University Press.
Shani, U., Tsur, Y., Zemel, A., & Zilberman, D. (2009). Irrigation production functions with water-capital substitution. *Agricultural Economics.* 40 (1), 55–66.
Siebert, S., & Döll, P. (2010). Quantifying blue and green virtual water contents in global crop production as well as potential production losses without irrigation. *Journal of Hydrology.* 384 (3), 198–217.

Siebert, S., Kummu, M., Porkka, M., Döll, P., Ramankutty, N., & Scanlon, B. R. (2015). A global data set of the extent of irrigated land from 1900 to 2005. *Hydrology and Earth System Sciences.* 19 (3), 1521–45.

Spiritos, E., & Lipchin, C. (2013). Desalination in Israel. In N. Becker (ed.), *Water Policy in Israel: Context, Issues, and Opinions* (pp. 101–23). Netherlands: Springer.

Sunding, D. (2000). The price of water: Market-based strategies are needed to cope with scarcity. *California Agriculture.* 54 (2), 56–63.

Taylor, R. & Zilberman, D. (2017). The diffusion of drip irrigation: the case of California. *Applied Economic Perspectives and Policy.* doi: 10.1093/aepp/ppw026

Tsur, Y. (2009). On the economics of water allocation and pricing. *Annual Review of Resource Economics.* 1 (1), 513–36.

Valipour, M., Ahmadi, M. Z., Raeini-Sarjaz, M., Sefidkouhi, M. A. G., Shahnazari, A., Fazlola, R., & Darzi-Naftchali, A. (2015). Agricultural water management in the world during past half century. *Archives of Agronomy and Soil Science.* 61 (5), 657–78.

Warner, J., Zeitoun, M., & Mirumachi, N. (2014). Part 3. Transboundary governance. In R. Q. Grafton, P. Wyrwoll, C. White, & D. Allendes (eds.), *Global Water: Issues and Insights* (pp. 47–84). Canberra, Australia: ANU Press.

Wichelns, D. (1999). An economic model of waterlogging and salinization in arid regions. *Ecological Economics.* 30 (3), 475–91.

World Health Organization. (2009). *Global health risks: Mortality and burden of disease attributable to selected major risks.* Available from: http://www.who.int/healthinfo/global_burden_disease/GlobalHealthRisks_report_full.pdf [Accessed May 27, 2017].

Xie, Y., & Zilberman, D. (2016). Theoretical implications of institutional, environmental, and technological changes for capacity choices of water projects. *Water Resources and Economics.* 13, 19–29.

Zhao, J., & Zilberman, D. (1999). Irreversibility and restoration in natural resource development. *Oxford Economic Papers.* 51 (3), 559–73.

Zilberman, D. (2015). Editorial—The Economics of Climate Change and Water: An Introduction to the Special Issue. *Water Economics and Policy.* 1 (3), 1502001. http://dx.doi.org/10.1142/S2382624X15020014

Zilberman, D., Dinar, A., MacDougall, N., Khanna, M., Brown, C., & Castillo, F. (2011). Individual and institutional responses to the drought: The case of California agriculture. *Journal of Contemporary Water Research and Education.* 121 (1), 3.

Zivin, J. G., & Zilberman, D. (2002). Optimal environmental health regulations with heterogeneous populations: Treatment versus “tagging.” *Journal of Environmental Economics and Management.* 43 (3), 455–76.

3

Innovation in Plant Breeding for a Sustainable Supply of High-Quality Plant Raw Materials for the Food Industry

If you think in terms of a year, plant a seed; if in terms of ten years, plant trees; if in terms of 100 years, teach the people.

—Confucius

3.1 Introduction

"Plant breeding is the development of new varieties with new properties, enabling the company who places such varieties on the market to obtain or increase its market share. The development of a new variety or new breeding technique requires much time, effort and money" (Niels et al., 2009). Plant breeders are in process of continuous innovation. They play an important role in a many socioeconomic objectives, such as food needs and security, energy production, environment, quality of soils, health, and quality in food chain in the context of sustainability. Innovation in plant breeding is built on a lot of know-how, new technologies, and capital in transversal management. It also requires the access to genetic resources, which is a prerequisite to any development of new plant varieties.

Competition and profitability of the seed industry is necessary for the sustainable evolution of the whole agri-food chain. Farmers are the first to benefit from the innovation of seed industry, especially when facing new environmental challenges as a result of pests and diseases. However, the development of new varieties should also better meet the needs of the processing industry and consumers. But these needs between stakeholders are often disconnected.

The challenge of this review is to study the relationship between seed industry, farmers, and the actors of the food chain by proposing some examples of different species that can represent various business models. So the first aim is to compare the research-and-development (R&D) investment of the seed industry to the value and area of production. The second aim is to have an exhaustive overview about available techniques and strategies in plant

Megatrends in Food and Agriculture: Technology, Water Use and Nutrition, First Edition.
Helmut Traitler, Michel Dubois, Keith Heikes, Vincent Pétiard and David Zilberman.

breeding. The third aim is to understand how the food chain operates from seeds to consumers. The final aim is to propose some recommendations for all actors of the agri-food chain.

The structure of this chapter is as follows: after a short introduction, section 3.2 is a study of the future agriculture challenges; section 3.3 is a presentation of plant breeding and its technical limitations and possibilities; section 3.4 presents the analysis of megatrends after providing the methodology of gathering data about the main cultivated species, their areas, and their production, analysis of investment in seed R&D, and analysis of deviations/distortions between R&D investments and production volumes and values. These trends provide the basis for formulating the recommendations. Finally, section 3.5 concludes the discussion and offers possible areas for future research.

3.2 Challenges for Future Agricultures and Food Industries

3.2.1 Strongly Growing Food Needs

The challenge for agriculture production in the 21st century is to produce sufficient high-quality food while limiting ecological impact. From 7.4 billion in 2016, the population is expected to increase to 8.5 billion in 2030 and exceed 9 billion in 2050, an increase of about 25% compared to 2016 population (World Population Prospect, 2012). The urban population, which reached 50% of the world's population in 2011, is expected to reach about 70% of the world's population by 2050, with large cities accounting for more than half of the world's population. This growing of "megacities" is generally done to the detriment of the best cultivable lands.

The total cultivated area, today about 1600 million hectares, will grow very little. Indeed, according to Food and Agriculture Organization (FAO; 2016a, 2016b, 2016d) assessment, gains in arable land will be offset by surface losses induced by overall economic development, urban growth, and desertification. The predictable decline in the cultivated area per inhabitant will therefore be of the same order of magnitude as population growth. This leads to a necessary increase in average yields in the same proportions, for the same result in terms of food.

The demand's growth for meat further increases the need for animal feed and land to produce it. According to the FAO (2016c), annual meat production will have to increase by more than 200 million tons to reach 470 million tons. Indeed, the economic development and growth of cities is accompanied by an increasing need of meat supply. However, the feed-conversion ratio[1] of meat consumption requires a significant increase in crop production. This feed-conversion

1 Is a measure of an animal efficiency in converting feed mass into increases of the desired output.

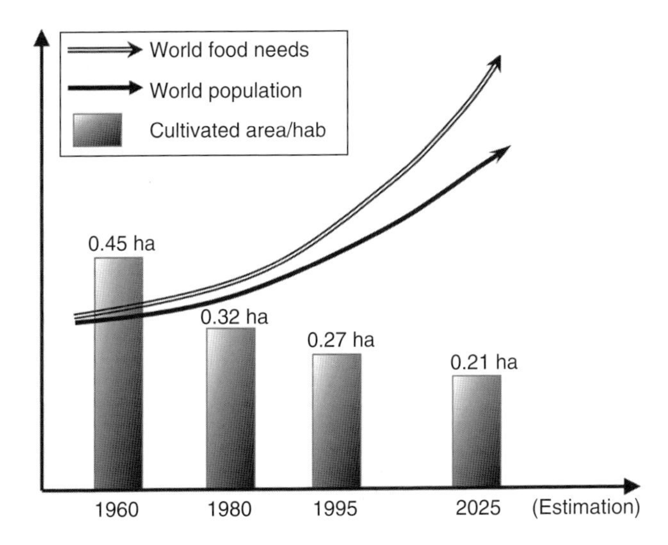

Figure 3.1 How to answer to a growth of population without a growth of cultivated lands. *Source:* FAO Statistics.

ratio of 1.7 for pisciculture is close to 2 for poultry production, around 3.5 for pork, and more than 7 for cattle. For the production of milk or beef from field crops (maize, soybean), this ratio may be as high as 10. For example, it is generally considered that one hectare (ha) of soybean could feed about 17 to 20 persons for 1 year when consuming beans while it is not more than 2,5 when consuming cattle products. Thus, by integrating these different parameters, according to the FAO (2016c), agricultural production, quantitatively, will have to increase by nearly 70% for the year 2050 compared to 2016. The annual production of cereals for food will have to be about 3 billion tons, which is close to the 2.1 billion for 2016 and is illustrated in Figure 3.1.

Figure 3.1 shows the growth in demand, coupled with stagnation in the growth of cultivated areas, leads to a decrease in the productive area per inhabitant. Based on field crops (excluding permanent grassland), in the year 2050 each hectare will have to feed 5.8 inhabitants, or each inhabitant will have, on average, 0.17 ha. If one calculates in "cereal equivalent," based on an average yield of 5 tons/ha (well below the *current* European average yield) and annual caloric requirements equivalent to 400 kg of cereals per year, one hectare could feed 12.5 people/year, or 0.08 ha/person is needed, provided a fully vegetarian diet. On the average, if the meat supply, consisting of a balanced mix of poultry, pigs, and ruminants, reaches 50% of the caloric intake, we just meet the projected data. In other words, the objective of an average global yield of 5 tons/ha is a goal that looks correct and will have, at minimum, to be met.

The African continent, especially in sub-Saharan Africa, has the lowest average yields. This would be the place where yield gains should be greatest because it is also the continent with the highest population growth for the next 20 years. An evolution sounds to be initiated. The Alliance for a Green Revolution in Africa (AGRA; 2015) has shown that 80 small and medium-sized African seed companies from 16 different countries have produced more than 80,000 tons of certified seed in 2014. For 2015, 18.2 million small farmers were able to benefit from certified seeds out of 9.2 million hectares. The goal is to double the income of at least 30 million smallholder farmers in Africa by 2020. Up to now, large seed producers are not present for such farmers.

World average cereal yields, calculated by the World Bank, are 3.9 tons/ha for 2013 (World Bank, 2016). FAO data (see Table 3A.1) show that, in 2014, maize is the only cereal crop with an average yield of more than 5 tons/ha.

The calculation of the average yield does not take into account the inputs used for cultivation, and particularly fertilizers and phytosanitary products. For instance, average yields of cereals exceed 8 tonnes/ha in Western Europe, but they are about half of them for organic cereals (France AgriMer, 2012). In other words, at the same technical level to date, the inputs "weigh" for 50% of the output. This issue needs to be analyzed.

3.2.2 Energy Issues

3.2.2.1 Value of Agricultural Energy Production

It seems that the uses of agriculture for nonfood purposes (fuel mainly, but also green chemistry, pharmacy) will grow. This is what emerges from the trends of the last 20 years as shown by the increase of nonfood crops. In 2011, world production of ethanol was 1.439 million barrels of oil per day (boe) or 72 million tons of oil equivalent (toe) per year, and that of biodiester of 403 thousand boe or 20 million toe, totaling 92 million toe (Souza et al., 2015).

The agriculture cannot replace the use of fossil fuels because the annual consumption of fossil energy is evaluated at 9 billion tons. On the basis of 3 billion tons of cereals (which is the target for 2050), this means that total agricultural production cannot reach the equivalent of 1 billion toe. The energy value of one ton of cereals is lower than that of 0.3 tons of oil. There is a scale issue. The need for fossil energy is about 10 times higher than agricultural energy production. One can imagine development of green chemistry, but certainly not the replacement of fossil energy. May be green energy could take a more important place when the technology will master the transformation of by-products of agriculture in energy.

No effective plant-breeding tool will multiply yields by 10! Solving the problem of the consumption of fossil energy, the stock of which will eventually be depleted, cannot reasonably be expected from agriculture.

3.2.2.2 Energy Needs of Agriculture

There is a need to limit the energy consumption of agricultural production and thus to increase the efficiency of energy use. The cost of fertilizers is first of all an energy cost. This efficiency combines new agronomic methods with adapted varieties. So there is a need to adapt current procedures for cultivar evaluation, to promote the breeding of multiresistant cultivars for low-input systems (Loyce et al., 2012).

Lowering the energy cost of inputs in agriculture is a major objective recognized by FAO and described in its objectives: "producing more with less." The objective is to achieve this average yield of 5 tons of cereal equivalent per year, but with a concomitant reduction of inputs. The two paths to be pursued in parallel are therefore the new agronomic practices and the selection of varieties resistant to diseases and predators (Bharadway, 2016; Singh et al., 2016).

3.2.2.3 Environmental Issues

It has now been proven that global warming resulting from climate change has and will result in increased droughts, floods, tornadoes, typhoons and storms, and rising sea levels, with increasing salinity of deltas. To achieve climate adaptation, varieties resistant to abiotic stresses will be needed.

It has also been shown that the yield objective, based on the use of pesticides and fertilizers, has led to a stalemate because of its impact on the environment. We must now rely on the multifunctionality of the ecosystem as a whole (Thompson & Gonzalez, 2016). To increase multifunctionality, we have to increase the number of cultivated species in longer and more complex rotations, including inter-cropping and multiple cropping (Byrnes et al., 2014).

There is therefore a need for a larger number of cultivated species, probably new ones adapted to specific agronomic functions and also new varieties from currently cultivated species to adapt them to the demands of the new agronomic methods (Singh et al., 2015).

3.2.2.4 Quality of the Soils

To produce better, we know that it will be necessary to have a soil of good physical, chemical, and biological quality. Soils must be protected from pollution and erosion and a high level of physical, chemical, and biological soil quality must be maintained. Conservation agriculture systemically combines direct seeding, inter-cropping and long rotations, making it possible altogether to capture carbon (against climate change), to be more productive, to limit inputs and to improve soil quality (Johnson et al., 1997; Tóth, Stolbovoy & Montanarella, 2007; Dubois & Sauvée, 2016).

Again, there is a need for species and cultivars that are adapted to this new context, and therefore a need to adapt current procedures (agro-practice itineraries) for cultivar and species evaluation, which should be adapted to such systems.

3.2.2.5 Health Issues

The human being is omnivorous. It needs many vitamins, many mineral salts, and even essential amino acids and essential fatty acids. Its complex needs are satisfied by a large variety of biological products, diffused at the global level but also produced and distributed at the local level. This leads to a minimum cost of logistics and of energy.

In addition to food toxins from plant pathologies (mycotoxins), it may possibly have pesticide residues depending on the sensitivities of the cultivars to these pathologies. Obtaining resistant cultivars will reduce the diverse risks. On the other hand, the previous issues show the interest of crops associated with cereals and fixing nitrogen such as fabaceae (Corre-Hellou et al., 2009).

These associations of cereals and fabacae, already found in the antiquity, allow good dietary supplements to reduce the consumption of animal products through increasing use of vegetable proteins in food products (e.g., meat analogues, dessert cream, dairy product substitutes, etc.). Health concerns thus integrate environmental issues by opening up opportunities for the development of "flexitarism" and "vegetarianism."

It is a question of rebalancing the ratio of vegetable to animal proteins. Proteins from fabaceae will be supplied by cereal products, dried vegetables, and vegetable proteins derived from the fractionation of seeds, leaves, tubers and will be selected for their techno-functional or nutritional properties. Vegetable proteins have a great diversity of structures and properties according to their botanical origin, their location in the plant, and the technologies used. Broad progress is possible (Guéguen, Walrand & Bourgeois, 2016). The trend toward the consumption of plant proteins, possibly purified, and used in food compositions can be considered a strong trend because it corresponds as much to health concerns as to the aforementioned problems. This means that a specific plant breeding effort on the cultivated fabaceae will have to be put in place. It will have to integrate industrial uses and extraction techniques.

3.2.2.6 Quality Control and Traceability, as Requested by Consumers, Food Chains, and Nonfood Agro-Industries

Traceability and quality control of agri-food products is a strong societal demand. In the context of a growing diversification of agricultural production, it is necessary to anticipate a need for traceability, which will concern the cultivated species, cultivars, the raw material storage, and transformations throughout the industrial food chain.

Some vitamins, such as B_{12}, iron, essential amino acids, essential fatty acids, and vegetable proteins (Gueguen et al., 2016), must be expected from plant products. With the decline in meat consumption, in nearly all advanced industrial countries, of about 2% per year, the selection of new cultivars will have to be considered by integrating the entire processing in the food chain.

The nonfood industries must answer to specifications of any industry according to the process of Total Quality Management.

3.2.2.7 Synthesis

It will be necessary to produce more and better on smaller areas in relation to the growth of the population. Produce more quantitatively, in better quality and in a wide range of products, on limited areas to be protected. The production will be more diversified to find a global answer, even if, in fact, a good part of the answers will be local. Bharadway (2016) presented that fossil fuels, fertilizers, water, and chemical products are at their peak of use, but this situation will not remain linear in the future. He deduced that the amount of arable land for crop cultivation is decreasing as a result of urbanization, salinization, desertification, and environmental degradation.

This synthesis proposed a change in plant breeding goals that need an improved rationalization between yield, areas, crops and farmers' practices, inputs coupled with high-quality food, and environment protection. It's consistent with a sustainable crop production. However, besides the fact that an increasing number of countries will be largely dependent on food imports, by impossibility of quantitative self-sufficiency (e.g., United Kingdom, Mediterranean Africa, part of the Middle East, Singapore, Japan, South Korea, Philippines), needs for food diversity will contribute to global food trade. An increased demand can be expected in terms of numbers of species and numbers of cultivars adapted to different agronomic and food demands. This delicate appropriateness will return in large part to the stakeholders in plant breeding.

3.3 Genetic-Based Techniques for Plant Breeding in the Context of Agricultural Production

3.3.1 Genetic Innovation and Agronomic Practices

With regard to the challenges faced by agriculture for fulfilling the needs of the 2050 world, policy makers may consider different approaches, and they probably will have to establish priorities.

These approaches are mainly based on the genetic knowledge and agronomic practices. Genetics of the crops is a prerequisite to significant quantitative or qualitative improvement of a crop production that is the result of interactions between genetic and environment plus agronomic practices. Only a part of genetic improvements can be useful if growing these crops with unadapted agronomics practices. However, some may be, without changing culture practices, resistant to some disease and may save the crop of certain productions. The best agronomic practices may also need adapted varieties. And each specific soil-climate condition needs specific varieties. These two key approaches should be considered simultaneously and adapted accordingly to the crop and the economic and sociological environment. Figure 3.2 depicts the evolution of corn yield in the United States since 1960.

It was generally recognized that genetic innovation has accounted for about 50% of the improvement of the crop yield during the last 50 or 80 years, the other 50% being as a result agro practices including agro chemicals for crop protection, fertilizers, irrigation, and mechanization (see Fig. 3.2). However, it is

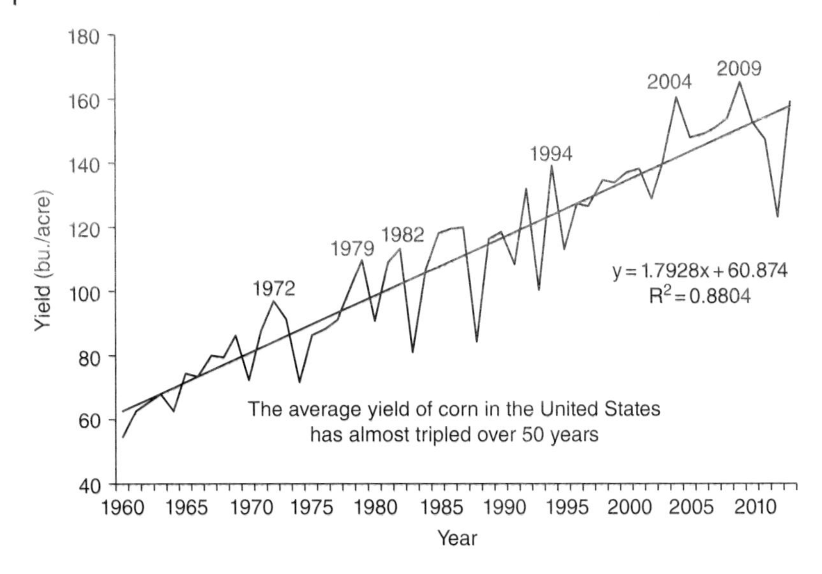

Figure 3.2 Evolution of corn yield in the United States since 1960. *Source:* farmdocdaily. illinois, July 9, 2014; The 2014 U.S. Average Corn Yield: Big or Really Big? Scott Irwin and Darrel Good, Department of Agricultural and Consumer Economics, http://farmdocdaily.illinois.edu/2014/07/2014-us-average-corn-yield-big-or-really-big.html

now understood that in the future genetics will account for 70% or 80% of the future improvements of crop production because of the limitation of agro chemicals and possibly in mechanization, even if precision agriculture will make it increasingly better to customize the crops and their production system. And if some new agronomic practices, like conservative agriculture, no-till, double- or intermediate-culture, could undoubtedly increase yield, improved varieties in these conditions will be even better. Agro performance traits are currently called "input traits," and quality traits are called "output traits". The demand for plant raw materials better specified to their final use ("decommoditization") will also require genetic innovation on the pathways permitting the crop to produce (or not produce) some specific biochemicals and secondary metabolites. For example, tomato for ketchup might not be similar to tomato for sauces or dehydrated soups and certainly not from fresh tomato. Genetic innovation will therefore become increasingly more important either for productivity or quality and safety of the produced raw materials.

Again, one should not deny the importance of agronomic practices, but for the aforementioned reasons, we have decided to focus this chapter on genetic and plant breeding.

From the early ages of agriculture, humans had the habit of domesticating and then selecting the plants they were growing. Science-driven plant breeding took off with genetics and more recently with cell and molecular biology. It is interesting to note that until the 1950s or 1960s plant breeding and the development of new varieties were mainly done in the universities and public institutes, and today, although a real industry, it is focused on the crops having the most seed value.

Plant breeding requires diverse competencies including genetics, agronomy, statistics, and cell and molecular biology. It was originally based on the so-called *phenotypic observations* in the fields that are determined by the combination of genotype and environment (generally called "gxe"). It is now possible to base decisions on both phenotype and genotype, which is clearly making the guidance of the process more reliable. Nevertheless, plant breeders have to keep in mind that the process must deliver a variety to be grown in a field (or a greenhouse), and has to be attractive for the farmers and the users or consumers. It has also to be produced in sufficient quantities in as short as possible time, called the "seed increase" phase. An outstanding variety that could not be rapidly produced and distributed to farmers is simply useless. For example, coffee breeders were facing this situation when producing the first hybrids of Arabica in Kenya or Ethiopia (Riuri 11). Producing hybrid seeds by hand pollination was easy for a small number of seeds but rapidly became impossible and unreliable for large production.

3.3.2 The Process of Plant Breeding and Its Main Limitations

Figure 3.3 describes the whole process that is quite similar for whatever the crop to be considered. There are many differences depending on the reproduction system, but in the end it is aimed at recombining the best traits according

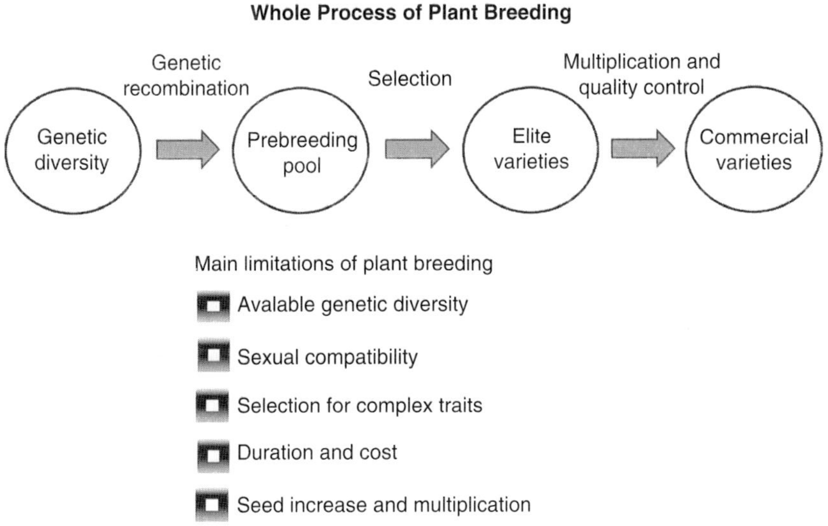

Whole Process of Plant Breeding

Genetic diversity → Genetic recombination → Prebreeding pool → Selection → Elite varieties → Multiplication and quality control → Commercial varieties

Main limitations of plant breeding

- Avalable genetic diversity
- Sexual compatibility
- Selection for complex traits
- Duration and cost
- Seed increase and multiplication

Figure 3.3 Plant breeding: The process and its limitations.

to the objectives in a given variety and to multiply it for making it commercially and rapidly available to the farmers.

3.3.2.1 Strategic and Technical Use of Available Genetic Diversity

It is useless to say that if not having genetic information encoding for a given trait in the genetic diversity it is impossible to get it in a new variety by conventional breeding. Plant breeders have always looked for new traits in natural diversity of the considered species or crossable relative species. In that sense, the collection, identification, and preservation of natural genetic diversity became a priority for securing the future of the crop when it will sooner or later be facing new environmental or qualitative challenges. The first approach of breeders is always to look at their breeding material, available collections, and check whether this new trait they are looking for might be present, and if so, to use it as a donor for the creation of new varieties. However, that has required time to master new technologies for facilitating this use of natural diversity, technologies that are now available at a continuously decreasing cost.

First necessary techniques to manage: It has to be collected and preserved. That is quite easy for many "orthodox" species because their seeds could be stored, in some cases, for long periods under controlled conditions. For others, "semi-recalcitrant or recalcitrant," living plants have to be maintained in the fields with all associated risks because their seeds cannot be stored (immediate germination as a result of no survival after dehydration) or even they do not produce seeds. For these last ones, scientists have developed methods of cryopreservation of different organs in liquid nitrogen securing genetic diversity. That has been useful for maintaining collections of many tropical species.

Secondly, we still have to optimize the cost for this preservation or conservation. We therefore need to identify the most genetically original material and not maintain everything that might be costly for recalcitrant species. Recent developments of DNA fingerprinting or genotyping by sequencing (GBS) allow identification of core collections that may represent 95% of the total genetic diversity. These so-called *optimized core collections* could then be shared in various breeding centers to be used by different breeders.

Third step: How to choose the individuals to cross? If having hundreds of different individuals (or accessions) in the core collection, the breeder has to select those that have the highest likelihood of having the trait they are looking for. When looking for a disease resistance, that is quite obvious by submitting all accessions under the pressure of this disease and selecting those that looks to be tolerant or resistant. When looking to complex quantitative traits, it is more difficult, and it is only with the support of DNA analysis that the breeder could identify the best donors to be crossed with a variety to be improved.

Fourth specific limit: how to obtain interspecific hybrids? The breeder has to get hybrid seeds from the crosses performed between the donor and the variety to improve. It may be quite easy and immediate, but also when looking to relative species possibly quite distant from the cultivated species, it may be more difficult. When facing this difficulty, scientists are often using a technology called "embryo rescue," which means extracting the young embryo that could not properly develop as a normal seed and developing it *in vitro* on an artificial medium under aseptic conditions.

3.3.2.2 New Sources of Genetic Diversity

An interesting trait existing in one given species cannot be transferred by conventional species to another species if no cross is possible or no result of the cross is obtained even with embryo rescue. Scientists have therefore looked for new techniques to create some genetic diversity that they could not find in the nature or in the collections.

The first technique plant breeders have considered was the mutagenesis provoked either by irradiation or chemicals such as ethyl methane sulfonate. That has not really been successful, not so much because of the efficiency of the treatment but more as a result of the difficulty to identify positive mutations happening randomly in a large number of treated plants.

The second technique is genetic engineering used for the creation of genetically modified organisms (GMOs). This technique that was rapidly developing early in the 1980s for the production of pharma products (e.g., insulin) or enzymes (e.g., chymosin) in microorganisms was for the first time successful on a plant (tobacco) in 1983 in the United States and Europe. It developed quite rapidly and the first commercial GMO variety was a tomato released in 1996 by Calgene and modified for turning off two enzymes involved in the softening of the fruit. Then most seed companies begun to apply this technique mainly for creating herbicide-, insect-, and virus-resistant crops such as maize, soybean, and cotton. Minor commercial achievements have been made for quality traits such as fatty acid composition of Canola and starch of potato. However, the seed industry has been facing a huge reluctance of consumers, nongovernmental organizations, and some authorities with regard to GMO crops because they are considered "non-natural products" and are required to be specifically labeled as such in numerous countries. Nevertheless, in 2015 about 180 million hectares has have been planted with GMO crops. From a more technical point of view, one limitation to the development of GMOs is that many important traits are not determined by a single or even a small number of genes, and in many cases, they are not yet known. From a commercial point of view, the regulatory costs for environmental and food safety became so important that this technology cannot be considered for minor crops, even if some success have been noticed on cabbages and other vegetables that have not yet been significantly released on the market.

The third technique, *tilling,* is a less random mutagenesis associating mutagenesis and high-throughput screening of the mutants based on DNA sequencing of the target gene. Those plants where the sequence of the target gene has been modified are developed and phenotyped for observing the effect of the mutation possibly after one sexual generation to have the mutant gene double recessive. That has led to new products improved for certain biochemical traits again in the area of fatty acids composition. However it is facing the same technical limitation that GMOs have about the number and knowledge of genes to be targeted. On the contrary, it is not submitted to the same regulatory and image constraints because it is not based on a DNA recombinant technique and leads to products that could have happen through mutations in the nature. It has also to be noted that hidden mutations might have been provoked on other genes than the targeted ones, leading to unexpected modification of important traits.

The fourth and most recent technique is called *gene edition,* or cluster regularly interspaced short palindromic repeats (CRISPR), and it targets a specific gene to be turned off. It is certainly the most advanced technique for creating a gene-targeted diversity, but the regulatory status is debated in many countries and whether the products will be considered as a GMO crops is still unclear. Here the gene encoding for the trait to be modified wanting to be edited needs also to be known.

It should also be noted that most of these new DNA-based technologies are being applied on isolated cells or tissues, and the regeneration of a whole plant from them has to be properly mastered via cell biology. This may still be a limitation to these technologies for some species.

Another approach, not based on DNA, was developed in the 1970s for increasing diversity. It was called "protoplast fusion." It is the fusion of single cells of which the cell wall has been enzymatically degraded for permitting their fusion either with chemicals or electrically opposed charges. Originally the idea was to combine nucleus genetic information between different species, but it only worked for species that are crossable and it was not as interesting as the breeders expected. It was, for example, possible to produce plants from Tomato + Potato cells, but these plants were chimeric and sterile most certainly as a result of the incompatibility of the two sets of chromosomes to pair in an equilibrate manner during mitosis and creating cells missing some chromosomes and others with extra ones. However, another application of protoplast fusion has been shown to be important for breeders. It permitted the transfer of genetic information contained into the cytoplasm, especially that of mitochondria, that may encode for the "cytoplasmic male sterility". This transfer has been used for producing hybrids of main field crops (e.g., winter oil seed rape, etc.) or vegetables (e.g., carrot, cabbages, etc.).

3.3.2.3 Selection of Quantitative and Complex Traits

Plant breeding always begins with the observation of agronomic, biochemical, and quality traits that is called *phenotyping*; it also ends by a selection or screening of specific combination. In theory, the number of plants to be screened is as high the probability to detect the right genetic combination and the ideal plant one is looking for. Although some traits may be quite easy to screen, other ones can be more difficult or costly to quantify. For example, about sensory quality of strawberries or tomatoes, it has been often claimed by the consumers that today's varieties do not have the quality of old varieties they used to eat in the garden of their grandparents, and that is true. It is certainly not because the breeders positively selected varieties having a neutral sensory profile. They simply could not select for positive sensory attributes because either they did not know them or they could not screen for them. When not screening for some specific characteristics, the breeders will logically progressively lose, by genetic drift, the genes encoding for them. One good example of this was the progressive loss of bakability of soft wheat varieties in France during 1970s and 1980s when focusing more on yield and animal feed. That imposed the import of hard red spring Canadian wheat to reequilibrate the quality of flour for production of bread.

Breeders are open to screen for any trait that makes sense for the farmers or the consumers, but they need reliable screening tools for doing so. Scientists have looked for biochemical markers or DNA markers that allow not only to screen but also to make sure of the presence of the required genes into the genetic background of the plant they are selecting. That is certainly even more important when considering perennial crops where the trait of interest may be expressed only in adult plants after maybe 5 or 7 years. Any early detectable DNA or biochemical marker that might be applied at the nursery stage of the tree allows a breeder to discard those plants that are not promising and should therefore not be pushed further for field evaluation.

Another problem faced by the plant breeders is the selection of quantitative traits such as yield that are not determined by a single gene but in most cases by numerous ones. Starting in the 1980s, breeders looked for what has been called quantitative trait loci (QTL). They discovered that it is possible to detect the areas of the genome (QTL) that are significantly associated to a quantitative trait. They superpose and compare phenotyping and genotyping "pictures" of a population of plants made from a cross between two parents differing for the trait. These areas maybe more or less large, depending on the density and distribution of the markers disseminated on the genetic map, and therefore it is correct to talk about gene or trait association and not correlation because they may be not directly involved in the determination of the quantitative trait. It is possible to detect the gene from the QTL directly involved in the trait by moving step by step, increasing the marker density of this zone, and checking for their association to the trait. Definitive proof of involvement of this gene from a QTL in the trait could also be established by transferring it to a plant that did not have it, showing the consecutive improvement of the trait. However, most of this work on QTL has not been done for the discovery of new genes and "native traits" but more for guiding the breeders with these markers through strategies that have been called marker assisted selection (MAS) and marker assisted recurrent selection (MARS). Another great benefit of this approach based on QTL is to differentiate plants that have the same phenotypic value but for different genotypic reasons, from those that have similar value as a result of the same genetics. The first ones should be considered for further breeding on this trait, and the second ones would not offer great perspectives to the breeders.

Recently the development of Genome Wide Selection, originally developed for cattle breeding, allows breeders to predict with good reliability the expected performances (hybrid vigor, etc.) of the created plants before scaling up and diversifying in many locations. Only the promising ones will be scaled up, saving time for crops where it is run step by step with intermediary seed increase (e.g., potato, fruit trees, etc.).

3.3.2.4 Duration and Cost

As it should have been understood from the developments discussed, plant breeding requires successive generations of crosses or multiplication to get to a new variety. The duration of a project is generally longer than 5 years, 7 to 10 years is common, and often with perennial crops between 10 and 20 years. It obviously depends on the duration of the generation time (seeds to seeds) and the number of generation to be completed for reaching the objective. Plant breeding is therefore organized as a pipeline to regularly release novel varieties. Breeders and scientists are also looking for any time-saving that is possible. Different strategies and innovations have thus occurred during the last 20 years.

For example, they use any possible climatic condition for speeding up the process. Growing plants alternatively in North and South hemispheres makes possible two generations per year of maize. Using greenhouses for fast growing and multiplication of some vegetables allows for saving time in the whole process. Multiplying with tissue culture plants that are not in sufficient number for moving to the following stage of the program is also used for saving time.

One of the most important technologies developed for saving time in breeding is the production of di-haploid (DH) plant lines. For many crops, either self- or cross-pollinated, breeders are looking for pure lines, meaning plants that have two sets of chromosomes strictly identical (homozygous), that are also called *fixed lines* because their progeny from selfing is 100% similar to the mother plant. Hybrids from two pure lines are also 100% homogeneous, and therefore, either it is for creating new self-pollinated varieties or new hybrids; breeders often have to go through pure fixed lines. These pure lines are currently obtained via successive selfing generation that could take years and years.

First in the mid-1960s some scientists succeeded in regenerating tobacco plants from pollen grains. Developed from a sexual cell, they had only one set of chromosome, which after doubling it with a natural alkaloid named *colchicine*, they obtained a 100% pure line and possibly saving five or six generations of selfing. This strategy has then been enlarged to many crops, using either pollen or ovule, but also interspecific crosses or irradiated pollen that permit the induction of the development of the ovule without effective fecondation by the pollen cell. DH technologies are now routinely used for many cereals (e.g., maize, wheat, barley) but also other field crops (e.g., rape seed and sugar beet) and vegetables (e.g., cabbages, pepper, melon, cucumber, etc.)

A large part of the cost of a plant-breeding project comes from field trialing at the successive stages of the process. Any reliable prediction tool that may decrease the number of plants to be tried or the number of locations and plots where to try them is a source of cost savings. New algorithms based on genomic selection or previously tested environments are becoming integrated in the whole breeding strategy for optimizing the use of the number of plots authorized to a breeder according to budget.

3.3.2.5 Seed Increase and Production

For any new variety, the seed increase and production is the last challenge the plant breeder is facing. Commercial release should be as fast as possible for limiting the risk of a new variety coming from the competition to surpass the one to be commercially launched. Competition in plant breeding is such that some vegetable seed companies could lose up to 30% of their production because the produced varieties became obsolete on the market during the time required for scaling up their production.

For clonal varieties propagated by horticultural methods or tubers, new *in vitro* tissue culture techniques have sped up the production and market release and for some of them eliminating virus (e.g., potato, raspberries) or other diseases that were currently disseminated through suckers (e.g., banana) or other graften sticks (e.g., cacao). More recently the tissue-culture regeneration through somatic embryogenesis allowed to propagate elite individuals of oil palm, date palm, Arabica hybrids, cacao, and forestry that could not be produced in large number, making their genetic benefits available to the planters. They could be key for rapidly renewing plantations with a new clone resistance to a disease (e.g., date palm resistant to Bayoud, oil palm resistant to Ganoderma).

Finally, quality and purity control of commercial batches of seeds or clones can be done at an affordable cost by the analysis of specific DNA markers that reinforced these controls and traceability down to the shelf.

3.3.3 Preliminary Conclusions

The preceding description of the main challenges faced by the plant breeders and the different techniques developed to turn them around, or limit their constraints, should lead decision makers to understand that innovation in plant genetics is not much limited by the technology. A lot of tools in conventional breeding, cell biology, and molecular biology have been developed and validated on different crops.

All these tools can be used successfully to meet the challenges of the future of agriculture. The breeders should now select among a diversified toolbox the most adapted ones according to the crop, the objectives, and the expected timing for developing a new variety. But, to our opinion, the food industry is facing a new paradigm for the supply of specialty crops that are not considered by the seed industry very often called "orphan crops." There are different reasons why conventional breeding or new technics are not developed for all these species, including some regulatory problems and resulting higher cost. For some of these crops that are often creating more value than commodities, the food industry will have to invent new business models, sharing value up to the seed or planting material for motivating private entities to invest on these orphan crops by transferring adapted technologies already validated on main field crops or vegetables.

It is not anymore a matter of technology, it is a matter of transfer, which may need regulatory decisions but also, for the food industry, strategic decisions. This will be discussed and analyzed in the following section.

3.4 Trends: Shift in Allocation of Resources to Global Needs?

3.4.1 Methodology

All data are extracted from FAO statistics (2016a), for the year 2014 and for 173 products. The 101 main productions were recovered, from which the surfaces and volumes produced by species were extracted. The total area under consideration represents 1.33 billion hectares, or 83% of the world's cultivated area. The 41 products corresponding to the production of staple foods were extracted (see Table 3A.1). They correspond to all the agricultural products aimed at satisfying the plant-based diet, that is to say the caloric, protein, and fatty acids. They cover 1.15 billion hectares. From this data, the dry matter production is calculated, which is converted into Kcal, taking into account the nature of the products. Yield can be evaluated in Kcal/ha.

The total Kcal produced by these 41 productions is thus estimated at 12.3×10^{15} Kcal. By considering an average need of 2800 Kcal per day/person and an average loss of 35%, that should allow to feed 7.8 billion people, which sounds consistent with the current global situation, knowing that 12% of production of sugar cane and about the same proportion of maize is used in bioethanol. There is no evidence that this is a major trend because the challenges for the future will be to produce enough. The 132 productions discarded from the preceding calculations mainly concern fresh fruits and vegetables, aromatic plants, and

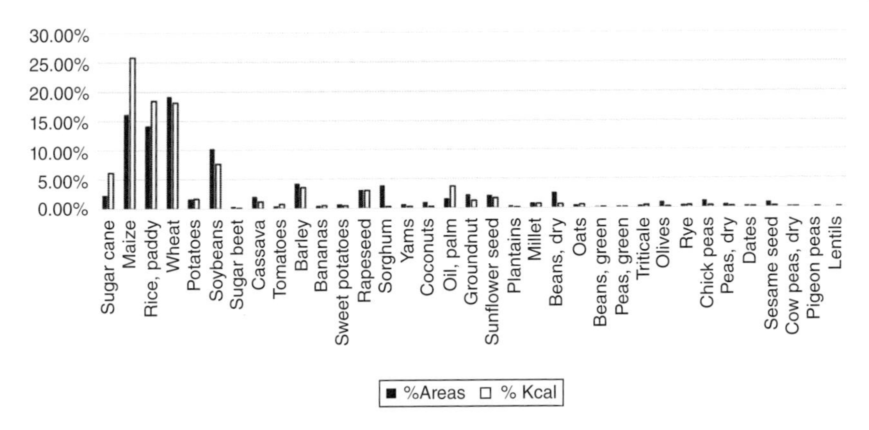

Figure 3.4 Average caloric consumption of different crops.

various local productions with a low overall impact. They certainly contribute to less than 20% of the world's caloric intake, while being essential (e.g., vitamins, minerals) and nevertheless cover more than 28% of the cultivated area.

All calculations are carried out from the cultivated areas, which *de facto* excludes pasture and consequently the part of animal production obtained from them. In addition, a large proportion of crop production is used for livestock. This creates a significant bias in global quantitative terms because a large part of crop production (mainly maize, soybean and wheat) is used also for livestock that has not been considered in the described calculation of caloric production. Nevertheless, it enables to roughly evaluate the contributions of each type of production to the average caloric intake of consumers. This is illustrated in Figure 3.4.

This figure is extracted from Table 3A.1 to show the major importance of the three main cereals. They account for 62.6% of the considered cultivated area. All cereals combined accounted for 68.6% of caloric intakes. Tubers and bananas 4.26% and the fabaceae, which have often been mentioned above, represent only 9.47%, with soybean alone representing 80% of all of them. Alfalfa, not presented here with 33 million hectares, only adds 3%, to be used in animal feed for dairy farms. This weakness of fabaceae has an important disadvantage; it does not promote R & D in plant improvement for protein production.

3.4.2 Analysis of Investment in Seed Research and Development

Using data from the professional world (e.g., turnover, average R&D rate, market for different crops), it is possible to estimate the distribution of R&D effort per large group of species. These results make possible to study the relationship between agriculture production and the R&D investment for genetic innovation and plant breeding as illustrated in Table 3.1.

Table 3.1 Investment of seed industry in R&D (US$).

Turnover of seed industry	$60 billion
R&D of seed industry	$7.8 billion
Of which Field crops	$6.24 billion
Corn	$2.5 billion
Soybean + Rapeseed	$2.5 billion
Other field crops	$1.24 billion
Vegetable and ornamentals	$1.56 billion
Of which Tomato	$0.7 billion

The global world turnover of the seed industry is not high compared to the value of world agriculture commercial production. It represents a little less than 4% of the total of plant commercial production value for food and feed. But it can surpass 10% of the production value for hybrid plant production. In that case, the gain for the farmer, thanks to a better yield, is generally higher. The higher the gain for the farmer, the more he is interested to buy the seeds.

The R&D investment with regard to the turnover of seed industry is remarkable. Only pharmacy and biotechnology have such high R&D expenses. These are innovation-driven industries. However, this magnitude has to be qualified also because it is focused on a reduced number of species. Of the $7.8 billion affected to private R&D, 6.24 (82%) are for field crops, which is more important than their representation in the total production calculated on the acreage, in volume or even in value. In particular, maize sees an over-allocation of R&D compared to its share in the overall production, but the discrepancy between affected R&D and production is even stronger for soybeans and oilseed rape. Similarly, tomatoes with 45% of R&D devoted to vegetables and ornamental plants, is out of proportion to its share in total production.

Table 3.1 allows to show the concentration of R&D efforts on a limited number of species; it evidences that many species can be considered as "orphans" and are of low interest to seed companies (e.g., oats, rye, cassava, coffee, cocoa, and many vegetables).

Moreover, in relation to biotechnology and the high cost of developing and releasing a genetically modified cultivar (due more to regulation than to R&D costs), the globalization of the seed industry has accelerated for 20 years (Niels et al., 2009). Now, the top five companies in the sector account for 52% of the seed market, and the next five, just 10%.[2] More recent evolution has shown an even increased globalization with the creation of no more than three companies dominating this market: Dupont, Bayer, and Chem China (Syngenta).

2 Phil Howard (https://msu.edu/~howardp/seedindustry.html)

This analysis about plant breeding goals entails an optimization among yield, acreage, caloric intakes, investment in R&D, coupled with high-quality food without environmental degradation. There are different approaches to enhance the sustainability of agricultural farms (Bharadway, 2016). The following analysis will show that, because of the discrepancy between the various determining factors, the stakeholders of agri-food chains are far away from sustainability.

3.4.3 Analysis of Deviations and Distortions of R&D Investments and Production Volumes

To understand the evolution of the seed sector, it is necessary to understand how the food chains operate, from seeds to consumers, as illustrated in Figure 3.5.

Farmers get inputs from their suppliers (e.g., seeds, crop protection, fertilizers and other technological components) to sell their production to primary processors, whose products can be used in secondary processing and distributors, from whom consumers buy the products they need. That is not say that they could sell fresh produces to the retailers or in a more direct way to consumers (see Fig. 3.5).

For the concern of information exchange, the arrows between stakeholders are two ways. Depending on what consumers buy, or what they ask for, the information goes back to the distributor, and hopefully to the producers. If consumers demand were only correlated with the price they are willing to pay, consumers would be able to dictate what they want. But they are often in contradiction, and the consumer population is diverse. So, different channels can be set up according to the product demand. But among products of similar perceived quality, consumers, as a population, are generally consistent; they choose the cheapest product. In other words, no food industry or retailers for private labels can raise prices of products for reasons that are unrecognized by consumers as worthless.

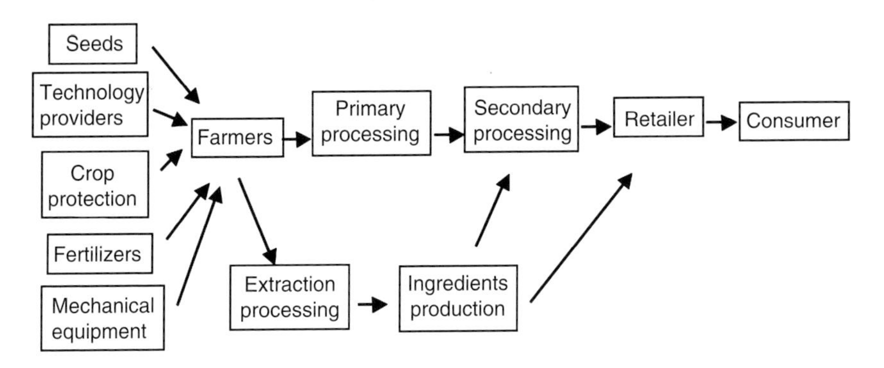

Figure 3.5 A schematic flow chart of the agro food chain.

There are heterogeneous entities along the food chain, with the consequence of loss of contact between the agricultural upstream and the consumers according to Maynard and colleagues (2016). When distributing the same products retailers can only differentiate on price and services. In this way, different distribution strategies can be developed, all the more based on price because volumes are large and the diversity of products is broad, or all the more based on the quality that the range of products targeted is specific. This can lead to distinct channels (the organic segment). Food manufacturers are looking to build a positive perception of their products to value them independently of the distributor and compete with private labels. The stronger and unique the image the more they can impose their price to the retailers. This image can be supported as much by a stable and appreciated quality than by a communication on the relation with the upstream part of the chain (i.e., traceability, sustainability, etc.). The upstream manufacturers, and in particular all those who produce ingredients, seek to meet business-to-business marketing specifications, that is to say techno-economic. Thus, the differentiation in terms of value is built according to the logic of evolution of any industrial product from the raw material to the concept of functional ingredient. Their goal is to provide solutions to their customers.

Farmers, even if wishing to differentiate themselves by changing their products as a response to a specific or local demand cannot all move in that same direction. A "production basin" is made up of many farmers, all competing, and where differentiation will be almost impossible. But the logic of "production basin" is necessary for logistical reasons. As, in a basin, they could not decide the selling price of their products, the only remaining solution for them is to control the production cost. But if farmers cannot control the price of their products and also the produced volumes that are depending on climatic conditions, at the end they have little potential to influence net income. Hence, they have to manage volatility in volumes and in prices. Singh and colleagues (2016) have shown that farmers may be the passive receivers of any technology or information or be the active partners in research action with breeders. They suggested that improved lentil varieties and integrated crop management technologies resulted in higher gain both in yield and net return. Other technological components like fertilizer management, irrigation, and pest and disease control contribute to increase in yield and net returns.

There are different innovation levels along the agri-food chains, but there is a lack of coordination between actors that is one reason of the quasi absence of innovative supply chain (Meynard et al., 2016). Niels et al. (2009) presented business models in the breeding sector:

- Local and traditional breeding companies that integrate variety development, production, marketing of seeds, and planting materials.
- Local and traditional breeding companies in home country but licensing their varieties to third parties in other countries.

- Traditional breeding companies that are also integrating biotechnology in their breeding programs.
- Companies that combine biotechnology activities with the production and marketing of seed and also licensing technologies or traits to other seed companies.

We can add another one: biotechnology companies that are focusing on income from services in R&D to seed companies and on licensing income, that are often called *technology providers*.

With all these different models can we imagine the possible strategies conducted by these competitive groups of companies? These strategies are connected to the short-term demands of farmers and not to the needs of the processors or final consumers market.

This strategy analysis does not take into account:

- Technical feasibility for improving yields or quality production (i.e., hybrids production and high hybrid vigor effect, efficient solution by biotechnology) and specific needs for agriculture such as disease resistance or abiotic stress resistance for a specific species, such as maize versus wheat;
- Specific difficulties in genetics resources and seed production;
- Needs of food industry and consumers that are not connected to the "seed profitability loop."

Taking into account these conditions, the relationship of farmers to seed companies can be described according to Figure 3.6. We can raise some business models through this figure.

The trend is that the turnover of the seed industry for a species is directly depending on the benefits and final net income of the farmer. The more value generated for the farmer by the variety, the more expensive the seeds can be. This positive feedback leads to investment and incessant growth of the species, which gradually takes "all the place." We can call it the *seed profitability loop*; we consider this operating mode as the *first business model* that is the traditional and dominant model today.

This is even accentuated by the possibilities of downstream development provided by the fact of supplying raw materials at a reduced price. Thus, we can now find new chips made of maize flour to the detriment of the potato (e.g., the differentiation of Lay's (Pepsico®) with its new product Fritos original corn Chips). That means the downstream industry unconsciously adapts itself to this business model and does not ask for improvement of the species originally used. In simple words, if the industry can make the final product better or cheaper with maize than with another raw material it will go for it. Two species are in the process of "taking all the stake" in seed investment: maize and tomato. Does this mean that they are the most promising species for the future of processors and consumers?

Analysis of 24 crop species seems not to support this conclusion and suggest more analysis to explicit a different business model (see Table 3B.1). Indeed,

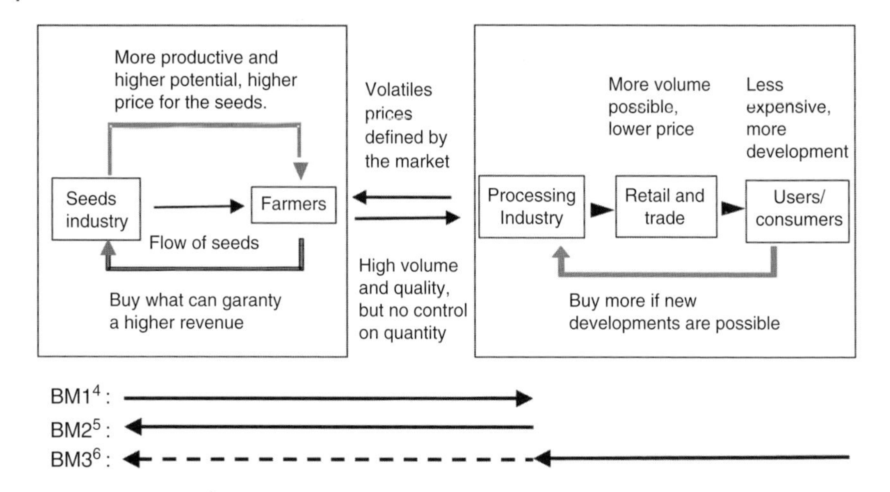

Figure 3.6 Relationship among seed industry, farmers, and the actors in the food chain.
[4]Business Model 1: From Seed Industry to Processing Industry
[5]Business Model 2: From Processing Industry to Seed Industry
[6]Business Model 3: From Consumers to Processing Industry. Dotted arrow to show the Integrated BM from consumers to Seed Industry.

among these apple, papaya, eggplant, and green bean are in the lead of the race for yield growth (which doubles), without any increase of the growing acreage. The lowest yields were for peas and cocoa (negative trend); orange, millet, cassava, barley, oats, and olives are a little better. Wheat, rice, potatoes, kiwi, but also tomato, are rather below average. In those above average, there is certainly corn, but also asparagus, banana, and coffee, but the latter is the only one without increase of area. Several different reasons could explain a slow growth yield: the achievement of significant yields relative to the potential of the plant, such as rice, wheat, orange, potato. These species resist to the domination of the first business model and suggest an impact of processing industry on plant breeding. The knowledge of plant breeding in these species reveals that on potatoes, wheat, orange (generally *citrus*), and rice, plant breeding actions have been performed to propose cultivars for improvement of the downstream industry. One could conclude that the needs of downstream industry might have impulse long-term R&D because they knew that large possibilities did exist on consumer markets.

Unfortunately that is probably not the real situation for most of these crops. The Table 3B.1, comparing the evolution of production and cultivated acreage between 1990 and 2014, is reflecting a global evolution between two given years and may not be directly representative of the trend. In addition it reflects the global increase of productivity that might be as a result as much to crop protection than to genetic innovation. Taking the example of coffee, it considers both Arabica and Robusta species and the positive evolution of average yield/ha is mainly because the increase of importance of Robusta (better yield than Arabica), during the 1990s in the global production. Vietnam, the main Robusta producer, became the number-2 coffee producer in the world during this period.

However, if this situation came to happen we could consider it as an emergence of a *second business model*: the downstream, industry provoking an evolution of the seed profitability loop. A better understanding about how this work needs more precise analysis.

There is a specific analysis about the species with very weak R&D investment (cocoa, cassava, oats, olive, millet, barley, peas, coffee). Peas and cocoa are typical examples, in this selected group of the cultivated species, of growth driven by demand, with low R&D investment for various reasons: surface is still limited and lack of involvement of downstream industries. Coffee is in the same situation with two specificities: on the one hand, particularly low yields, and on the other hand, the high value of its production and its importance in international trade. This might lead to a *third business model*, by which the consumers demands for traceability and quality would be taken into account by retailers and processing industries, but, that did not change the seed profitability loop. In this model, the risk is that the cost of raw material increases, because of the decrease of cultivated areas. An *integrated third model* could be that the processing industries, even in competition, could find a way for specific plant breeding.

One of the problems in this kind of vertical integration of breeding by the downstream industry is the following: what kind of business model could create value for the industry or the retailers without creating strong image risks on the end products? Could the company take some industrial property rights and value on the created varieties without being accused of squeezing poor growers, especially when talking about tropical crops produced by small holders of emerging countries? The only way that has yet been found is to create a consortium of food companies running precompetitive R&D that would benefit the whole community. However, it is quite difficult to organize and manage, and in most cases, it relies on unstable resources.

This analysis revealed also that nearly 60% of the R&D of the seed industry is focused on crops for animal feed, because this is where the seed profitability loop is optimized. It concerns commodities for which only the cost of production comes into account. The processing tomato may know this benefit, thanks to the industry of tomato sauce that is vertically integrated through contract farming because of the need for a regular supply of fresh fruits every day during season. This is probably the reason for its lesser advantage than for maize and soybeans (the latter dominates R&D on fabaceae). It is therefore understandable that the seed industry is ultimately mainly controlled by American and European groups. The construction of significant research potential has taken place through this process, which requires high consumption of meat produced from the cheapest feed. So one could claim that commodity production for industry is an adaptation of downstream industry according to the *first business model.*

What would be the consequences of such plant-breeding situation in the long run? Analysis of the evolution of maize and soybeans suggests a rather worrying scenario. The increase in synergy between farmers and seed producers, thanks to higher yields and lower production costs, results in a high overall growth of these crops and their gradual domination of a small number of cultivated species. Large-scale crops would move toward monoculture, and other species could simply be abandoned by the farmers.

Indeed, on a world scale, it becomes unreasonable to put a hundred times more investment on maize than on cassava, for example, even if the strategic and financial choices of the breeding companies look reasonable from a pure financial point of view. We can conclude that R&D investment is pulled by the seed profitability loop and really not by any optimization in affecting resources on the basis of global alimentation needs, value of the final products, and no more in a sustainable approach.

What is an "orphan species"? It is a species known and even appreciated, but that has not entered the seed profitability loop described for various but generally well-identified technical and business reasons (e.g., genetics, biology of reproduction, protectability, size and value of seed market). This may be a species identified as important for nutritional, agronomic, ecological or environmental, economical and societal reasons, but it was not possible to construct this loop of mutual benefit between seed producers and their customers. This would require prior support from consumers and the agri-food industry, such as what appears to be taking place for certain vegetables for frozen and canned products where integrated models, are progressively set up with exclusive varieties customized for the processors.

Table 3B.1 also makes it possible to conclude that the improved yields and production costs offered by plant breeding tools have also benefited other species not included in the profitability loop. For 7 species, yields have increased by more than 50% and for another 10 by 20% to 35%, but these averages mask considerable variability depending on the region, either for agronomic reasons, or access to certified seed. Technologies developed thanks to R & D investments made on the dozen species that account for 90% of the turnover in the seed industry could be transferred to orphan species, but that would require to create value on seeds or plants of these species and to build favorable conditions at the regulatory and organizational levels in the different local contexts.

The distortion between R&D in plant breeding and global needs, as well quantitative than qualitative, in food production, can be well understood. Whatever are the strategies of seed industries, they are driven by the seed profitability loop and not by a potential interest for the food chains and the consumers. This profitability loop is built on three major characteristics: size and renewal rate of the market, possibility to improve yields by protected varieties (attractivity for the farmers), and technical feasibility for fast improvement. It has no connection with environmental needs or nutritional needs.

3.5 A First Set of Conclusions and Recommendations

Agriculture will have to adapt its productions to climate change, but it can also mitigate its effects, and plant breeding must play an important role in association with the new systems of production (i.e., agro-ecology, conservative agricultures). This is already largely recognized.

The first immediate conclusion is that increasing the production of biofuels represents a major risk to long-term food security without resolving otherwise the global issue of energy (i.e., production and consumption). The replacement of fossil energy by biomass has a limited impact, as even the net primary production cannot reach 10% of energy value of fossil energy already used. Therefore, policies that encourage the use of biofuels from potential food production should be reconsidered cautiously. Except if new development would allow to use by products of agriculture, it would be much better not to involve too much, on a long-term vision, agriculture production for energy. Policies that favor the breeding in most species, supporting the needs of consumers/citizens and food industry would be more adapted.

The second conclusion is that cattle and dairy production should not compete too much for croplands because the energy transformation rate is high, and fabaceae products could be much better for nutrition purposes. Apart for soybean breeding that has never been really oriented to make substitutes of animal-based food products, nearly all fabaceae are orphan species for breeding programs.

The third conclusion is that the fast evolutions of technologies useful for plant breeding have broadened the possibilities for crop improvement. Whatever the biology of reproduction of the plant species, a plant-breeding program can be designed for improving yield, disease resistance, adaptation to abiotic or biotic stresses, adaptation to new systems of production, including less inputs, environment protection, and energy consumption. We have now the technical possibilities to find genetic answers to the challenges of the future, in interaction with a lot of innovation in NTIC, robotic, and new agricultural systems. The large range of technologies offers the breeders a choice of strategies depending on the aims and on the genetic traits to be combined, but also depending on national or international regulations about the different technologies. It is not the purpose of this chapter to argue about transgenic or gene-edited plants, or about the recognition of the different used methodologies for obtaining new combinations of traits. But it has to be considered that for any powerful technique, heavy and expensive regulatory controls or studies for authorization already led to an amazing globalization of seed companies. Any low-cost powerful technique will help any small or local company to focus on an orphan species.

With all we know today about genetic recombination, the recommendation could be not to unduly increase the cost for launching a new variety onto the market. Any unnecessary added cost for launching a new variety onto a market is an impediment for small breeders to access to the market.

In developing countries, about 80% of the necessary increase in production will come from yield increase, and probably less than 20% could come from an expansion of arable land. The growth rate of yield for the crops on which breeding companies invest massively declines regularly despite increasing biggest R&D investments, and the strategic choices of the seed companies may therefore have negative effects in the medium and long terms. The recommendations in these conditions are to help, by political and regulatory actions, the emergence of small producers of certified seeds for varieties adapted to the local conditions. Once these productions are set up, there are possibilities for the breeders to increase the value of production of farmers and to create some new seed profitability loops. Investment in agricultural R&D has been neglected in most GDP countries. Because R&D in developing countries are dominated by the public sector, additional investment will have to come from public money, unless breeding companies will change their strategy and offer new opportunities for local breeders with possibly new local regulations.

The fourth conclusion is that, although billions of dollars are invested into R&D by seed companies, there is an excessive concentration of investment on a few numbers of species that are now close from their maximum potential of production. The low R&D investment on some important species for food, as, for instance, cassava, sorghum or any other orphan species, although important in international trade, nutritional, social and environmental values, creates a real disadvantage for producing them and therefore for some countries to import other crops while the local ones might be used for the same food purpose. It is really possible that each dollar invested in orphan crops could offer better benefits for food security, nutritional, and environmental impacts. It depends of new business models to be invented for providing a fair return on investment to the breeding companies that would decide to consider them.

The question becomes: how to build new business models according to the real needs of the society in the next 30 to 50 years? *Business model 1*, which is built only on that we named the seed profitability loop, is not sustainable, if being the only one driving plant-breeding activity. The *business model 2* looks to have a much better effect, although it is not sufficient to save the orphan species, the production of which could decrease inexorably if a third new model does not emerge. It is interesting to understand that the business model 2 has been developed according to different criteria depending on the concerned species (e.g., public R&D, foundations, downstream industries or seed companies), only in developed or middle developed countries. The potential involvement of the downstream stakeholders of the chain and food industry is also dependent on the crop. Like it was for many years with dairy products, the food industry may be vertically integrated (contract farming) for crops that have to be managed within few hours after harvest as it is for frozen and canned vegetables, tomato paste, or for specific products (baby foods). In this business model, plant breeding may be driven by the specifications of the processor in a fair value added sharing model with the partner from the seed industry. However, as for one given crop, the customizing to one single food processor needs may not justify to enter into breeding of a new species that is limited to crops where seed industry already has an activity driven by the seed profitability loop.

The commitment of downstream industries in plant breeding is limited to the crops where the industry has clearly identified risks on the supply chain or traits and specifications of interest for them, but also a dynamic and responsible marketing strategy of the derived products. However, there are examples where it is not sufficient. For instance, tequila production, oatmeal, and quinoa-based dishes, until now have not yet lead to significant plant-breeding programs, although it would be possible. The major issue is to become fully conscious about risks and benefits for a product based on a given crop that could not be easily substituted by another plant raw material and that is not in a dynamic process of improvement. In a competitive market, the upstream involvement of food industry should not create a competitive disadvantage on costs. For increasing interaction between plant breeding and downstream industries, one prerequisite would be to permit, through regulation, agreements between competitors for joint involvement in upstream R&D (i.e., plant breeding). This is particularly important when the final consumer is active nearly everywhere in the world and the agriculture (e.g., coffee, tea, cocoa, vanilla) is only in countries with low labor cost. Low labor cost of agriculture and high value of the final product, associated with competing multinational food companies, is a system that prevents genetic improvement because it supports short-term vision and only technical innovation in processing and final products according to short-term consumer insights. The impediment could come from a necessary agreement between competitors that might have an impact, on short term, in the final price of their products.

Many possibilities do exist. One would be a real and fair sharing of the downstream added value with the upstream by subsidizing the purchase of improved plants or seeds by the producer. This can also come from a gain in productivity and profitability for producers that will allow them to self-finance more expensive plants or seeds. This may first require a loan on future crop to initiate a virtuous circle. For some species, this may justify direct verticalization or crop contracting. In this case, end producers engages themselves directly, in-house or outsourced, in the supply of its plant raw material.

The last recommendation would be to have a general pragmatic position on a case-by-case basis, in the food chains, as well as for plant species or accessible technologies and when needed to involve regulation authorities.

3.6 Summary and Major Learning

- For quite diverse demographic, environmental, societal reasons food production might be facing shortage of plant raw materials in the next 20 to 30 years
- The productivity of agriculture will therefore have to meet these challenges, especially if nonfood demands of agriculture products continue to increase.
- It is generally admitted that most of these gains of productivity will come from genetic improvement of the cultivated varieties.
- The development of new agronomic practices such as precision agriculture, inter-cropping, and others will also require the development of adapted varieties.
- Plant breeders have developed and validated numerous conventional and biotechnological (molecular and cell biology) techniques on main field crops and vegetables. They now have a quite diversified toolbox to tap in according to the crop, the objectives, and the breeding.
- The economics of seed industry has led it to focus on a small number of crops where seed sales is highly competitive and profitable: 33% of private R&D expenses in field crops are dedicated to maize alone and 50% of private R&D expenses in vegetables are dedicated to tomato alone.
- This situation has led to the creation of numerous orphan crops, not receiving significant genetic innovation that are quantitatively and qualitatively strategic for many food products.
- This situation is increasing the production and income gaps for the farmers between modern and orphan crops, leading to a reduction of the diversity of cultivated species but also to big risks of shortage.
- Food industry will have to take or reinforce consortium initiatives as the Sustainable Agriculture Initiative (SAIN) or World Coffee Research and will also have to be investing in new business models where part of the added value will be effectively shared upstream, not only with the farmers, but also to their seed or plant suppliers.

3.7 Appendix Tables

Table 3A.1 Areas, production, and calorie intake.

Product	Quantity	Acreage (ha)	Yield (t/ha)	Areas (%)	Dry matter	Millions Kcal	Kcal (%)
Sugar cane	1 884 246 253	27 124 723	69,47	2.36	188 424 625	753 698 501	6.13
Maize	1 037 791 518	184 800 969	5,62	16.08	882 122 790	3 175 642 045	25.82
Rice, paddy	741 477 711	162 716 862	4,56	14.16	630 256 054	2 268 921 796	18.44
Wheat	729 012 175	220 417 745	3,31	19.18	619 660 349	2 230 777 256	18.13
Potatoes	381 682 144	19 098 328	19,99	1.66	57 252 322	206 108 358	1.68
Soybeans	306 519 256	117 549 053	2,61	10.23	260 541 368	937 948 923	7.62
Sugar beet	269 714 066	4 471 580	60,32	0.39	40 457 110	16 182 844	0.13
Cassava	268 277 743	23 867 002	11,24	2.08	40 241 661	144 869 981	1.18
Tomatoes	170 750 767	5 023 810	33,99	0.44	25 612 615	92 205 414	0.75
Barley	144 489 996	49 426 652	2,92	4.30	122 816 497	442 139 388	3.59
Bananas	114 130 151	5 393 811	21,16	0.47	17 119 523	61 630 282	0.50
Sweet potato	106 601 602	8 352 323	12,76	0.73	15 990 240	57 564 865	0.47
Rape Seed	73 800 809	36 117 722	2,04	3.14	62 730 688	376 384 126	3.06
Sorghum	68 938 587	44 958 726	1,53	3.91	10 340 788	37 226 837	0.30
Yam	68 132 129	7 755 803	8,78	0.67	10 219 819	36 791 350	0.30
Coconut	60 511 756	11 939 801	5,07	1.04	9 076 763	32 676 348	0.27
Oil, palm	57 328 872	18 697 276	3,07	1.63	57 328 872	458 630 976	3.73
Groundnut	43 915 365	26 541 660	1,65	2.31	26 349 219	158 095 314	1.29
Sunflower seed	41 422 310	25 203 554	1,64	2.19	35 208 964	211 253 781	1.72
Plantains	30 667 662	4 495 057	6,82	0.39	4 600 149	16 560 537	0.13
Millet	28 384 668	9 591 795	2,96	0.83	24 126 968	86 857 084	0.71

(Continued)

Table 3A.1 (Continued)

Product	Quantity	Acreage (ha)	Yield (t/ha)	Areas (%)	Dry matter	Millions Kcal	Kcal (%)
Beans, dry	26 529 580	30 612 842	0,87	2.66	22 550 143	81 180 515	0.66
Oats	22 721 702	5 298 873	4,29	0.46	19 313 447	69 528 408	0.57
Beans, green	21 720 588	1 527 613	14,22	0.13	3 258 088	11 729 118	0.10
Peas, green	17 426 421	2 356 340	7,40	0.20	2 613 963	9 410 267	0.08
Triticale	16 953 565	4 135 952	4,10	0.36	14 410 530	51 877 909	0.42
Olives	15 401 707	11 000 000	1,40	0.96	2 772 307	22 178 458	0.18
Rye	15 242 551	5 306 288	2,87	0.46	12 956 168	46 642 206	0.38
Chick peas	13 730 998	13 981 218	0,98	1.22	11 671 348	42 016 854	0.34
Peas, dry	11 186 123	6 931 941	1,61	0.60	9 508 205	34 229 536	0.28
Dates	7 600 315	3 500 000	2,17	0.30	6 080 252	21 888 907	0.18
Sesame seed	6 235 530	10 819 558	0,58	0.94	5 300 201	31 801 203	0.26
Cow peas, dry	5 589 216	2 178 613	2,57	0.19	4 750 834	17 103 001	0.14
Pigeon peas	4 890 099	826 523	5,92	0.07	4 156 584	14 963 703	0.12
Lentils	4 827 122	622 427	7,76	0.05	4 103 054	14 770 993	0.12
Beans, dry	4 139 972	2 150 905	1,92	0.19	3 518 976	12 668 314	0.10
Chestnut	2 051 564	530 809	3,86	0.05	1 641 251	5 908 504	0.05
Buckwheat	1 924 082	2 011 289	0,96	0.17	1 635 470	5 887 691	05
Lupins	1 014 022	31 432 088	0,03	2.73	861 919	3 102 907	0.03
Hazelnuts	713 451	462 489	1,54	0.04	428 071	1 541 054	0.01
Quinoa	192 818	195 342	0,99	0.02	163 895	590 023	0.00
Brazil nuts	109 300	11 300	9,67	0.00	65 580	236 088	0.00
Total		**1 149 436 662**				**12 301 421 666**	**100**

Source: ©FAO, FAOSTAT 2014

Table 3B.1 Area, yield and production from 1990 to 2014 for 24 cultivated species.

		Area harvested (ha)	Yield (t/ha)	Production (t)	Area increasing	Yield increasing	Production increasing
Apple	1990	5 146 764	7,98	41 047 874			
	2014	5 051 851	16,75	84 630 275	−1.8%	110.0%	106.2%
Asparagus	1990	580 105	3,47	2 012 830			
	2014	1 451 123	5,40	7 830 219	150.1%	55.5%	289.0%
Bananas	1990	3 774 758	13,23	49 932 878			
	2014	5 393 811	21,16	114 130 151	42.9%	60.0%	128.6%
Barley	1990	74 186 774	2,41	178 937 965			
	2014	49 426 652	2,92	144 489 996	−33.4%	21.2%	−19.3%
Beans, green	1990	938 951	6,30	5 912 848			
	2014	1 527 613	14,22	21 720 588	62.7%	125.8%	267.3%
Cassava	1990	15 185 262	10,03	152 243 051			
	2014	23 867 002	11,24	268 277 743	57.2%	12.1%	76.2%
Cocoa	1990	5 712 868	0,44	2 531 907			
	2014	10 434 201	0,43	4 450 263	82.6%	−3.8%	75.8%
Coffee	1990	11 250 724	0,54	6 062 766			
	2014	10 485 408	0,84	8 790 005	−6.8%	55.6%	45.0%
Eggplants	1990	836 066	13,59	11 359 951			
	2014	1 870 728	26,83	50 193 117	123.8%	97.5%	341.8%
Kiwi fruit	1990	66 715	12,64	843 102			
	2014	219 134	15,73	3 447 604	228.5%	24.5%	308.9%
Lentils	1990	3 219 508	0,80	2 563 048			
	2014	4 524 043	1,07	4 827 122	40.5%	34.0%	88.3%
Maize	1990	131 038 436	3,69	483 623 773			
	2014	184 800 969	5,62	1 037 791 518	41.0%	52.2%	114.6%

(*Continued*)

Table 3B.1 (Continued)

		Area harvested (ha)	Yield (t/ha)	Production (t)	Area increasing	Yield increasing	Production increasing
Millet	1990	37 496 606	0,80	30 004 192			
	2014	31 432 088	0,90	28 384 668	−16.2%	12.9%	−5.4%
Oats	1990	20 691 016	1,93	39 938 278			
	2014	9 591 795	2,37	22 721 702	−53.6%	22.7%	−43.1%
Olives	1990	7 402 116	1,22	9 023 959			
	2014	10 272 547	1,50	15 401 707	38.8%	23.0%	70.7%
Oranges	1990	3 175 061	15,66	49 707 067			
	2014	3 885 966	18,23	70 856 360	22.4%	16.5%	42.5%
Papayas	1990	232 096	14,18	3 290 566			
	2014	411 355	30,80	12 671 038	77.2%	117.3%	285.1%
Peas, dry	1990	8 677 343	1,92	16 638 494			
	2014	6 931 941	1,61	11 186 123	−20.1%	−15.8%	−32.8%
Peas, green	1990	973 182	7,70	7 491 306			
	2014	2 356 340	7,40	17 426 421	142.1%	−3.9%	132.6%
Potatoes	1990	17 659 630	15,11	266 827 112			
	2014	19 098 328	19,99	381 682 144	8.1%	32.3%	43.0%
Rice, paddy	1990	146 987 918	3,53	518 579 272			
	2014	162 716 862	4,56	741 477 711	10.7%	29.2%	43.0%
Soybean	1990	57 207 200	1,90	108 455 175			
	2014	117 549 053	2,61	306 519 256	105.5%	37.5%	182.6%
Tomato	1990	2 901 330	26,31	76 328 887			
	2014	5 023 810	33,99	170 750 767	73.2%	29.2%	123.7%
Wheat	1990	230 750 672	2,56	591 324 680			
	2014	220 417 745	3,31	729 012 175	4.5%	29.1%	23.3%

Source: ©FAO, FAOSTAT 1990–2014

References

AGRA. (2015). Programs Performance Progress Reports 2007–2015. Available from: https://agra.org/wp-content/uploads/2016/04/2015-Progress-Report.pdf [Accessed June 3, 2017].

Bharadway, D. N. (2016). Sustainable agriculture and plant breeding. *Advances in Plant Breeding Strategies: Agronomic, Abiotic and Biotic Stress Traits.* 2, 3–34.

Byrnes, J. E. K., Gamfeldt, L., Isbelle, F., Lefcheck, J. S., Griffin J. N., Hector, A., Cardinale, B. J., Hooper, D. U., Dee, L. E., & Duffy, J. E. (2014). Investigating the relationship between biodiversity and ecosystem multifunctionality: Challenges and solutions. *Methods in Ecology and Evolution.* 5 (2), 111–24

Corre-Hellou, G., Faure, M., Launay, M., Brisson, N., & Crozat, Y. (2009). Adaptation of the STICS intercrop model to simulate crop growth and N accumulation in pea–barley intercrops. *Field Crops Research.* 113, 72–81.

Dubois. M., J., F., & Sauvée, L. (Eds.). (2016). *Évolution agrotechnique contemporaine—Les transformations de la culture technique agricole.* UTBM.

FAO. (2016a). Available from: http://www.fao.org/faostat/en/#data/QC [Accessed June 3, 2017].

FAO. (2016b). Produce more with Less. Available from: http://www.fao.org/ag/save-and-grow/fr/accueil/index.html [Accessed June 3, 2017].

FAO. (2016c). Available from: http://www.fao.org/fileadmin/templates/wsfs/docs/Issues_papers/Issues_papers_FR/Comment_nourrir_le_monde_en_2050.pdf [Accessed June 3, 2017].

FAO. (2016d). Available from: http://www.fao.org/world-food-day/2016/theme/en/ [Accessed June 3, 2017].

France AgriMer. (2012). Variétés et rendement des céréales biologiques 2011. Available from: http://www.franceagrimer.fr/content/download/18021/142430/file/11%20-%20Etude%20FAM%20-%20Vari%25C3%25A9t%25C3%25A9s%20et%20rendements%20c%25C3%25A9r%25C3%25A9ales%20bio%20r%25C3%25A9c%202011.pdf [Accessed June 3, 2017].

Guéguen, J., Walrand, S., & Bourgeois, O. (2016). Plant proteins: Context and potentialities for human food. *Cahiers de Nutrition et de diététique.* 51 (4), 177–85.

Johnson, D. L., Ambrose, S. H., Bassett, T. J., Bowen, M. L., Crummey, D. E., Isaacson, J.S., Johnson, D. N., Lamb, P., Saul, M., & Winter-Nelson, A. E. (1997). Meanings of environmental terms. *Journal of Environmental Quality.* 26, 581–89.

Loyce, C., Meynards, J. M., Bouchard, C., Rolland, B., Lonnet, P., Bataillon, P., Bonnefoy, M., Charrier, X., Debote, B., Demarquet, T., Duperrier, B., Félix, I., Heddadj, D., Leblanc, O., Leleu, M., Mangin, P., Méausoone, M., & Doussinault, G. (2012). Growing winter wheat cultivars under different management intensities in France: A multicriteria assessment based on economic, energetic and environmental indicators. *Field Crops Research.* 125, 167–78.

Meynard, J.-M., Jeuffroy, M.-L., Le Bail, M., Lefèvre, A., Magrini, M.-B., & Michon C. (2016). Desining coupled innovations for the sustainability transition of agrifood systems. *Agricultural Systems.*

Niels, L., Hans, D., Geertrui Van, O., Hans, R., Anthony, A., Derek, E., Annemiek, N. (2009). *Breeding business. The future of plant breeding in the light of developments in patent rights and plant breeder's rights.* Centre for Genetic Resources, the Netherland (CGN), Wageningen University and Research Centre.

Singh, A. K., Singh, A. K., Misha, A., Singh, L., & Dubey, S. K. (2016). Improving lentil (*Lens culinaris*) productivity and profitability through farmer participatory action research in India. *Indian Journal of Agricultural Sciences.* 86 (10), 1286–92.

Singh, R. P. Prasad, P. V. V., & Reddy, K. R. (2015). Climate change: Implications for stakeholders in genetic resources and seed sector. *Advances in Agronomy.* 129, 117–80.

In Souza, G. M., Victoria R., Joly, C., & Verdade L. (Eds.). (2015). *Bioenergy & Sustainability: Bridging the Gaps.* Paris: SCOPE (Scientific Committee on Problems of the Environment).

Thompson, P. L., & Gonzalez, A. (2016). Ecosystem multifunctionality in metacommunities. *Ecology.* 97 (10), 2867–79.

Tóth, G., Stolbovoy, V., & Montanarella, L. (2007). Soil quality and sustainability evaluation—An integrated approach to support soil-related policies of the European Union. EUR 22721 EN. Luxembourg: Office for Official Publications of the European Communities.

World Bank. (2016). Cereal yield (kg per hectare). Available from: http://data.worldbank.org/indicator/AG.YLD.CREL.KG [Accessed June 5, 2017].

World Population Prospect. (2012). Available from: https://esa.un.org/unpd/wpp/publications/files/key_findings_wpp_2015.pdf [Accessed June 3, 2017].

4

The Agriculture of Animals: Animal Proteins of the Future as Valuable and Sustainable Sources for the Food Industry

Some animals utter a loud cry. Some are silent, and others have a voice, which in some cases may be expressed by a word; in others, it cannot. There are also noisy animals and silent animals, musical and unmusical kinds, but they are mostly noisy about the breeding season.

—Aristotle

4.1 Livestock and Animal Husbandry

A large part of agriculture is production from livestock, commonly called animal husbandry. To better understand the terminology livestock production, we reference the following:

> Domesticated livestock have played a pivotal role in the development of human civilizations around the world and continues to be an integral part of human culture, society, and the local and global economy. Domestic livestock has contributed to the rise of human societies and civilizations by increasing the amount of food and nutrition available to people in four ways: by providing sources of meat, milk, and fertilizer, and by pulling plows. Throughout history livestock have also provided leather, wool, other raw materials, and transport. Livestock furnish high quality protein and energy foods, and function as part of integrated, renewable systems of plant and animal agriculture. The digestive systems of ruminant animals such as cattle, sheep, goats, llamas, and camels are specially adapted to convert plant materials that humans cannot utilize into proteins of high biological availability to humans. ("Livestock Production," n.d.)

Megatrends in Food and Agriculture: Technology, Water Use and Nutrition, First Edition.
Helmut Traitler, Michel Dubois, Keith Heikes, Vincent Pétiard and David Zilberman.

To make sure of our understanding we can look at how the term *livestock* is defined.

> Livestock are domesticated animals raised in an agricultural setting to produce commodities such as food, fiber, and labor. The term is often used to refer solely to those raised for food, and sometimes only farmed ruminants, such as cattle and goats. ("Livestock," n.d.)

Animal husbandry takes the definition a step further as we see here.

> Animal husbandry is the management and care of farm animals by human beings, in which genetic qualities and behavior, considered to be advantageous to humans, are further developed. The term can refer to the practice of selectively breeding and raising livestock to promote desirable traits in animals for utility, sport, pleasure, or research. ("Animal husbandry," n.d.)

The production of meat, milk, and fiber by animals has progressed along with civilization and changed from nonspecialized production (e.g., the backyard cow, chicken, and sow) to the present situation of highly specialized farms with high levels of production from one species, at least in developed countries. This is obviously not the case around the world because there is a huge range of animal production systems ranging from small-holders with one cow or goat to farms and ranches with thousands of cattle or pigs in one production system.

As global income has risen in the past several decades, demand for animal products has increased. Despite for calls for "Meatless Mondays" and a vocal segment of the population that is either vegan or vegetarian, the demand for products derived from livestock has grown and the trend continues.

4.1.1 How We Got to Now

Although history from deep in the past is always interesting, we will briefly focus more on the past century and half and the developments that have led to today's animal agriculture. In 1862 90% of the U.S. population was farmers (today that is less than 2%), and a similar percentage was true throughout the world. There was little mechanization, therefore labor was provided by humans or draft animals. This meant that farm size was necessarily small to be able to handle all the chores and allow the draft animals to eat and rest before working again the next day.

The invention of mechanical equipment and rural electrification greatly changed the need for labor on farms, and along with it, was cause for a massive migration from farms to urban areas. This had a huge impact on the production of crops, which changed the path of production from animals as well. Now, crops could be grown in larger and larger quantities on farms, and the way to care for animals was also changed drastically.

Instead of needing to use large amounts of feed for draft animals such as horses and mules, that feed could now be used for more specialty production for meat, milk, and eggs. Electricity changed the face of dairy production, with dairy farmers previously being limited by how many cows they could milk by hand. That certainly is not a problem in today's production environment.

Breeds of animals have been developed primarily based on the geographic region where they originated. Thus, you can easily guess where the Guernsey cow, Hereford bull, and Yorkshire pig were developed. As transport of animals increased, breeds were spread far and wide, and today in some species only a few breeds are dominant. Very few of these are "pure" though, with only a small percentage of farms maintaining "purebred" status and not using genetics that might come from outside of the breed. More discussion on breeding will come in later in the chapter.

Around the world, farm size continues to grow in both hectares and animal numbers. A large farm is always subject to debate in terms of actual scale, but the trend has been underway for many years and does not appear to be slowing. A few years back, I met with a group in South Africa and later in the week with a group from Northern Ireland. Although the definition of what was a large farm changed significantly from one venue to the next, the conversation about the farms was exactly the same!

This does not mean there will be no small farms; in fact statistics tell us that the number of small farms has been increasing in some areas. However, it is realistic to predict that large farms will increasingly produce a higher percentage of the world's total production. In another section farm size and the potential impact in developing countries will be discussed.

With 1.3 billion people working in the livestock production and marketing chain around the globe, many people rely on livestock to provide them food or a livelihood. The importance of animal agriculture's impact on the global economy should not be underestimated!

4.2 Animals: A Source of High-Quality Proteins

Animals provide proteins that are the basis for many diets around the world. This is a simple statement but an important one to remember as we consider global diets and how they might develop in the future. Consumption of animal products has contributed to better nutrition in many countries.

According to the FAO (2009),

> at the global level, livestock contribute 15 percent of total food energy and 25 percent of dietary protein. Products from livestock provide essential micronutrients that are not easily obtained from plant-based foods. Livestock contribute 40 percent of the global value of agricultural output and support the livelihoods and food security of almost a billion people. The livestock sector is one of the fastest growing parts of the agricultural economy, driven by income growth and supported by technological and structural change. The growth and transformation of the sector offer opportunities for agricultural development, poverty reduction and food security gains, but the rapid pace of change risks marginalizing smallholders, and systemic risks to the environment and human health must be addressed to ensure sustainability.

The World Health Organization (n.d.) concludes that nutritional needs for high quality proteins of a major part of the world population can best be satisfied through livestock products. Livestock products can also supply a wide variety of crucial micronutrients such as iron and zinc as well as vitamin A. Although livestock products in developing countries are mainly consumed for their nutritional value and taste, excessive consumption of these food sources may result in critically high amount of fats.

Doing a search for "animal protein" or something similar is an interesting exercise. The results show pages that range from complete support of including animal proteins in the diet to the complete opposite with strong positions about excluding animal protein. The inquirer can basically confirm his or her perspective. What there is agreement on is that animal proteins are of high quality and similar to proteins that exist in humans and are considered in most instances to be a good source of dietary protein.

One of the challenges that exist in today's world is to sort out what is fact and what is fiction. Anyone with a little technical know-how can write a blog and post it on their Web site, and with the right optimization tools can have the page show up early in a Web search. But doing so does not make the page accurate, and some of the worst offenders are also some of the most popular pages! Nevertheless, it seems clear that a diet that includes animal protein in some form is beneficial to the consumer, and even though there will be people with a variety of reasons who choose to have a diet that excludes some or all animal proteins, the clear majority of the world will continue to choose to eat animal proteins.

4.3 Animal Protein Demand in Emerging Markets

By 2050 an expanded world population will be consuming two thirds more animal protein than it does today, bringing new strains to bear on the planet's natural resources, according to FAO (2011). Populations and income growth are fueling an ongoing trend toward greater per capita consumption of animal protein in developing countries. Meat consumption is projected to rise nearly 73 percent by 2050; and dairy consumption will grow 58 percent over current levels.

Between 1970 and 2013 global livestock production almost tripled. The growth was not being driven by developed countries, but rather by rapid expansion of meat, milk, and egg production in the developing world. World Bank statistics for the period of 2000 to 2013 show their livestock production index for the world to increase from 90.2 to 117.1 (2004–2006 = 100), with low-income countries increasing from 83.8 to 123.5, low-middle-income countries from 83.4 to 125.1, high-middle-income countries 85.5 to 125.1, and high-income countries showing an increase from 98.6 to 104.4. The trends are not only driven by more population growth in emerging markets but by a growing desire for animal products as consumers have more income available (World Bank, n.d.).

To add a little more understanding, the following can be summarized from a World Health Organization (n.d.) report.

> Due to the pressure from a combination of population growth, growing incomes as well as growing urbanization the sector of livestock is growing at its fastest rate yet. This simply reflects the importantly growing demand for high value proteins from animals. While in the years around 1998 218 million tons of animal proteins were produced, it is expected that this number rises to 376 million tons in 2030.
>
> Not surprisingly, increasing income translates to increased consumption of animal proteins. A rather important decline of animal protein prices in recent years led to a further increase of consumption of such proteins especially in developing countries. As suggested above, growing urbanization is one of the most important drivers in this development, largely based on improved supply chain quality. It can be observed that populations in cities have easier access to a more diverse diet that is richer in proteins and fat from animal sources.

To further illustrate the change in use of animal products, the Table 4.1 provides projected global consumption trends for meat and dairy products through 2050.

According to Population Pyramid (n.d.), the global population is predicted to be 9.725 billion in 2050. As the population continues to expand one key question is whether the production and consumption trends established in the past decades will continue as they have.

Population growth, urbanization, and higher levels of income growth are all expected to continue their current trends, which points to the continued increase in demand for animal protein. And these trends are more extreme when looking at developing economies. Thus, although it is difficult to predict the future, it is a relatively safe bet that the desire for animal protein continues to increase.

Table 4.1 Projected meat and dairy consumption.

Projected total consumption of meat and dairy products ***(million tons)***					
	2010	**2020**	**2030**	**2050**	**2050/2010 % increase**
World					
All meat	268.7	319.3	380.8	463.8	173%
Bovine meat	67.3	77.3	88.9	106.3	158%
Ovine meat	13.2	15.7	18.5	23.5	178%
Pig meat	102.3	115.3	129.9	140.7	137%
Poultry meat	85.9	111.0	143.5	193.3	225%
Dairy not butter	657.3	755.4	868.1	1,038.4	158%
Developing countries					
All meat	158.3	200.8	256.1	330.4	209%
Bovine meat	35.1	43.6	54.2	70.2	200%
Ovine meat	10.1	12.5	15.6	20.6	204%
Pig meat	62.8	74.3	88.0	99.2	158%
Poultry meat	50.4	70.4	98.3	140.4	279%
Dairy not butter	296.2	379.2	485.3	640.9	216%

FAO 2011.

One thing is certain in looking to the future. Livestock production is globally quite diverse and will continue to be—ranging from intense large-scale management systems designed to maximize output and productivity to small holders operating at the lowest possible input and cost level. Depending on the geographical, economic, and political constraints, the system will vary from country to country, region to region, and even within a more localized area.

With that said, more and more of the total production around the developed world will come from medium to large farming operations as economies of scale come in to play more and more. This is not to say that smaller farms and production units will not have a place; they will if they are able to successfully market their products and capture more added value that goes directly to the producer. However, the trend that has been occurring for many years will continue. That is, mechanization, technology, and production systems that increase efficiency will drive farms to larger size and human labor will be replaced with mechanical labor.

Pigs and poultry have reached large-scale operations in much of the world, and cattle have tended to be less intensive in many parts of the globe. Efficiencies of scale have allowed this to happen and it will continue. The developing world will see the same trends in scale, which no doubt will raise questions around many social issues.

The dairy industry provides a good example of the wide range in size and scale of farms around the world. Small farms (less than 10 cows) make up 96% of the farms and have 57% of total dairy cow population. Medium-sized farms (11 to 100 cows) are 3.7% of the farms and own 27% of the cows, and large farms (>100 cows) are 0.3% of the global dairy farms, own 16% of the cows, but produce more than 40% of global production (International Farm Comparison Network, 2016).

Animal production (both farming and processing) employs a high percentage of the workforce in developing countries, and for them livestock is not only a way to eat, it can also be their bank account and is used as a risk-management tool. More than half of the people in Africa depend on agriculture for their living. So not only is the production of livestock important for developing countries from the standpoint of supplying needed nutrition, but it also provides employment in these economies.

Many opportunities are available for livestock production to increase in developing countries, but with that come challenges as well. Overall productivity of animals is quite low, and to increase output efficiency must be increased. There will be competing interests for resources such as land and water, as well as increasing scrutiny regarding health and welfare of animals and the need to adapt to a changing climate. All of this points to a need, maybe even a demand, that productivity must change. This can be accomplished but will not happen easily or without planning.

To increase efficiency numerous areas must be addressed at both a farm level and from a societal level. Included in a list would be improved feeding and nutrition, better veterinary care and disease control, breeding and genetics, animal welfare, application of technology, and governmental policies toward livestock farming. Later in this chapter we will address some of these issues but to address all would require multiple chapters or more probably an entire book.

4.4 Optimal Animal Welfare: Sustainable, Humane, and Healthy

For animal production to continue its expansion the production must be sustainable, humane, and healthy. This should be the goal of every farm, and realistically must be achieved if a farming enterprise is to be successful. But we must say that trying to provide a definition of these that all agree on can at times prove to be difficult.

Sustainability can be measured both as environmental sustainability and economic sustainability. To be successful for anyone raising livestock they must be able to do both. Humane seems to be a simple question, but again can create some discussion on exactly what is humane animal treatment. Healthy seems to be the easiest of these three points to agree on, but possibly not always depending on what defines a healthy animal. Most farms will tell you that their goal is to treat their animals in a manner that keeps them healthy and is humane so that they are productive and can contribute to the farm's economic progress. This might not be the case with every farm, but after a lifetime of working with farmers I am sure that is the sentiment in almost all cases. An animal that is sick or having physical issues is not productive and therefore costly to the farm.

I mention economics right away when introducing this topic for a reason; many people do not want to recognize that animal production is a business. For some it is a small business, and for some a large business, but in any case, it is what farmers depend on for their livelihood. So, this is where economic sustainability enters the picture. Decisions must be made on the healthy treatment of animals and also on the economic sustainability of a farm.

Much can and will be debated about how animals should be raised, how they should be housed, what amount of space is needed, and a myriad of other questions. I have seen animals housed in many types of systems from highly intensive to pastoral settings, and when properly managed,they can provide conditions that allow for good animal welfare.

Intensive farming systems are often criticized as being environmentally unsustainable. However, this is countered by increasing efficiency of animals in intensive systems as compared to extensive systems. The example of dairy production in the United States provides a good snapshot of this. According to the US Department of Agriculture (2002), in 1950 there were 22 million dairy cows in the United States that produced 116 billion pounds of milk. By 1998, the number of cows had decreased to 9.1 million and production has increased to 157 billion pounds, with the average cow producing 3.25 times more milk than in 1950. In 2016 the trend remained the same with 9.3 million cows producing a total of 216 billion pounds. The increased yield per cow increased 32% from 1998 to 2016, and when comparing 2016 to 1950 production per cow had increased 430% (US Department of Agriculture, NASS, n.d.).

Beef cattle, pigs, and poultry have similar stories to tell on increasing efficiency, but this is also a sustainability story. If livestock production is to continue to increase to meet global demand, the number of animals needed at current production levels cannot be supported. Each animal requires a certain amount of resources for basic metabolism. After that requirement is met, resources are used for production, whether meat, milk, or eggs. So, although a low-input system requires fewer resources per animal, to have equal production as a higher input system requires more *total* resources.

4.4.1 Animal Production Increase

For animal production to increase in the future, efficiency and better use of resources must occur. We cannot expect that we will obtain two-thirds more production by 2050 by increasing the number of animals by that amount. With no more land to find and increasing pressure from population growth to use the land having two-thirds more animals simply will not happen!

There will be many arguments about the statement just made, and much research will be undertaken to prove it one way or another. That is as it should be and will undoubtedly uncover new ways at looking at maximizing the limited resources we have.

Opportunities do and will exist in the future to make better use of resources, and research needs to continue in this area. As we have more animals, and they are more concentrated either on farms or near urban areas, the disposal of waste becomes a bigger issue. Animals have been identified as a major source of greenhouse gas emissions and that will need to be addressed. Some of this will be done through changes in management practices, which reduce the use of energy, through changing feedstuffs to those with a lower carbon footprint, and breeding programs that focus more on environment-related traits, to name a few possibilities.

In the United States, when everyone had a few cows, nobody really cared how the manure was disposed of and where, even though the cattle population was much larger as described previously. Now, with large number of animals in a single operation the volume of waste is massive, and proper disposal draws a lot of discussion.

One option that has emerged to help solve this problem is the use of anaerobic digesters. These can be installed on a farm and used to create energy from the methane gas that is created by the manure, in turn reducing the electricity cost for the farm and at the same time lessening the environmental impact. Though not widespread, use of digesters is increasing and holds the potential for reducing emissions. The by-product from digesters can still be used as fertilizer, which reduces the need for other inputs on land used for crops. As with any technology, the cost will be a key factor in the widespread adaptation. At present, few farms can afford to buy one, and must have a large enough scale of operations to keep the unit running efficiently. If (maybe when) a digester is economically viable and can be run on smaller farming operations, there will be widespread utilization.

To be sure, there is and will be a place for all types of management systems in the future —extensive grazing, free-range, organic, intensive, and so on. Much of this will be driven by local conditions, economics, and political considerations. A farmer operating in the Eastern Horn of Africa will not have to same situation as one in the Netherlands, California, or Japan. That is why it is difficult—no impossible—to describe what will be used throughout the world, as we will not have one-size-fits-all systems.

One of the paradoxes with a discussion on the humane treatment of animals is related to the discussion on efficiency. We want efficient animals, and we also want and need to treat them humanely. I use the word *efficient* from an economic viewpoint that we want to use environmental and economic resources efficiently, which typically means maximizing the producing units through better feed conversion, maximizing output, or some other related measurement. However, maximizing output also puts stress on animals, which can lead to higher rates of disease or death loss. There will be increasing pressure from the public to maximize the welfare of animals.

Much like a discussion on sustainability, discussion of humane treatment of animals has several sides. To some, the only humane treatment is to allow an animal to roam free and not ever be confined. There will be production systems that use this approach successfully. However, if we are to increase production to the extent predicted and needed, using intensive systems will be a necessity.

A current example of the differences in opinion of what is humane involves the area that I work in, that is artificial insemination of cattle. There have been some recent advertisements by an animal rights organization that the use of artificial insemination for cows was inhumane and was essentially raping cows. Yet if I look at the other side and think of the increased safety to both cows and humans by not having a bull present that could attack at any time, the argument does not seem to hold water and in fact seems to be absurd. The definition of humane treatment will continue to evolve, and it will be important for agricultural producers to balance the demand from consumers to treat animals while educating them on what is actually best practice for animal health and welfare.

The health of animals seems to be a more straightforward discussion, and it is hard to argue against having healthy animals. Again, this becomes a matter of what is the best way to achieve that objective. Technology will no doubt play a part in this as animal-monitoring systems in intensive management systems will become more important in the early detection and treatment of diseases. Breeding will also play a part in being able to genetically change the susceptibility to insects or diseases and therefore prevent illness from occurring.

An acknowledged risk with more intense livestock production is the potential for widespread disease outbreaks that can be transmitted through an entire herd or flock quickly. The movement of people and goods across long distances adds to the problem in that a disease on one continent could quickly be spread to another continent and animal population. Because of this, risk-mitigation measures must be in place to help reduce this possibility at the farm level as well as governmental level.

4.5 Animal-Breeding Programs

Selective breeding to improve animals has occurred for many years, but only in the past century has it been systemized and organized to the point that rapid genetic progress could occur. As we said previously, breeds of livestock developed over time to fit the specific geographic region where they originated. As people moved to new areas and even new continents, they took some of their stock with them as they established their new farm.

Farmers typically would breed their "best to the best," but if the parents were not highly selected the opportunity simply did not exist except in the case of a somewhat lucky recombination of genes in the mating. This can and did happen, but it is not a good way to expect to make genetic improvement.

When artificial insemination came on the scene, it was mainly used in cattle to control the spread of diseases along with making better use of superior animals. But the tools did not yet exist to do the computations that would allow the identification of elite genetics. Farms did maintain production information that helped them better manage their animals, and that became the basis for being able to determine which animals were the best from a genetic standpoint. Computers, with their ability to rapidly process large amounts of data, allowed the identification of superior animals that could then be used for successive generations of improved animals. More recently, the evolution (or maybe revolution) of genomics has increased the ability to select breeding animals more intensely and thus has increased the rate of genetic gain across species.

Genetic improvement is driven by the ability to calculate breeding values and to then use a program that allows the genetics of superior animals to be intensely used. Although we should avoid getting in to all the minute details of breeding value calculation, some background is important to understand what might happen in the future.

There are three basic elements that make up the calculation of a breeding value today: the pedigree, the animal and its offspring's performance (if any), and the genomic breeding value estimate. All of them are used to predict an animal genetic merit, and the genetic merit will change over time as information is added.

The pedigree is simple; it is the family tree of the animal. Performance of the animal and its relative provides the clue that it will be a superior animal genetically and historically has been the best predictor available to use in predicting the elite genetic animals of the future. With that said, a pedigree still has a high degree of uncertainty in predicting the performance of offspring and thus can only be used as a screening tool to help breed the next generations of improved animals.

The performance of an animal itself is the next piece of the puzzle. This is how much milk a dairy cow produces, the weight gain of animals being grown for meat, the number of eggs produced, and so on. This is the phenotype of the animal and is the basis for all genetic evaluations. There are multiple phenotypic measurements recorded depending on the species, and breeding values are then calculated for each of the traits measured. For reference, a dairy animal in the United States would today have more than 40 individual breeding values calculated. The production of breeding values requires a tremendous amount of computational resources and again is only possible because of the tremendous increase in capacity in the computers used today.

Although it sounds simple, in the case of a dairy bull it is not so because a dairy bull does not produce milk. So, the only way to measure his breeding value is to obtain phenotypic information on his offspring. Because of biological limitations, the animal is then almost 5 years old before the "true" breeding value estimate is known. The same would be true to a certain extent in other species, but the difference is the length of time to obtain phenotypic data on offspring.

Accurate phenotypic information, and in large amounts, is imperative to be able to calculate genetic rankings. If not available, breeding values are less accurate and lead to genetic rankings that do not allow the best animals to be selected. There will be more discussion on the importance of phenotypes later in this chapter.

4.5.1 Genomic Breeding of Animals

Finally, a more recent development in animal breeding is the use of genomic or DNA breeding values, which came about in the mid-2000s. By using the genetic profile of an animal, predictions can be made on an animal's genetic value at a very young age—essentially at birth. This has revolutionized the animal-breeding industry with the ability to rapidly identify superior animals and use them in breeding programs as soon as physiologically possible to greatly speed up genetic progress.

In the past 50 years, much of the focus on breeding programs has been to increase weight gain or some other production metric based on output. There is variation on this across species and across geographical areas, but the basis for such selection is for a simple reason. This selection has been based on what is easy to measure. To measure a kilogram or pound of weight gain can be done without any special tools and typically has the greatest impact on the payment a farmer receives. Other attributes that are more difficult to measure or economically value take a back seat to the easier and more direct methods.

The dairy industry in the United States is a good example of what happens when the breeding focus is primarily on production. With the drive to increase milk, protein, and fat production came a decrease in fertility. This was not intended, but the "unintended consequences" always must be looked out for. Because fertility has a negative correlation with production and there was not a genetic evaluation for fertility, selection for the production traits caused a corresponding decrease in fertility. Researchers made a concerted effort to

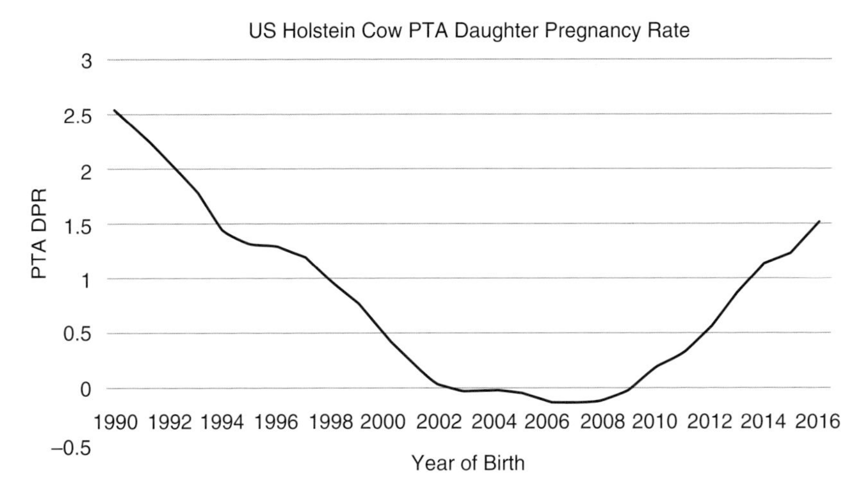

Figure 4.1 Holstein cow pregnancy rate. From Robert Fourdraine, Cooperative Resources International, International Center for Biotechnology Mt Horeb, WI, USA.

create a genetic evaluation for fertility and it was released in 2003. Sires being used at the time were mostly negative for fertility, but dairy farmers' selection immediately changed to put emphasis on increasing the fertility level of their herds. The result was a reversal of the genetic trend, as seen in Figure 4.1. Remember that genetic trends take time to change, and this is especially true in species with long generation interval like bovines. This is a clear example of being able to change a trait once data is available.

Selection in dairy animals has changed with more focus now placed on traits related to health or overall economic efficiency rather than a narrower focus on production output. Single-trait focus has been the norm, but as more and more traits are, and will be identified, the need to develop multitrait economic indexes becomes more important than ever. A multitrait index, as the name implies, incorporates many breeding values that by themselves have small impacts and should not be focused on individually, but taken together can move the genetics of a group of animals toward a desired goal.

Pigs and poultry have had more focus on multitrait improvement than dairy animals, and will continue to do so in the future. As we look forward, there will be a growing emphasis placed on breeding a healthier, more efficient animal. Although the move for efficiency has been going on for some time, the measurement of it will become increasingly fine-tuned and the definition will quite possibly change.

Many strategies can and will be employed to change the genetics of animals. Some are low tech and can be easily implemented, others exist now but are relatively high cost, and yet others are still at some point of the idea phase and still need to be moved to a practical and usable strategy.

Developing countries genetic progress in cattle has been limited because of several factors. First and foremost is the lack of a structured data collection program that then prohibits any type of selection program to be used within country. There is much concern in developing countries about retaining genetic diversity and use of domestic breeds that have been bred for local conditions such as heat tolerance or insect resistance. This concern is valid and needs to be considered in long-term plans. However, without information to develop a good way to identify superior animals of these local breeds little progress will be made in them, and they stand to get passed by as farmers use a higher level of imported genetics.

The use of gender sorted or sexed semen for use in artificial insemination is increasing in bovines as well as other starting to develop in other species. Acceptance and usage varies widely by country depending on economic conditions, but holds the possibility to allow for the speed of genetic progress to increase. For bovines, a producer can obtain female offspring from animals that have strong maternal traits they want to retain in the herd and use male semen to produce calves that will be used for beef. In swine-producing females is desirable for slaughter also, so few males would be produced.

Going one step further than sexed semen, the use of embryos will likely become commonplace in many dairy and beef operations as technology improves and costs are reduced. Producers will be able to purchase a genetic package in an embryo that provides the traits that fits their goals and objectives. In cattle, this will be similar to what happens in pigs and poultry where a specific line of genetics is used. This allows more collaboration between the producer and the processor to deliver a food product that has less variability and meets the specifications of the processor.

A word of caution on this: A common thought in some developing countries is to purchase and implant embryos and all problems will be solved because the most modern genetics are being used. But this ignores two important points. First, introducing a high performing animal into a herd that has previously had low genetic potential creates big management problems. The ability to feed and manage a high-performance animal is not acquired overnight, and this is a disaster waiting to happen. The ability to manage the animal must be there. Second, often overlooked is the need for "localization" or the need to handle stressful conditions that likely were not present in the environment the embryos came from. If special conditions like high heat and humidity exist or endemic disease, the imported embryo likely will not be adapted to handle the conditions.

As discussed previously, more emphasis will be placed on creating animals that have a smaller environmental footprint. This will occur by breeding animals that are more efficient in turning the food they eat into animal products, and it is likely that there will also be emphasis on reducing methane gas emissions from ruminants. Feed efficiency, which has seen tremendous improvement in poultry and pigs and to a lesser extent in beef cattle, will be more directly used in dairy animals. With the drive to increase total production, dairy cows have become too large. The increased size, which equates with more feed used for maintenance, will need to be reversed.

Another broad trend that will occur is the move to breeding healthier, more robust animals. Not only will this be driven by consumer demand, but a strong case can also be made about the economies that will occur both at the farm level and societal level. Reducing treatments used on animals will become increasingly important, as will breeding animals that are resistant to insects and other pests. Current breeding methods can be used to accomplish this, but the rapid gain will come as greater use of DNA analysis or genomics occurs.

The use of genomics is unfolding and is nothing short of revolutionary. The ability to now predict the genetic value of an animal at birth now lets us greatly speed up genetic progress. Previously we said that historically selection has been for primarily production traits, and often at the expense of health or fitness traits that could not be measured, or at least could not be measured in time to have an impact on making genetic progress. That has now changed because we can determine the genetics for many traits, including health and fitness, early in the animal's life.

Historically most of genetic progress in animal breeding has come from males because of the accuracy of their genetic evaluations and their ability to have many offspring. Now females can also be evaluated with a high degree of accuracy at an early age, and with reproduction techniques like in-vitro production of embryos, we can also produce more offspring than ever before.

For a farm, it will be possible to do a genomic test on all animals born and then make determinations on how to use them later in their life. Today it is possible to select the highest genetic merit animals for production and some fitness traits, but in the future, a selection might be based on how an animal will respond to a feed or health protocol. If it is known what the response will be, management can be developed around maximizing the response expected.

The next steps in breeding program changes are being developed but at this point not at the point of commercial availability to take to the farm. They have the potential to drive genetic progress even faster if able to be fully taken advantage of.

Gene editing is the first to come to mind and is eagerly awaited. Different than genetically modified organisms, using gene editing speeds up the process of selection that occurs in breeding programs today and allows the genome of the target animal to be changed. The possibilities for this technology are endless, and this is where the greatest progress could be made in regard to health traits. Some animals have a genetic resistance to certain diseases, and if this can be identified and transferred broadly into the population, the implications for having healthier animals is great.

An example of this is occurring now in bovines with the ability to create hornless or "polled" animals. This is naturally occurring, but to create animals by traditional methods will take 10 to 15 generations to have animals that are competitive genetically. Polled animals are a positive for animal welfare, as they then do not have to go through the stress of being de-horned. It is easy to imagine that this could apply to insect resistance also.

Gene editing opens the ability to move genes from breed to breed among the same species. This is just what would happen with the polled gene mentioned and could additionally be used to take advantage of other breed traits that have genetic advantages. It was previously mentioned the need to take advantage of breeds in developing countries that were accustomed to the local conditions. This technology opens the door to be able to do this and use the advantageous traits from another population, in ways similar to what has happened with plant breeding.

Epigenetics, the study of heritable changes in phenotype that does not involve changes in the underlying DNA sequence, is another area that is being explored and will provide opportunities in the future. Much is to be learned about epigenetics and how the practice can be used to turn genes on and off.

4.6 The Use of Big Data for Management and Genetic Evaluations

Let's face it, it's a data driven world. Data is used in most of the things we touch today to drive decisions that impact the products we buy and the way they are marketed and sold to us. In today's world, the term *big data* is used extensively and is used for a vast amount of marketing purposes as well as research. In the world of animal agriculture, big data has been around for some time but is certainly being used on a much broader basis than in the past, and the extent that it will be used in the future will increase substantially. The ability to capture information is the key to this and will be so in the future as new methods are developed to digitize information and turn that information into something useful.

At the farm level, producers have long measured the output of their animals to determine which one was better. This has been done both formally through a systemized method of recording the information or through more informal methods as simple as "the eyeball" or looking at the animal. It is through the former that evaluations can be made and turned into useful data on a larger scale. This is phenotypic information—a phenotype being the expression of an animal's genotype or genetic makeup in combination with the environmental factors that influence the animal.

For the purpose of this chapter there are two main uses of the data captured about the performance of animals. The first is as a useful tool for managing, and the second is to use in genetic evaluations for the purpose of changing the genetic makeup of the population. There are other obvious uses (financial analysis is the first that comes to mind), but this chapter will be limited in scope to the use of data in management and genetics.

Previously it was mentioned that selection for many years has been driven by a desire to increase production of a single trait such as a kilogram of meat or milk produced, and the reason being that this was easy to measure. Our ability to gather more sophisticated measurements has been limited, and even when captured it was not possible to compile a data set that would have broad use. The game has now changed with the ability to digitize almost anything.

The list of traits that have historically been measured is lengthy and has increased dramatically over time. An example is that US genetic evaluations have gone from only predicting changes in milk and fat production to a long list of evaluated traits as seen in Table 4.2. Some of the traits listed also have subtraits calculated and thus the number of options in making genetic decisions is almost limitless. In the latest catalog of the dairy sires GENEX has available, there were 66 individual or multitrait evaluations listed, which reinforces the idea of using an index for making genetic decisions rather than selecting on individual traits.

For beef animals, the story is similar but not as extreme. An Angus sire would have 24 traits or indexes listed; easier to make selections than a dairy animal but still a lot of information! Other species have less public genetic evaluations available, but would have the same types of analyses done to predict the performance of offspring. All of this is possible through the collection of phenotypic data along with the use of pedigree and genomic information as described.

One of the common phrases often quoted in my work is "If you don't measure it you can't manage it," and it is certainly true. A great example of this developed in the dairy industry in the United States with fertility late in the 20th century, as shown in the previous section of this chapter. Only by having measurements on a large population and then using the information in breeding values was the industry able to reverse the trend.

Gathering phenotypic data has historically been a challenge because the ability to take measurements has been limited by having farmers, an employee, or someone hired from the outside visit the farm and physically record a measurement. There is no doubt that this has amassed a large set of historical information over the past century but has also presented a challenge in how to manage all the data collected and how farmers should use the information to help them better manage their herd. With page after page of information that is not clearly summarized with key performance indicators, the data provides little value for a manager! Today the options for farm managers to clearly have information at their fingertips to aid in making decisions are greatly expanded and the promise for the future is a bright one.

As we move to the future the ability to capture data automatically will increase drastically. Already on-farm systems gather much more information

Table 4.2 Comprehensive list of relevant traits.

Year	Evaluation changed	No./year	Release months	Traits included
1936	Bull (new)	1	Summer (exact month varied)	Milk, fat
1940	Bull	**12**	**Each month** (bulls with many additional daughters since last evaluation re-evaluated)	Milk, fat
1960	Bull	**4**	**February, May, August, November**	Milk, fat
1964	**Cow**	**2**	**January, July**	Milk, fat
1967	Bull/cow	**3**	January, **May, September**	Milk, fat
1977	Bull	3	January, May, September	Milk, fat, **protein**
1978	Bull/cow	**2**	January, **July**	Milk, fat, protein (bulls only), **type**[1]
1985	Cow	2	January, July	Milk, fat, **protein**, type[1]
1994	Bull	2	January, July	Milk, fat, protein, **somatic cell score (SCS), productive life (PL)**, type[1]
1995	Cow	2	January, July	Milk, fat, protein, **SCS, PL**, type[1]
1997	Bull/cow	4	**February, May, August, November**	Milk, fat, protein, SCS, PL, type[1]
2002	Bull	4	February, May, August, November	Milk, fat, protein, SCS, PL, **calving ease (CE)**, type[1]
2003	Bull/cow	4	February, May, August, November	Milk, fat, protein, SCS, PL, CE (bulls only), **daughter pregnancy rate (DPR)**, type[1]
2006	Bull	4	February, May, August, November	Milk, fat, protein, SCS, PL, CE (bulls only), **stillbirth (SB)**, DPR, type[1]
2008	Bull/cow	**3**	January, **April, August**	Milk, fat, protein, SCS, PL, CE (bulls only), SB (bulls only), DPR, type[1]
2010	Bull/cow	3	April, August, **December**	Milk, fat, protein, SCS, PL, CE (bulls only), SB (bulls only), DPR, **cow conception rate (CCR), heifer conception rate (HCR)**, type[1]

[1] USDA calculates type evaluations for breeds other than Holstein; Holstein type evaluations calculated by Holstein Association USA.
Source: Council on Dairy Cattle Breeding

than we imagined possible 30 years ago. An example is robotic milking systems that are rapidly growing in popularity. Not only do they collect information that has traditionally been measured such as milk weights, but they also measure length of time a cow takes to milk, mastitis, and similar parameters. This can then be used to determine if a cow is becoming sick, is not eating, or a number of physiological issues, which she may be encountering. This early warning system can be a tip off for some type of treatment to prevent a more serious problem and holds large potential for the reduction of antibiotic use or other treatments in animals as well as increasing life span because of animals being healthier.

The list of possibilities for sensors to be used in herd management is extensive. Integration of them on dairy farms will provide opportunities to capture and use the following:

- Collect milk weights and composition of the milk
- Hormone levels for the detection of pregnancy or estrus
- Estrus activity
- Body temperature
- Rumination activity
- Movement (or lack of, therefore indicating health issues)
- Infections
- Blood glucose
- Calving

The list could be much longer, and for other species comparable measurements could be listed that measure productivity or health. To be used, sensors will need to be accurate and provide measurements that are reliable. They must also be fully integrated across other systems on the farm so that data transfer is seamless, allowing management decisions to be easily made.

Technology will continue to decrease in price, making the availability and use of sensors on farms a commonplace occurrence. Not only will sensors aid greatly in farm management, but they also will become the collection point for data, which can then be summarized and used for genetic modeling and predictions.

Although I would expect that there will be widespread use of sensors on many farms, data used for genetic purposes will increasingly come from a limited number of farms that have large numbers and have many data collection points. For pigs and poultry, this is the case now and will become more so for bovines. One of the reasons for this will be an increasing awareness of the need for data privacy and concerns about its use if shared on a broad basis, as well as the development of proprietary traits and breeding values by breeding organizations.

The use of sensors to capture phenotypic information will greatly expand the opportunities for new genetic evaluations. Again, this will not necessarily be the traditional measurement of traits that are relatively easy to measure, but traits that are harder to capture and require numerous animals to capture enough information for a genetic evaluation. One could logically ask the question of when is there too much information? As pointed out previously, the amount of information available today seems like it might be overload, and yet the direction is that there will or should be more available.

The key point here is that, while more and more data is being collected and traits evaluated, the use of traits as individual selection points will decrease as more use is made of indexes that include multiple traits, are more focused than they previously have been on areas of low heritability, and will increase the overall health and welfare of animals which not only has value to the producer but to the public as well.

The lack of a structured data-collection program in developing countries was discussed as an impediment to making genetic progress in local breeds. Collection of phenotypic data is the backbone for any genetic evaluation system, and is necessary if rapid genetic progress is expected to be made. Without it, hoping to make steady progress in the genetics is merely a dream. There must be an organized, systematic approach to data collection along with a sound animal identification program in place. The genetic diversity deserves preservation, but it will not happen without proper strategies put in place to identify the best animals. With proper preservation, there is great opportunity for cross-breeding and use of the local genetics for their environmental adaptation. Perhaps some of the new technologies discussed in this section will allow data collection programs to be established that help achieve this.

4.7 Summary and Major Learning

- Livestock and animal agriculture have a major impact on the global economy. Animals are the largest users of land in the world, and animal agriculture employs 1.3 billion people. It is easy to overlook the fact that in developing countries ownership and raising animals can be the main livelihood for many people.
- Animals provide a substantial amount of dietary protein consumed globally, and it is a high-quality protein that provides essential micronutrients.
- The world population is expected to reach 9.725 billion by 2050. Consumption of meat is expected to grow by 73 percent and 58 percent for dairy products during that time.
- Global livestock production is growing rapidly with faster growth in developing countries than developed countries. This trend will continue, with the driving forces in developing countries being population growth, income growth, and urbanization.
- The way livestock is raised around the world is quite diverse, from large-scale intense operations to many small holders. This will continue, although a larger and larger share of production will take place on large-scale farms.

- Production in developing countries will increase, but there is opportunity as productivity and efficiency is low compared to the developed world. The competition for scarce resources will demand that productivity in developing countries must increase.
- Everyone wants animals that are healthy, treated humanely, and do not harm the environment while at the same time contribute to the economic wellbeing of the farm. Farmers want to and will treat their animals well, but sometimes production methods are not understood. This in turn causes friction between the farmer and the consumer.
- Intensive farming is efficient, not only from an economic point of view but also when viewed from the standpoint of how resources are used. Less total resources per unit of output are used in an intensive management system versus an extensive management system.
- Technology will emerge that can be used on farms to reduce the amount of methane gas released into the atmosphere. If economically viable they will see widespread use.
- Although there will be more production in intensive management systems, production will occur in a wide range of methods as a result of local conditions affected by economics, climate, politics, tradition, and a variety of reasons.
- Efficiency of animal production must increase if the projected increase in production is to happen. There is not enough land to support two-thirds more animals.
- Humane treatment of animals is important and will become increasingly more important in the consumers view.
- Animal breeding will focus more on total animal wellbeing and economics rather than so much focus on individually measured production traits.
- The ability to do a better job of selecting for health and wellness traits will emerge and become a major part of breeding strategies.
- Breeding for efficiency of production in the use of resources will also occur. This will reduce the environmental impact of animals, but there will also be the possibility to breed and manage for lower methane gas emissions. Smaller dairy animals will be bred that will reduce their environmental footprint.
- Genomics will play a major role in breeding with the emergence of the ability to have genetic evaluations for health and fitness traits early in an animal's life. This will speed up genetic progress and also reduce treatments that are required for sick animals.
- Other areas of technology such as gene editing and the study of epigenetics are emerging and hold the potential for making a large impact on the production and well-being of animals.
- Data is important in animal agriculture and is needed for both managing a herd and to be used in the calculation of genetic evaluations.
- Phenotypic data is the basis for all genetic evaluations and has been collected for many years in developed countries. Previous focus has been on

production focused traits while in the future more traits will be measured that are related to animal health and well-being.
- Good data is important for a farm manager and must be available in a way that is easily accessible and can be used for making fast management decisions.
- Emerging technologies like sensors will be commonly used on farms to collect much more data than is now being gathered. Sensors will be integrated across the farm and will provide the farm manager with accurate details regarding the animals on the farm so they can maximize production and minimize negative health events.
- In developing countries, lack of a data collection system is a large problem and decreases the opportunity for the use of local breeds in modern breeding programs. Although local breeds offer genetic diversity as well as disease resistance and heat tolerance, there will be little genetic progress as long as a method to collect phenotypic data does not exist.

References

"Animal husbandry." (n.d.). Available at https://en.wikipedia.org/wiki/Animal_husbandry [Accessed March 6, 2017].

FAO. (2009). Livestock in the Balance. Available at http://www.fao.org/docrep/012/i0680e/i0680e01.pdf [Accessed March 6, 2017].

FAO. (2011). World Livestock 2011—Livestock in food security. Available at http://www.fao.org/docrep/014/i2373e/i2373e.pdf [Accessed June 3, 2017].

International Farm Comparison Network. (2016). Dairy Farm Structure for today, trends and drivers for future. Presentation for 2016 IFCN Supporter Conference, September 14–16, 2016.

"Livestock." (n.d.). Available at https://en.wikipedia.org/wiki/Livestock [Accessed March 6, 2017].

"Livestock Production." (n.d.). Encyclopedia of Food and Culture. Available at http://www.encyclopedia.com [Accessed March 6, 2017].

Population Pyramid (n.d.). Population pyramids of the world. Available at https://populationpyramid.net/world/2050/ [Accessed June 3, 2017].

U.S. Department of Agriculture. (2002). The changing landscape of U.S. milk production. Available from https://www.ers.usda.gov/webdocs/publications/47162/17864_sb978_1_.pdf?v=41056 [Accessed June 12, 2017].

US Department of Agriculture, NASS. (n.d.). Available from https://www.nass.usda.gov/

World Bank. (n.d.). Livestock production index. Available from http://data.worldbank.org/indicator/AG.PRD.LVSK.XD [Accessed June 12, 2017].

World Health Organization. (n.d.). Global and regional food consumption patterns and trends. Available at http://www.who.int/nutrition/topics/3_foodconsumption/en/ [Accessed March 6, 2017].

Part 2

The Future of the Food Industry

5

The Food Trends—The New Food—Enough Food?

Great things are not accomplished by those who yield to trends and fads and popular opinion.

—Jack Kerouac

5.1 Historical Food Trends: From Then to Now

The preceding chapters have dealt with the specifics of agriculture as well as the role of water and water management in the entire equation of food from the land. This chapter sets out to look back and at the same time to project to the future. It's almost a situation similar to the one in "back to the future," without the use of a time machine but rather the attempt to achieve this by looking into the long-term perspectives of food and food trends in both directions, backward as well as forward.

5.1.1 Food and Beverages during the Period of Classical Greece

Let me discuss and describe typical meal components, and some food and beverage related fashions, for instance in Athens during the fifth century or so BC. Just as a reminder, this is the age of the historians Herodotus and Thucydides, dramatists such as Sophocles, Euripides, and Aeschylus, and philosopher, Socrates. These are the times of "classical" Greece, and after many years of destructive wars against the Persians, the city of Athens was rebuilt again under their leader Pericles, including the great Parthenon and other architectural masterpieces, which can still be seen to this day.

And yes, people had to eat and drink in those days, a fact that is of course obvious, but never really much mentioned in the typical history books. Let me get to this in some detail. So what did Greeks eat in those days? Most of the knowledge that we have on these matters stems from literary and artistic evidence. Greeks had three to four meals a day, so no real difference to our

Megatrends in Food and Agriculture: Technology, Water Use and Nutrition, First Edition.
Helmut Traitler, Michel Dubois, Keith Heikes, Vincent Pétiard and David Zilberman.

mainstream food habits. Their breakfast mostly consisted of barley bread dipped in wine, complemented by figs or olives or both. On frying pans they prepared some kind of breakfast pancakes made with wheat flour, olive oil, honey, and curdled milk. Other pancake variants used spelt.

Early light lunch was eaten at around noon and an additional light meal was sometimes consumed in the late afternoon. Dinner was the most important meal of the day, typically taken at nightfall. Unlike in most of today's cultures, men and women took their meals separately, which, in many cases in the more affluent households would be served by slaves. Today we would call them "waiters" and give them a tip.

One of the social dining highlights was the so-called "symposium," a kind of banquet but more often an after-dinner drinking bout. So first food was consumed, accompanied by wine, which was followed by the drinking part, typically accompanied by snacks such as chestnuts, beans, toasted wheat, and honey cakes. The goal was to have these snacks absorb alcohol and thereby increase the capability to consume wine.

So, what were the foods consumed by the contemporaries of Socrates and Pericles? Like in many cultures of our world today, cereals formed the staple diet, the two main grains being wheat and barley. Bread was either baked as flat bread or leavening agents such as alkali or wine yeast. Already in these days there was a distinction between rich and poor or those who could afford the more expensive, "nobler" bread made from bread wheat and those who could just pay for the coarser brown bread. This distinction has held for many centuries...until more recently when nutritionists told us that the coarser, complete grains are the healthier variant.

Barley was also used to make bread, although more difficult to bake good bread. Barley was often roasted before milling, thereby leading to a coarse flour, which was used to make "maza," the basic Greek dish in these days. Maza would be served cooked or raw, as a broth or made into dumplings or flatbreads, and sometimes augmented with cheese and honey. Such bread was often served with accompaniments, such as either meat or fish, fruit or vegetable.

Fruits and vegetables, next to bread were the most significant parts of the diet in ancient Greece, while meat consumption was much lower in these days compared to modern times. Important fruits in the old days were figs, raisins, and pomegranates. Dried figs were typically eaten as an appetizer or when drinking wine, and sometimes accompanied by grilled chestnuts, chickpeas, or beechnuts.

The most prominent vegetables in these days were lentils. Vegetables were eaten either as soups, boiled or mashed, seasoned with olive oil, vinegar, herbs, or fish broth. Mashed chickpeas—fava—are still to this day, a highly popular appetizer in a Greek taverna.

Although all or most of all these mentioned cereals, vegetables, and fruits, including honey and wine, were affordable to the large majority of the population in fifth-century Greece, fish and meat, pretty much like today, were more abundant in the kitchens of the wealthier. In the countryside, hunting (trapping) for birds and hare was popular, and some peasants had farmyards with chickens and geese. The wealthier landowners could raise goats, pigs, and sheep. Much of the meat was transformed into sausages, most likely to render it more stable for storage.

As far as recipes from those days are concerned, a Spartan one stands out: soup made from pigs' legs and blood, seasoned with salt and vinegar—the so-called "black soup." The dish was typically served with maza, figs and cheese, and sometimes accompanied by game and fish.

Fish and seafood products were typically consumed in the Greek islands but were more often transported to the Athenian markets, where one could typically find anchovies and sardines. Moreover, eels, conger-eels, and sea-perch were considered to be great delicacies.

Eggs and dairy products were also part of the diet of Greeks in those days during the classical period and quails and hens were bred for their eggs. Such eggs were cooked soft- or hard-boiled and consumed either as appetizers or desserts. Additionally, the various parts of eggs—whole, yolk, or white—were used as ingredients in food preparation, pretty much like today.

Milk was mainly consumed in the countryside, closer to where goats or especially sheep would be held in farmyards. Butter was known but not frequently used. Other dairy products were curdled milk, similar to cottage cheese, as well as probably to yogurt. Cheese of all kinds (goats, sheep) was a popular staple food, in both formats, soft as well as hard cheeses. Cheese was either eaten alone or in combination with honey or vegetables and was included in the preparation of many other dishes.

When it comes to beverages, despite the ancient Greek's love for wine (has not really changed since), water was the most widespread and popular beverage. Milk from goats was consumed but to a far lesser degree. As far as wine is concerned, it is believed that all three main variants of wine—white, rosé, and red—were produced and consumed. However, wine was typically mixed with water and sometimes honey was added to make it sweet.

Cereals such as barley mixed with water and herbs were used to produce a beverage called "kykeon," which was most likely fermented, at least that is the assumption here. It was widely used in the Eleusinian Celebrations and was considered to be stimulant, to say the least ("Ancient Greek cuisine, n.d.; History Guide, 2009).

5.1.2 Food and Beverages in the Roman Empire

Although there are certainly many similarities between food and beverage raw materials as well as recipes and consumption habits between classical Greece and the heydays of the Roman Empire, there are many subtle differences and sophistications. Let me randomly select the first century AD as the period for which I attempt to describe the major food and beverage habits and discuss a possible evolution in not only habits but also food or nutrition-related trends during this period. We might expect a continuation of certain food trends as we have seen them in classical Greece. Much can certainly be attributed to the fact that both Greece and Rome are Mediterranean countries and therefore have similar food and beverage raw materials at their disposal. I do realize that the expansion of the Roman Empire during its high time went far beyond the Mediterranean but to simplify my task, I shall stay in that region.

The first century was the century that lasted from the year 1 to 100 according to the Julian calendar. It is often written as the first century AD and is considered to be part of the Classical Roman era. During this period Europe, North Africa, and the Near East fell under increasing domination by the Roman Empire, which continued expanding, most notably conquering Britain under the emperor Claudius. The reforms introduced by Augustus during his long reign stabilized the empire after the turmoil of the previous century's civil wars. Later in the century, the Julio-Claudian dynasty, which had been founded by emperor Augustus, came to an end with the suicide of Nero in 68 AD. There followed the famous Year of Four Emperors, a brief period of civil war and instability, which was finally brought to an end by Vespasian, ninth Roman emperor and founder of the Flavian dynasty. The Roman Empire generally experienced a period of prosperity and dominance in this period, and the first century is remembered as part of the Empire's golden age.

The first century saw the appearance of Christianity, following the life of Jesus of Nazareth in the Roman province of Palestine ("1st century, n.d.). So, what happened agriculture and food wise in the Roman Empire during this century? Pretty much like the Greeks, Romans had eventually a habit of two breakfasts, one at dawn and a second one, *prandium*, around lunchtime. In late morning they ate a late lunch, like they broke fasting twice, over a fairly short period of time during the morning hours. Their evening meal, *cena*, over time grew larger and became more diverse under the influence of Greek culture and increased trade. It has to be added, that the wealthier Romans in these days tried to finish all their business activities rather early during the day so that this cena could already begin at maybe 2 PM, often lasting until late at night. This sounds like a really long meal.

Not everyone could of course afford such a long period for eating cena nor the richness of such a meal. The working classes would eat much simpler, and a typical cena would consist of a kind of porridge, or rather pottage, the so-called *puls*. Puls is a pottage (a thick soup or stew) made from farro grains (spelt, emmer, and einkorn), boiled in water, and flavored with salt. It was a staple dish in Ancient Rome. The dish was considered the indigenous food of the Ancient Romans and played a role in archaic religious rituals. The basic grain pottage could be elaborated with vegetables, meat, cheese, or herbs to produce dishes similar to today's polenta or risotto ("Puls (food)," n.d.).

Depending on social status, typical foods eaten by the less fortunate during these meals were salted, flat, round leaves made from the wheat-like cereal emmer (*Triticum dicoccon*). Later on wheat was introduced to make bread. Richer families, so-called upper classes, would eat eggs, cheese, honey, fruits, drink milk, and wheat increasingly replaced emmer. This is a recurring observation both in classical Greek as well as ancient Roman times; the "lower grade" cereals such as barley or emmer were replaced by the more highly regarded wheat. Currently, there is a rather strong movement to again use these "old" grains, especially emmer in today's food mix.

Other food ingredients, typically consumed by the more affluent Romans, were bread dipped in wine and accompanied by olives, cheeses, and grapes. Interestingly enough, when my wife and I ate a late afternoon snack in a taverna on a Greek island, our favorite snack combination was: bread, cheese, and olives, the only difference being that we skipped the dipping and drank the wine straight away.

Moreover, wild boar, beef, sausages, pork, lamb, duck, goose, chickens, small birds, fish, and shellfish were also popular in those days, especially among those folks who could afford it. By the way, hunting for small birds, especially for migratory birds is still a popular pastime and sport in Italy to this day ("Ancient Roman cuisine," n.d.).

Patrick Faas's book *Around the Roman Table: Food and Feasting in Ancient Rome* (2003) has more than 150 recipes from the ancient Roman times that were adapted to today's cuisine. It lists salads, soft-boiled eggs in pine-nut sauce, lentils with coriander, roast wild boar, ostrich ragout, roast tuna, fried veal escalope with raisins, nut tart, and many more. It reflects very well the ingredients diverse and rich Roman cuisine, which is a tradition that is held up to this day in traditional, modern Italian cuisine.

5.1.3 Food in Medieval Times in Central Europe

I vividly remember a scientific conference on the topic of essential fatty acids back in the early 1980s. The conference venue was at the London Zoo and one of the co-organizers was Michael Crawford. He was, and probably still is, quite a character and was already then one of the leaders in the field of certain long chain fatty acids such as docosahexanoic acid and brain development in mammals. While the conference was held at the zoo, the final get together and special dinner took place in the Museum of London in the London Wall. The conference venue was already quite outlandish but the museum even beat that. Not only was the dinner held there, but the various courses of the dinner also reflected different time periods and were set in the corresponding environment.

What really personally impressed me though was not the setting and all the wonderful pieces and sceneries we were surrounded by but the appetizer, the starter we were served. It was a recipe from the late 13th century, and if my recollection does not fail me it was some kind of paté or terrine, made from then-popular vegetables and meats such as lentils, cabbage, and lamb. I might be off with the recipe a bit, but it was quite awesome to think that the recipe had been created and was popular some 700 years earlier. I worked for a Swiss food company and was thinking that the confederation of Switzerland was founded at around the same time, and like the country, the recipe was old, too. I do not recall that I was especially struck and impressed by the taste of the starter but the idea stuck with me. We had several courses with recipes from more recent times and ultimately, I recall, we were served a dish that contained modern day (1980s) nutritionally balanced ingredients.

The industry of agriculture and farming in those middle ages was, at least in central Europe, very much built on the structure of the society, that is, nobles and peasants, artisans and civil servant, soldiers and priest, slaves (serfs) or free people. These were the times when the German painter Albrecht Dürer had written that one should not work longer than 4 hours per day and when the nobles could charge the un-free and they had to pay their taxes or their crop shares directly to nobles and church. Land was mostly owned by either the church or the nobles. Farmers had to lease their piece of land and typically applied the technique of crop rotation, mostly because they did not know how to fertilize the land. So, for instance, wheat was grown in the first year, followed by barley, and then that piece of land rested for a year.

Villages or manors were mostly owned by nobles, and land that was not in use to grow crops, the so-called commons, were used by serfs or peasants to let their animals, typically cows or sheep, graze. Farming in these times was rather simple and farm tools such as wooden plows could only scratch the surface and harvesting of grain was done with a sickle and grass was mowed with a scythe ("European farming during Middle Ages to 1800," 2016).

So, how and what did people eat in the Middle Ages? Well, the answer is simply: basically all what we eat today, prepared from slightly different, less "pure" or refined ingredients and prepared with less-sophisticated cooking and baking tools. Recipes can be found for bread, meat, pottage, fowl and poultry, fish, vegetables, sauces, salads, desserts as well as cakes.

Bread was typically made from wheat often together with honey or rye, the latter for the less wealthy. Ale-barm (the scum or foam) was used for raising the dough. The most popular way of baking bread was done by the peasants. They would take their grain and grind it by hand in a wooden mortar or stone trough. The so ground grain would then be mixed with water and baked to unleavened bread, so-called *oatcakes*.

Meat recipes were often based on poultry or fowl. "Blawmanger" (or blanc-manger, eating white) was a popular meat recipe using rice and minced chicken.

Ground almonds were sometimes added. The white color of all ingredients explains the origin of the word. Today's dessert "blancmange," popular in the United Kingdom has only a faint resemblance with the original recipe of blawmanger.

Pottages, similar to antique times, were very popular in medieval times. Recipes for pottages were mainly based on vegetable and stock, cooked in an earthenware pot or cast-iron cauldron, in a kind of vegetable soup, probably similar to today's minestrone in the Italian cuisine.

Fowl recipes were typically very varied and made use of every type of fowls there were to be found, with probably the greatest variety in medieval Britain. Dishes ranged from chicken to dressed peacock, not forgetting pheasant, partridge, duck, and goose usually stuffed with herbs like parsley and sage. Depending on the season, grapes, again with herbs might have been used for stuffing chicken.

Fish was less abundant in medieval central Europe and Britain than meat fowl and vegetables. Fresh water fish, especially salmon and trout, were most popular and were often prepared and served with dedicated sauces such as the one made by blending white wine, vinegar, rye breadcrumbs, and water and spiced with cinnamon, pepper and onions.

Vegetable recipes were mostly popular with the poor and vegetables were really their staple food. Recipes are numerous and carrots of different colors, peas, fava beans and chick peas were common and the cheapest source of proteins. Vegetables were preferably cooked.

Sauces, like in modern cuisine were a popular accompaniment to many medieval dishes and in those days were more wine or vinegar than milk and cream based. Thickening of sauces was achieved by the addition of breadcrumbs. Sometimes sauces were overpowering in their taste and would have the role to mask off-flavors from meat and fish.

Salad recipes were not quite the same in those days as we would typically find them in today's cookbooks or granny's personal recipe. A medieval salad mainly consisted of herbs and flowers seasoned with oil and vinegar. Most popular ingredients were primrose, sweet violet, mint, and parsley. Other plants used were borage, fennel, garlic, leeks, lettuce, rosemary, sage, and spinach. Salads were already then considered a meal with health benefits.

Desserts already played an important role in medieval times and were made from eggs, milk, cream, sometimes almond milk, often enriched with honey to add sweetness and flavor. Pine nuts were often used in such desserts. It is interesting to realize that almond milk was already popular and its present day popularity is just a confirmation of a long-standing tradition.

Cakes existed already then but they are not to be compared with modern cakes of any kind. One of the major reasons was the absence of sugar and cake products were rather bland and simple. The closest relative today is probably a biscuit or cookie, however with very little or no sweetness. Medieval cakes were often spruced up with spices such as mace or clove to make them more tasty and interesting to eat.

As far as beverages were concerned, every potable liquid seemed to be fermented because water was generally considered not pure, although boiling of water would have rendered it inoffensive. So, basically everything was more or less fermented, and alcohol (ethanol) was the recurring theme. Even fruit juices such as pear, blackberry, wild plums, and berries came fermented into some form of wine, liqueur or mead. Wine, such as claret was often spiced with honey, cinnamon and other ingredients.

It is interesting to point out that water in many Mediterranean countries had long been "disinfected," or rather rendered potable, by mixing approximately 25% or so of anis-based liquors such as raki, ouzo, pastis, arak, sambuca, and even older recipes, such as tsipouro, which dates back to the monks on Mount Athos in the 14th century. While central Europe fermented berries and fruits to make microbiologically safe beverages, the European south turned toward higher alcoholic concentrations.

Given the long list of different recipes briefly listed it can safely be said that variety was already an important element in the diet of people of the middle ages. It must also be added that wealth was a critical factor in this and variety and nutrient mix increased with increasing wealth of the individual. Without being too daring or proposing something outrageous, this is exactly the situation that we find in today's cultures and societies around the globe.

5.1.4 From European Renaissance and Enlightenment to the First Industrial Revolution

I want to speed up the history excursion a bit and therefore will attempt to span the long period of the 15th to the end of the early 20th centuries as far as agricultural development and typical food trends are concerned. I do realize that I jump over many periods including baroque times, Napoleonic, and post-Napoleonic times. I also do admit that this is rather Europe-centric and the Americas and Asia are pretty much left out if it, except for the new crops that originated in these places and came, through the ages of seafaring, including global trade, colonialization, and discovery to central Europe and the British isles.

As far as agriculture was concerned, the trend toward safe (meaning available) and sustainable crops continued with new agricultural products being added, such as the potato. Field crops included wheat, rye, barley, and oats; they were used for bread and animal fodder. Peas, beans, and vetches became common already from the 13th century onward as food and as a fodder crop for animals; as an additional benefit, it also had nitrogen-fixation fertilizing properties. Crop yields peaked in the 13th century and stayed more or less steady until the 18th century. Though the limitations of medieval farming were once thought to have provided a ceiling for the population growth in the Middle Ages, recent studies have shown that the technology of medieval agriculture was always sufficient for the needs of the people under normal circumstances, and that it was only during exceptionally harsh times that the needs of the population could not be met ("Economic history of Europe—Agriculture," n.d.).

With the onset of the industrial revolution and especially with the invention of such devices as the steam engine, tools became available to improve agricultural output quite substantially and this certainly triggered an agricultural revolution. Gregory Clark's "The Agricultural Revolution and the Industrial Revolution: England, 1500–1912" (2002) suggested that "agricultural revolutions," as a kind of counterpart to the often cited "industrial revolution" have taken place in Britain over a period of several centuries between the mid-16th and the mid-19th centuries. And he ends by suggesting that such agricultural revolutions were based on improvements of agricultural practices, either by increasing yields or by intensifying the use of land. So what did people eat during these decades and centuries and did their food-related and nutritional habits change and to what degree? Let me use Italy as an example for this transition period between medieval and modern times.

> During the Renaissance, Italy had the most skilled, well known and creative cooks in Europe. They took Italian fine dining to new levels of refinement and prestige. Large, elaborate banquets were served in the dining rooms of the dukes and princes who governed the many small states throughout Italy.
>
> Many of the Medieval flavors and preparations were carried over to the Renaissance, like the generous use of spices, the addition of sugar to savory dishes, the widespread consumption of roasts, stuffed pastas, tarts and pies.
>
> The use of light sauces made of fruit or aromatic plants were mixed or thickened with the soft part of bread, grilled bread, flour, almonds or eggs. Sometimes, these sauces were flavored with acidic juices and mixed spices.
>
> During the Renaissance, people developed a great love for giblets and the innards of butchered animals, poultry and fish. In addition, you could find a large selection of stews, long pasta noodles, stuffed pasta and maccheroni. Milk and dairy products were used often: butter became as important as lard, heavy cream became popular and people began cooking all types of cheeses.
>
> Fruit and citrus were fundamental flavoring agents and fruit became a prominent part of the dishes served at the beginning of a meal. ("The food of the Renaissance," n.d.)

Here is one typical recipe "herb tart," cooked for pope Martin V by his personal cook and translated from Latin:

> Take some fine aromatic herbs, such as parsley, marjoram, rue, mint or sage and so on, and pound them in a mortar. Then take some raw egg and fresh cheese and mix with some raisins; add saffron, ginger and other sweet spices together with some fresh butter.
>
> Then make the dough; use it to line a greased pan, fill with the mixture and some more butter and cover with more dough. When it is cooked, sprinkle with sugar and whole pine nuts. And this will be superlative for courtiers and their wives. (Bonardi, 1995)

Other period recipes made use of well-known food ingredients such as pork, poultry, veal, Mediterranean vegetables and cheese such as mozzarella. Although there was certainly an agricultural revolution taking place after the renaissance period, not the same can be said when it comes to evolution of food and eating habits, and the most important event influencing our food habits was probably the invention of the industrial fridge during the first half of the 19th century.

Artificial refrigeration began in the mid-1750s and developed in the early 1800s. In 1834, the first working vapor-compression refrigeration system was built. The first commercial ice-making machine was invented in 1854. In 1913, refrigerators for home use were invented. In 1923 Frigidaire introduced the first self-contained unit. The introduction of Freon in the 1920s expanded the refrigerator market during the 1930s. Home freezers as separate compartments (larger than necessary just for ice cubes) were introduced in 1940. Frozen foods, previously a luxury item, became commonplace ("Refrigerator," n.d.).

Once food raw materials and thermally sensitive ingredients could safely be stored, this became a game changer to what was possible and what foods could be prepared and safely stored. Industrial food production on a large scale first became possible then. It was this invention, far more than any of the others that have been described up to this point, that changed how we looked at food, how we dealt with it, how it could become commercialized, and how it was possible that one of the biggest, if not the biggest, industries in our world—agriculture and food combined—could rise, almost worth the description "food revolution." It is estimated that worldwide approximately 1 billion people or 40% of the global work force work in agriculture, thereby it is the world's largest job provider (Momagri, n.d).

The numbers for the food industry are equally impressive, yet much lower. The 10 largest global food companies employ approximately 1.3 million people, and the best "guestimate" is that the total number of people working in the packaged food industry might be several millions up to probably 10 million on

FOOD THROUGH THE AGES

Major Food Components: A Selection

Classical Greece	Ancient Rome	Medieval Central Europe	Renaissance Italy
❑ Barley ❑ Beans ❑ Chestnuts ❑ Figs ❑ Olives ❑ Lentils ❑ Fruits ❑ Fish ❑ Meats ❑ Cheese ❑ Eggs ❑ Milk ❑ Honey ❑ WINE	❑ Grains (barley, emmer) ❑ Bread ❑ Eggs ❑ Cheese ❑ Fruits ❑ Milk ❑ Olives ❑ Grapes ❑ Meats ❑ Poultry ❑ Fish ❑ WINE	❑ Bread ❑ Meat ❑ Pottages ❑ Fowl ❑ Fish ❑ Vegetables ❑ Sauces ❑ Salads ❑ Desserts ❑ Cakes ❑ FERMENTED beverages	❑ Innards ❑ Pasta noodles ❑ Milk ❑ Butter ❑ Pork ❑ Poultry ❑ Veal ❑ Vegetables ❑ Cheese ❑ Mozzarella ❑ Aromatic plants ❑ Refined recipes ❑ WINE

Figure 5.1 Food through the ages: An attempt to paint a food picture.

a global basis. I am not including restaurants and fast food businesses in that number, or those who work in packaging machines and materials or other manufacturing line equipment; this is strictly food and beverage manufacturing. Based on the number for agriculture mentioned, this translates into approximately 4% of the total global workforce work in food and beverage manufacturing. Even if I am off by 50%, this is still a rather important number. I do emphasize both of these numbers because they reflect the absolute importance of these industries on a global scale, not so much for the money they may generate but for the incredibly high number of work places they offer.

Let Figure 5.1 shows a typical selection of major food components and trends in those days. Please note that wine and fermented beverages represent an important part of this history. Let me, however get to the last section of this short historical discourse of this chapter, namely food and food habits in the 20th century.

5.1.5 Food in the 20th Century: The Real Food Revolution

I call this "the real food revolution" because I strongly believe that the founding of science of nutrition was really the starting point of this revolution. And personally I do link this first founding stone of nutrition science to the discovery that certain fatty acids in our food are really "essential fatty acids," or in other words, molecules that the body needs for good functioning but cannot make on its own.

I do accept that this is a personal view of someone who has worked many years in lipid sciences and had the chance to work in areas such as discovery of new sources of functional and essential Ω6 as well as Ω3 polyunsaturated fatty acids. So yes, it's personal and the history of nutritional science dates back much further than the 1930s when Burr and Burr described the existence and role of essential fatty acids for the first time. I insist on this feat because it hardly finds a mention in the "official" nutrition history records.

As early as around 400 BC the Greek physician Hippocrates of Kos supposedly said: "Let your food be your medicine and your medicine be your food." Foods in these days were widely used as cosmetics or even in wound healing. Liver juice containing higher levels of vitamin A were squeezed onto the eye in cases of eye diseases. People had anecdotal evidence without scientific proof yet. However, it can be described as nutritional and metabolic functionality, wisely applied.

Much later during the mid-1750s, scurvy, a disease found among sailors on long voyages, could be cured or prevented by adding citrus fruits such as lime or lemons to the diet. Yes, we learned much later in the 1940s, when vitamin C was discovered that these fruits contained such a valuable and essential ingredient. It can safely be said that even without knowing, our elders possessed nutritional knowledge obtained through many generations of trial and error and more or less successful results.

Antoine Lavoisier is often called the "father of nutrition and chemistry" because he discovered the actual process by which food is metabolized already back in 1770. Seventy years later, the German Justus Liebig described the chemistry of carbohydrates, fats, and proteins and E. V. McCollum developed methods that allowed for the discovery of food nutrients in 1912, and in the same year Casimir Funk coined the term *vitamin* to describe a food ingredient with a vital role and contribution to our health and well-being. And I can go on with breakthrough findings, such as the discovery of essential amino acids as protein building blocks by William Rose back in the 1930s ("A history of nutrition," n.d.).

Much more happened during the second half of the 20th century in the area of nutrition, and some of the highlights are the discovery of hormones governing many or most bodily functions as well as many more micronutrients and their respective roles in our metabolism such as selenium, zinc, and iron. So much was going on in the area of nutrition research that I do not want to embark on any one of these in particular but rather want to point out that although nutrition research of a similar kind briefly described here is still ongoing and will *never* be finished. Nutrition research, even more so than other fields of research, is always ongoing, and many nutritionists who I had met during my many years of work in the food industry never had a conclusive opinion about anything they ever worked on!

To be fair, I was one of them when I worked in the area of lipid metabolism and lipid nutrition, and I know what I am (and was) talking about. Typical semi-conclusive phrases mostly sounded like "more research is required..." and, "although our findings partly support this particular claim..., it would be better to conduct more research before we can fully support the claim supposed to go on a particular product label."

Back to food and food trends of the 20th century and how these have influenced what we are eating today. When it comes to food trends during the 20th century, many of these were born from the reality of two World Wars that lasted "only" about 10 years in total but have influenced European societies in decades before, in between, and after the wars. Poverty and crisis—real ones—were almost the norm between the early 1910s and the late 1950s—so almost an entire half of a century. Missed opportunities, hunger, persecution, cruel genocides, and depression of any kind—personal as well as economic—were the reigning factors, and it took the Marshall Plan and the creation of European Economic Community (forerunner to the European Union) to stabilize the continent and give it the longest period of peace (yes, there were the Balkan wars in the early 1990s, and cynics will be plenty and mighty quick to point this out) and prosperity for many, not for all.

But then the utilitarian utopian idea of the "greatest happiness for the greatest number as the foundation of moral and legislation" proclaimed by Joseph Priestly late in the 18th century always was utopia, and many other utopian societies such as the late 19th-century California Colonies, for instance the especially short-lived Kaweah Colony located in the Sierras above Three Rivers, were never successful over any longer period of time. The Kaweah Colony only lasted for 6 years from 1886 until 1892! Don't get me wrong; I am all for utopia but with the clear vision and knowledge in the backs of our minds that we will never totally or even closely achieve it, but at the same time, it gives us sense and direction. This is not a history book, although it has just begun to sound like one; let me get back to food and nutrition, stuff that I know better.

As briefly mentioned, the 20th century, as far as food and nutrition are concerned, can safely be divided into two camps, both from a time period of view as well as from a geographic period of view, the latter on a more global basis, the former rather concerning Europe. The camps I am referring to are: affluent versus poor, or in food terms, proteins versus cheap carbohydrates—a distinction that can still be found to this very day and for which we do not have a real solution.

And so many things depend on finding a remedy to this discrepancy. Between the two World Wars and for some time after the Second World War, hunger, famishing, and deprivation were the norm. No diet was necessary to lose weight or to make one's body leaner and slimmer. Lean and slim was the norm and none of today's so popular diets or gyms were even heard of. I was born during these last years of food and resource scarcity after World War II, and the mantra of my parents was: finish your plate! And I was not alone in this; every kid of my age that I knew went through the same reminder by their parents.

Since probably the early 1960s, this era of scarcity was over in Europe and in the Western world; none of this was, and still is, the case in many places around the world: expensive proteins versus cheap carbohydrates and to some degree fats and oils. I have no real solution to this dilemma other than the industry of agriculture has to adapt in ways that will be difficult to accept at first but will eventually be easy. An increasing number of people, stagnant surface of arable land, and increasing scarcity of water in traditional agricultural geographies must lead to new thinking in all of these areas. These topics were discussed in the preceding chapters, and the findings and conclusions should be a valuable basis for necessary change.

A large portion of any change has to come from the individual and their eating habits. Nutrition science can and must help in this dilemma and must become more decisive and active and less ambiguous and nebulous in its recommendations. Maybe the biggest breakthrough in the science of nutrition during the 20th century was the discovery of the "essentiality" of certain nutrients, both macro as well as micro. And my favorite example of essential fatty acids is a prominent one. It was always clear that food and food components were essential in keeping us alive. However, certain nutrients are essential beyond energy, and their presence in our food actually makes our body function in the first place.

It can safely be said that the nutrition science's findings achieved during the 20th century are a confirmation of all food- and beverage-related know-how acquired over centuries and millennia and should make us wonder, how the elders knew what they knew and did what they did. The simple answer may be trial and error, and that is probably the one that comes closest to the truth. They had ample time, not in their own life times but accumulated over all these many years until today.

5.2 Present-Day Food Fashions and Trends: A Never-Ending Story

Because abundance has become so prominent in affluent societies, or at least to a large degree, deprivation and restriction had to be reinvented. And this reinvention meant devising new or forgotten ways of restricted food intake to reduce calories overall to eliminating certain food groups to specific timing of food intake or dissociation of food ingredients and their separate or consecutive consumption. As mentioned elsewhere, one can find more than 100 more or less different so-called diets when searching the Internet.

However, the numerous diets are maybe less important than the real, important, sensible, and sustainable trends—although it is a fact that at any given time a large portion of people in affluent societies follow these. More recent studies suggest that the trend for dieting is slowly declining, but it still affects a fairly large number of the population. In the United States, the NPD Group has been tracking dieting habits of Americans for almost three decades and found that the percentage of adults on a diet has substantially decreased by approximately one-third over these years, down to 20% of the adult population being on a diet (NPD, 2013).

At a proportion of 20%, it is easy to calculate that approximately 65 million Americans are on a diet at any given time, that's roughly the entire population of the United Kingdom! Or in simpler words: a lot. On the other hand, there is no direct relationship to improved public health and increased life expectations and slightly reduced numbers—since the late 1990s—of cardiovascular heart disease or stroke in these populations, and these declines are more linked to improved eating habits overall paired with increased awareness of physical fitness and other factors such as reduced smoking and more responsible drinking (CDC, n.d.).

5.2.1 Food and Nutrition Trends: A Story of Perception, Deception, and Beliefs

Much has been said and written about the major food and nutrition trends of our modern times, and everything seems to be quite a bit confusing. Although the overwhelming desire of most people in the so-called developed world is to lose body weight through one of the many diets—often but not always combined with physical activity—the overwhelming need in developing countries or poor societies is to get enough food to be able to survive. Often food is produced in areas where poverty levels are so high that locals cannot afford food, and it is mostly exported to more affluent countries. The real dilemma here is how to feed a growing world population in the most equitable way. I do realize that this almost sounds like Douglas Adams's "ultimate question of life, the universe and everything," yet "42" is really not the answer here. It's not that simple, but it is at the heart of today's food and nutrition trends, which are so diverse that there is no clear and simple answer.

So, to simplify the approach, I suggest that there are two major trends/

- Healthy food for those who can afford it.
- Enough food for those who have little to nothing.

Let me elaborate more on the former trend or rather family of underlying trends. Healthy food is a vague descriptor and can be more clearly defined or rather described by words such as: *equilibrated, balanced, natural, clean, safe, well-tested, organic, respectful to* nature and environment, *low carbon footprint*, and a few more, more technical such as *no GMOs* (genetically modified organisms), *"clean" label* (meaning short list of ingredients), no *or only a few E-numbers, recyclable packaging*, or *packaging materials from renewable resources*. All the latter, the more technical ones are only relevant for packaged, industrial food products.

And more recently, another trend has surfaced: reducing food waste, along the supply chain from field to fork and more importantly even "after fork," or "instead fork" (after consumption or instead of consumption). What I mean by that is simply that too much food is being thrown away either because products were too close to their expiration date or just too much was cooked and not consumed.

The FAO (2011) list the following facts regarding food waste

- Roughly one third of the food produced in the world for human consumption every year —approximately 1.3 billion tons—gets lost or wasted.
- Food losses and waste amounts to roughly US$680 billion in industrialized countries and US$310 billion in developing countries.
- Industrialized and developing countries dissipate roughly the same quantities of food—respectively 670 and 630 million tons.
- Fruits and vegetables, plus roots and tubers, have the highest wastage rates of any food.
- Global quantitative food losses and waste per year are roughly 30% for cereals, 40–50% for root crops, fruits and vegetables, 20% for oil seeds, meat and dairy plus 35% for fish.
- Every year, consumers in rich countries waste almost as much food (222 million tons) as the entire net food production of sub-Saharan Africa (230 million tons).
- The amount of food lost or wasted every year is equivalent to more than half of the world's annual cereals crop (2.3 billion tons in 2009/2010).
- Per capita waste by consumers is between 95 and 115 kg a year in Europe and North America, and consumers in sub-Saharan Africa, south and southastern Asia, each throw away only 6–11 kg a year.

Figure 5.2 shows the per capita food loss.

These are quite some impressive numbers and a simplistic conclusion would be, if more than half of the world's annual cereals crop were not entirely or not at all wasted, we might easily be able to feed a few more billion people on our planet. This is an oversimplification and not easy to achieve; however, often it's the simple solutions that actually work. The question is, however, how could this waste problem be solved so that everyone can share in the outcome of reducing food waste.

It is similar to the situation when it comes to water as a resource to animal and plant life; there are several solutions and much of this has been discussed in Chapter 2, but let me list a few items.

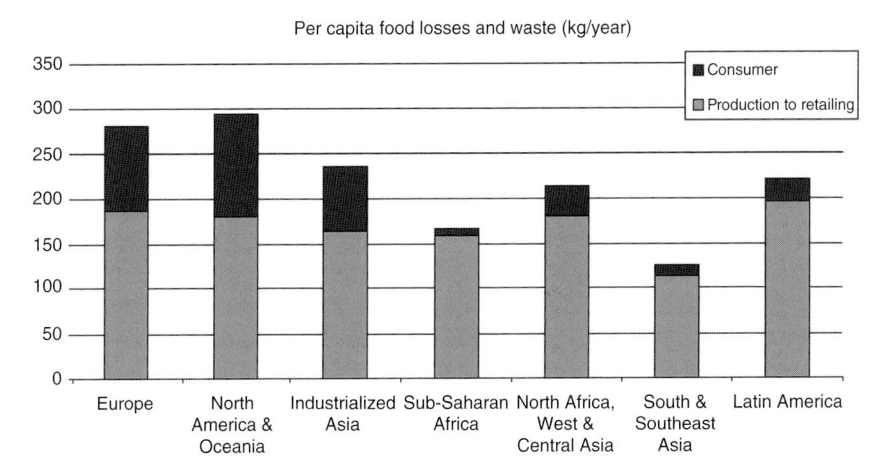

Figure 5.2 Per capita food losses and waste (kg/year). © FAO, 2011, Global food losses and food waste, presented at Interpack 2011, Düsseldorf, Germany. Available at http://www.fao.org/save-food/resources/keyfindings/en/ [Accessed August 20, 2016].

- Conservation
- Efficiency increase
- Fixing the leaks
- Optimize irrigation, and
- Minimize evaporation losses

They could be in any order and the similarities to reducing food waste are rather easy to see.

- Better preservation
- Increase production yields in agriculture and industrial food manufacture
- Improve supply chain (inbound and outbound)
- Optimize packaging of food products without increasing packaging materials, and
- Minimize manufactured food going to waste by increased consumer awareness and participation

Much of this is already done, but apparently not enough, otherwise we wouldn't see the food waste numbers we do. It appears that there is an abundance of food and food materials but as said many times before and by many prominent scholars and writers; the stuff just doesn't get to the people. It's the sad truth. My list should, however help quite a bit, so let us go into some more detail.

The important question that still remains: even if we would solve the logistics behind food waste and make more food and food products available to all, would there still be enough arable land and water to grow crops and breed and feed cattle? There are predictions that arable land because of climate change and weather as well as geophysical-related disasters will be less and less available for agriculture in the traditional farming areas around the globe. New ones may come up, however, and the question remains how to cultivate and use these without forcing or convincing people to move to these new locations. Well, some of this is already happening today, when many populations are almost forced out of their traditional habitats and moved into new areas, most often not welcome by those who already live there.

The UNHCR writes the following when it comes to natural disasters and their consequences.

> Some families and communities have already started to suffer from disasters and the consequences of climate change, forced to leave their homes in search of a new beginning.
>
> Scarce natural resources such as drinking water are likely to become even more limited. Many crops and some livestock are unlikely to survive in certain locations if conditions become too hot and dry, or too cold and wet. Food security, already a concern, will become even more challenging. (UNHCR, n.d.)

Just a word of clarification, in case it is not clear: *food security* simply means assurance that enough food is available for all, at all times, and everywhere.

When it comes to numbers of people displaced by natural disasters, the Internal Displacement Monitoring Center (2015) writes that since 2008 more than 26 million people per year have been displaced from their homes.

This book, unlike many other publications and political declarations before, attempts to find answers to these questions and issues by taking the underlying conditions as given and suggest solutions on that basis. I cannot promise that this book will "save the world," far from it, but the combined knowledge of its authors in fields, such as plant agriculture, animal husbandry, water management and food, nutrition, and technology should help to give good pointers.

5.3 New Food Sources: New Protein Sources

Previously in this book I have stressed the need for a more just distribution of food, let alone bringing enough food and food products to all. Although many in affluent societies try to reduce their overly rich food intake, most or all people in less fortunate societies simply don't have enough food to sustain energy and health in the long run. So the real challenge is food security, and the assurance that food is not only available where needed but also nutritionally balanced, affordable, safe and, tasty. After all, if it tastes bad, people are not inclined to eat their food, unless they have acquired the taste. Because there is no universally good taste, although many highly rated and often quoted chefs may want you to believe it, taste is not only local, but also learned, acquired, and ingrained on a family level.

5.3.1 Insects: A New Food Source?

Yes, taste is important, but security even more so. And yes, there are more and more food sources becoming available, or at least recognized as potential ones. Insects have become popular in the last couple of years and were hailed as a new source. Source they are, new they are not. Many populations in this world, such as for instance the Australian Aborigines eat insects, and insects are popular in Asian cuisine as well. You might not find grilled insects in your neighborhood Asian restaurant, but they are definitely part of that cuisine. And then there are the insects you eat inadvertently, without knowing that you actually consume them.

It might come as a surprise to you but many chocolates, especially manufactured in hotter climates, may contain small amount of insects, cockroaches and similar critters that crawled into the transportation bags of raw cacao beans or into the factory's storage silos. The US Food and Drug Administration (2000) has the following rules regarding what they call "insect filth."

> The following represents criteria for direct reference seizure to the Division of Compliance Management and Operations (HFC-210), and for direct citation by District Offices:
>
> Insect Filth:
>
> - The chocolate in six (6) 100 gram subsamples contains an average of 60 or more insect fragments per 100 grams, or
> - Any one subsample contains 90 or more insect fragments, even if the overall average of all the subsamples is less than 60.

In other words, this means that a 100-g chocolate bar may (and probably does) contain up to 59 insect fragments. Most often, such fragments cannot really be seen in the final chocolate product anymore because roller refining of the chocolate mass may have ground everything to indistinguishable fine particles of maybe around 10 micrometers.

When you develop allergies upon consuming chocolate, chances are that you are intolerant to a protein found in a critter that had clandestinely crawled into the jute bag and fed on cacao beans, at least until these beans were roasted at temperatures typically above 120 °C. Although these temperatures denature proteins nicely, they still may leave them with allergenic activity.

It could (and probably does) happen to coffee beans as well, however roasting temperature easily reach double (240 °C) in an industrial roaster so that any possible insect parts are literally roasted to death and only may add to the flavor, probably in good ways. I am not sarcastic, and I really do mean this.

I do hope that you still eat chocolate and drink coffee, and still consume your peanut butter or popcorns or similar pleasures. In one way or another, all food raw materials have been in some contact with insects and insects will always find a way in. So, what I am really saying is that we already eat insects, or at least "insect fragments" in our daily food and should not be shocked to learn about initiatives and development work when it comes to using insect protein specifically to enrich our food and especially enrich it with valuable proteins.

5.3.2 Increased Food Security through Exploiting New Protein Sources

In reality, these examples are just curiosities because it is estimated that 2 billion people consume insects and insect proteins as part of their traditional diet (van Huis, et al., 2013). This number accounts for approximately 30% of the world population. More than 1,900 insect species have reportedly been used as food and food source. Most commonly consumed insects are beetles, caterpillars and bees, wasps and ants. Grasshoppers, crickets, locust and other bugs and insects are also consumed.

An article by M. van der Spiegel and colleagues (2013) paints a similar picture. As the title of the article suggests, the authors weave a larger food net when it comes to the safety of novel protein sources and expand from insects to microalgae, seaweed, duckweed, and rapeseed, the latter not for its oil but for its protein part.

Much is moving; food regulations, especially in Europe are not up to speed und there is still much uncertainty as to how to proceed on many fronts from identification and breeding of most promising sources, technological challenges of adaptation and integration into the typical food manufacturing unit operations, assurance of food safety and absence of any toxins, to regulatory agreements. The article's major conclusions are rather straightforward and especially targeted to the European food market.

> Novel protein sources, like insects and algae, are expected to be increasingly used in Europe as replacers for animal-derived proteins. Technical and processing properties are being investigated but, to date, possible food safety hazards associated with the use of novel proteins in feed and food applications are hardly known. These hazards may include a range of contaminants, like heavy metals, mycotoxins, pesticide residues, as well as pathogens. Food business operators that wish to put on the European market products derived from novel proteins should comply with European legislation and possibly additional national legislation. To date, European law is not conclusive on several issues regarding the use of novel protein sources in feed and food products.

There are still many open questions as to the ability of "industrializing" insect breeding, feeding, growth, and "harvesting" especially when it comes to optimizing both quality and yield of edible proteins.

A fairly recent paper by Mark E. Lundy and Michael P. Parrella (2015) sheds some doubt, or at least tempers too high hopes and quick wins in this area, especially given the fact that for crickets, protein-conversion rates are not any better than for chicken and part of the authors' conclusion is simply if crickets will need to be fed high-quality organic diets so that they can grow to a worthwhile size to use them as an industrial source of proteins.

5.3.3 A "Crazy" Idea for Other Food Sources: Beyond Proteins

The importance of secure and safe and balanced protein sources was already emphasized quite a few times, not only in this chapter and its preceding sections, but in many more chapters. I realize that proteins are not the only important macro-ingredients in our food mix. Although carbohydrates are fairly easy to cultivate and harvest—and I use *fairly* in a rather nonchalant way—this is not the case when it comes to lipids, which is the fats and oils in our food. It could be a good opportunity and help to improve lipid security when we could expand to novel, yet rather old lipid sources such as tall oil, a component of wood, next to cellulose and lignin. Tall oil is typically used in industrial applications as a binder and glue component.

Under US Food and Drug Administration, Title 21 (2016), we find the following line.

> Sec. 172.862 Oleic acid derived from tall oil fatty acids.
>
> The food additive oleic acid derived from tall oil fatty acids may be safely used in food and as a component in the manufacture of food-grade additives in accordance with the following prescribed conditions:
>
> a) The additive consists of purified oleic acid separated from refined tall oil fatty acids.
> b) The additive meets the following specifications: ...
> c) It is used or intended for use as follows:
>   1) In foods as a lubricant, binder, and de-foaming agent in accordance with good manufacturing practice.
>   2) As a component in the manufacture of other food-grade additives.

Today tall oil and its purified component oleic acid C18:1 is only used as a food additive for specific applications such as the aforementioned lubricant, binder, or de-foaming agent.

Wouldn't it be appropriate to at least start discussing to use all major tall oil fatty acids, palmitic C16:0, oleic C18:1, and linoleic acid C18:2. Yes, it would require quite some substantial refining to get rid of the resin acids and other components such as fatty alcohols and a few highly interesting plant sterols, which are already used in food to replace—some suggest, to reduce—cholesterol.

The once-purified free fatty acids from tall oil could be esterified with glycerol, even in enzymatic and directed ways to produce highly desirable tailor-made, designed fats and oils with positive nutritional properties. Yes, I realize that much work, especially on the regulatory side, would be required, but the chemistry and manufacture are well known and should be pretty straightforward. I truly believe that this would be a really worthwhile project to pursue in any food R&D lab, academia, or industry.

5.4 Vegetarian Food and Its Potential Societal and Economic Impact

This is a touchy subject because by some it is almost seen as a religion, a strong belief in a cause ideally followed by everyone. I do not intend to add to any dispute here; however, I will express my opinion from a food, health, and nutrition point of view. A succinct overview article on the subject of vegetarianism in all or at least most of its formats was published by Tori Avey (2014). It looks back as far as ancient Greece and describes and comments on the various factions of vegetarians. The habit of consuming vegetarian diets is becoming increasingly popular; a fact that is reflected in the growing large variety of vegetarian food products in grocery stores. The article also suggests that the vegetarian and especially vegan diet needs to be containing all necessary food elements in order to live a healthy life.

She puts her finger on the real issue by saying "With proper attention to nutritional intake…". This is the real question and people who follow this, or any other diet, with or without meat, with or without fish or dairy products or eggs, have to make their own decisions as to what is good for them. When it comes to feeding toddlers and growing children during their years when they are having specific nutritional needs for their development, this becomes an entirely different story though especially in the case of veganism. From many personal discussions with vegans I always was left with the impression that vegan lifestyle was almost seen in a quasi-religious fashion, and I felt surrounded by missionaries. I do respect everyone's choice but have my slight doubts when it comes to enforce choices to everyone.

The BBC (2016) reported that a law on vegans and their young children is proposed to be voted on by the Italian parliament. It basically says that it would be illegal for parents to feed their young children a vegan diet. Although this proposed law might go a tat too far, it definitely sheds a bright light on the importance of a healthy and balanced diet for our children. If these very same children of vegans choose to adhere to a vegan life style once they become adolescent, so be it. Let everyone choose their most beloved food and food products as long as it is not detrimental to society and the health care system.

There is much to be said in favor of vegetarianism (not necessarily veganism) because it can help improve the balance of our food-related resources and have a positive impact on important environmental parameters such as emissions of greenhouse gases, methane and carbon dioxide most prominently. And this is and would be a good thing, at least to a certain degree. If we were, however, to follow the vegans' suggestions to only use food that grows in the soil, on shrubs and trees, and maybe in water, then eventually the following animals would not be any longer needed and therefore become extinct: cows, pigs, chickens, sheep and goats, buffalo, camel, and quite a few others that only or mostly exist to feed us and to supply clothing and similar materials. One could argue that these animals would become wild animals again and roam the "great plains" or the mountainsides of our planet. This is likely, but these animals would then compete for space, feed versus food, and still contribute to greenhouse gas emissions. The situation might become worse than it is today, because it may get out of hand. Consequently, if people would not consume any of these animal proteins any longer, the amount of arable land would have to be increased further to satisfy all the nutritional needs.

From a purely economic point of view, which of course has an impact on society and how it can survive, by letting all domestic food animals go free and unused, all of the jobs that are globally related to animal husbandry would be lost, and people would have to turn to other occupations. In times when more and more jobs become automated and robots and machines take over, this might seem like humankind's biggest nightmare. Today, agriculture employs approximately 1.3 billion people or around 40% of the total global workforce. In Africa more than 50% depend on agriculture and in Oceania (without Australia and New Zealand) this number is above 60% (Momagri, n.d.). A large portion of these 1.3 billion people depends on animal and animal husbandry-related jobs; they would all be gone or will eventually go away should vegans have their way. This is, beside total respect for everyone living the life they believe is best for them, a serious issue and should not be taken lightly.

The FAO ("Animal production," n.d.) writes the following, "Livestock contribute 40 % of the global value of agricultural output…"

This is really a big number because on top of this, it supplies food security and livelihood to approximately 1.3 billion people, which are families and businesses linked to animal agriculture. Yes, there is much to be improved and yes, we need to be careful as to how we deal with livestock and food animals, but the fact is that more people would die from hunger and extreme poverty in our world if all these jobs went extinct. This is something to think about and discuss calmly and attempt to find solutions that help both sides.

5.5 Urban Gardening and Urban Agriculture

In the last section of this chapter I want to analyze and discuss the role and potential contribution of a minority form of agriculture, minority in terms of surface, namely *urban gardening*. People in cities and townships, almost everywhere in this world, have always had the desire to have their own little plot of land next door, or elsewhere close by, and grow some vegetables or even hold chickens or rabbits. It is clear that arable land in an urban environment is rather limited, yet may have an important contribution especially to the diversity of plant agriculture.

Wikipedia introduces the topic's history with the following words:

> Community wastes were used in ancient Egypt to feed urban farming. In Machu Picchu, water was conserved and reused as part of the stepped architecture of the city, and vegetable beds were designed to gather sun in order to prolong the growing season. Allotment gardens came up in Germany in the early 19th century as a response to poverty and food insecurity. Victory gardens sprouted during WWI and WWII and were fruit, vegetable, and herb gardens in US, Canada, and UK. This effort was undertaken by citizens to reduce pressure on food production that was to support the war effort. Community gardening in most communities are open to the public and provide space for citizens to cultivate plants for food or recreation. ("Urban agriculture," n.d.)

Urban gardening's or urban agriculture's impact is both of economic as well as social nature, and it may add to food security in substantial ways, at least in the immediate vicinity of such urban gardens.

The allotment gardens ("Schrebergärten"), were the secret dreams of many families in postwar central Europe, and my family in Vienna was no exception. Unfortunately waiting lists to be allotted such a small plot of garden with a really small shed or tiny house on it were many years long, so no real chance to get one. Today, the situation is slightly different because people have discovered all sorts of unused "land" such as balconies, windowsills, or roof gardens in addition to small community plots in between city housing complexes.

The FAO considers "urban and peri-urban agriculture" (UPA) an increasingly important contribution to food security; however, it also mentions the risks as proper regulation is not in place everywhere. So, food security may be achieved, but total food safety not so quickly.

> Urban and peri-urban agriculture (UPA) can be defined as the growing of plants and the raising of animals within and around cities.
>
> Urban and peri-urban agriculture provides food products from different types of crops (grains, root crops, vegetables, mushrooms, fruits),

animals (poultry, rabbits, goats, sheep, cattle, pigs, guinea pigs, fish, etc.) as well as non-food products (e.g. aromatic and medicinal herbs, ornamental plants, tree products).

UPA includes trees managed for producing fruit and fuel-wood, as well as tree systems integrated and managed with crops (agro-forestry) and small-scale aquaculture.

UPA can make an important contribution to household food security, especially in times of crisis or food shortages.

Produce is either consumed by the producers, or sold in urban markets, such as the increasingly popular weekend farmers' markets found in many cities.

Because locally produced food requires less transportation and refrigeration, it can supply nearby markets with fresher and more nutritious products at competitive prices.

Consumers—especially low-income residents—enjoy easier access to fresh produce, greater choice and better prices.

Vegetables have a short production cycle; some can be harvested within 60 days of planting, so are well suited for urban farming.

Garden plots can be up to 15 times more productive than rural holdings. An area of just one square meter can provide 20 kg of food a year.

Urban vegetable growers spend less on transport, packaging and storage, and can sell directly through street food stands and market stalls. More income goes to them instead of middlemen.

Urban agriculture provides employment and incomes for poor women and other disadvantaged groups.

Horticulture can generate one job every 100 sq m garden in production, input supply, marketing and value-addition from producer to consumer.

The FAO (2016) also reports that today urban agriculture is practiced by 800 million people worldwide by growing vegetables or fruits and raise animals in cities and other urban environments.

There are two major takeaways from this information:

1) Some products can be harvested as early as 60 days after planting or seeding.
2) Urban garden plots can be up to 15 times more productive than rural holdings.

The first point would suggest that instead of one or maximum two harvests per year in traditional, large surface farming mode, there could be up to possibly four harvests (depending on climate and water availability), and the second point straightforwardly states that yield can be up to 15 times higher on a simple surface comparison. When combining these two effects, we might expect an overall efficiency increase of at least around 20 to a maximum of 60 times higher compared to large field growing. For an average of for instance

30 times higher efficiency, we would require a 30 times smaller field size in the urban gardening situation. In other words, one hectare of land (10,000 m^2) translates to approximately 300 m^2 of land in an urban garden. Could easily be the size of a nice community roof top garden or maybe 10 private balconies or one plot of communal land. And then there is vertical planting and growing, a trend that I first saw proposed in the Land Pavilion of Walt Disney's Epcot Center in Orlando, Florida, in the mid-1990s. Vertical gardening would add even more surface to the entire urban agriculture movement and should be keenly observed, followed, and applied wherever possible.

Regulations and application of basic food safety standards still need to be put into practice in many parts of the world, especially when it comes to commercializing the output of urban agriculture in bigger ways than at the local farmer's market. I do not say this because I love regulations—actually I rather detest them—but food, first and foremost, needs to be safe, and there is no compromising on this point.

5.5.1 The Urban Bee-Highways

Let me finish this chapter with a little side note concerning bees in urban environments. Although the diversity and even the presence of bees in large-scale agriculture is increasingly endangered, one can observe an increased presence of bees in urban environments. Cities and their plant diversity have become a safe ground for bees to live, work, and multiply and also do their work outside the city. In some places, cities have begun to what they call "bee highways," ensuring not only that bees can thrive and survive but that they have an incentive to traverse the city, literally from flower to flower, and do their pollination in agricultural areas outside the direct urban agricultural environment.

In 2015, Oslo in Norway was the first European city to have created such a bee highway (Guardian, 2015). There is a simple and straightforward rationale to these efforts, not only in Norway but also elsewhere: bees are considered to be increasingly endangered and warrant special attention. After all, the value of the work they do, pollination, is considered to be worth more than €150 billion per year on a worldwide basis.

When a good third of our agricultural food production needs bees and their pollination work, we better take the fate and well-being of bees seriously and make sure that they have the best environmental conditions so that they have a good "life–work balance." Bee-friendly designed urban gardens and urban habitats for bees are part of this important effort, especially in light of the fact that if there were no bees we would have to perform work that is valued an estimated €153 billion, without even knowing how to do it. The manufacturers of cotton swabs might become rich while we struggle to artificially pollinate, probably rather clumsily, our food crops. Think about it when you see the next bee flying toward you!

5.6 Summary and Major Learning

This was not an easy chapter because it spanned from a historic overview on food trends to pollinating bees. Nevertheless, the following topics were discussed and analyzed.

- Agriculture always was and always will be the twin sibling of food.
- A short voyage led us through the centuries and agriculture and food was described, analyzed and discussed from the times of classical Greece in the fifth century BC, through the times of the Roman Empire, the European Middle Ages, Renaissance, industrial revolution, the 20th century, to the present days.
- It was suggested that there were always two major food trends: enough and good, which could be translated to "food security" and "healthy and nutritious food."
- Food waste was identified as the number-one enemy to "food security" by losing roughly one-third of the world's food production every year, which amounts to approximately 1.3 billion tons of normally perfectly satisfactory and nutritious food material.
- The leaders in food waste are Europe and North America with a strong tendency of wasting more food between production and retailing than at the consumers' level.
- Several strategies were suggested to reduce food waste or to enhance food security such as better preservation, increased production yields, much improved supply chain, optimized and better adapted packaging, and last but not least, involve the consumers to a larger degree by increasing awareness of the issue.
- The question was raised and discussed if we were to solve all or most of the food waste-related issues, whether there would still be enough food and water to make enough food for a growing world population, especially in light of today's shifting populations because climate change and natural disasters. This book will not "save the world," but the combined knowledge of its authors attempts to give workable answers.
- To satisfy especially the need for proteins for a globally growing population, alternative food sources were discussed and analyzed and one popular, not really new source, are different types of insects. It was mentioned that today approximately 2 billion people consume insects and insect proteins as part of their traditional diet, or in other words almost 30% of the world population consumes insects one way or another.
- Industrialization of insect breeding, feeding, growth, and "harvesting" is still being explored, and legislation with regard to safe consumption of insect-based food products still remains to be adapted or expanded.
- As far as other alternative food sources are concerned, the potential usage of wood derived tall oil fatty acids could become an additional food source

provided technology is applied in affordable ways to render such fatty acids safe for human consumption as designed triglyceride lipids.

- Vegetarianism of the various sorts was discussed and analyzed and two major points were made: First, reduction of animal proteins in our diets is probably a good thing from both, a nutritional and environmental perspective and secondly, total veganism would wreak havoc to our agricultural communities because of the long-term loss of all (all!) domesticated food animals.
- To expand arable land for plant and to a smaller degree even for animal agriculture, the increasing popularity of urban agriculture, urban gardening was discussed and analyzed. FAO reports that approximately 800 million people practice urban gardening today.
- Increased harvest frequency as well as largely improved output per surface unit makes urban agriculture a viable and probably much-needed additional player in the quest to feed the growing world population.
- The role and fate of bees as the "pollinators-in-chief" were briefly discussed, and it was suggested that urban gardening habitats because of their greater diversity have an increasingly important role to play when it comes to the bees' well-being and long-term survival. It was briefly mentioned that on a global basis the work performed by bees in agriculture is estimated at a value of €153 billion.

References

"A history of nutrition." (n.d.). Available at http://www.nutritionbreakthroughs.com/html/a_history_of_nutrition.html [Accessed June 3, 2017].

"Ancient Greek cuisine." (n.d.). Available at https://en.wikipedia.org/wiki/Ancient_Greek_cuisine [Accessed June 3, 2017].

"Ancient Roman cuisines." (n.d.). https://en.wikipedia.org/wiki/Ancient_Roman_cuisine [Accessed June 3, 2017].

Avey, T. (2014). From Pythagorean to pescatarian—The evolution of vegetarianism. Available at http://www.pbs.org/food/the-history-kitchen/evolution-vegetarianism/ [Accessed June 3, 2017].

BBC. (2016). Italy proposal to jail vegans who impose diet on children. Available at http://www.bbc.com/news/world-europe-37034619 [Accessed June 3, 2017].

Bonardi, G. (1995). *Giovanni Bockenheym e la Cucina di Papa Martino V*. Milan: Mondadori.

Clark, G. (2002). The Agricultural Revolution and the Industrial Revolution: England, 1500–1912. Available at http://faculty.econ.ucdavis.edu/faculty/gclark/papers/prod2002.pdf [Accessed June 3, 2017].

CDC. (n.d.). Available at http://www.cdc.gov/dhdsp/ncvdss/index.htm [Accessed June 3, 2017].

"Economic History of Europe—Agriculture." (n.d.). Available at https://en.wikipedia.org/wiki/Economic_history_of_Europe#Agriculture [Accessed June 3, 2017].
"European farming during Middle Ages to 1800." (2016). Available at http://historylink101.com/lessons/farm-city/middle-ages.htm [Accessed June 3, 2017].
Faas, P. (2003). *Around the Roman Table: Food and Feasting in Ancient Rome.* Chicago: University of Chicago Press.
FAO. (2011). Global food losses and food waste, presented at Interpack 2011. Düsseldorf, Germany. Available at http://www.fao.org/save-food/resources/keyfindings/en/ [Accessed August 20, 2016].
FAO. (n.d.). Animal production. Available at http://www.fao.org/animal-production/en/ [Accessed June 3, 2017].
FAO. (2016). Urban agriculture. Available at http://www.fao.org/urban-agriculture/en/ [Accessed August 24, 2016].
"1st century." (n.d.). Available at https://en.wikipedia.org/wiki/1st_century [Accessed June 3, 2017].
Guardian (2015). Oslo creates world's first "highway" to protect endangered bees. Available at https://www.theguardian.com/environment/2015/jun/25/oslo-creates-worlds-first-highway-to-protect-endangered-bees [Accessed August 24, 2016].
History Guide. (2009). Lecture 7: Classical Greece. Available at http://www.historyguide.org/ancient/lecture7b.html [Accessed June 3, 2017].
Internal Displacement Monitoring Center. (2015). Global estimates 2015: People displaced by disasters. Available at http://www.internal-displacement.org/publications/2015/global-estimates-2015-people-displaced-by-disasters/ [Accessed June 3, 2017].
Lundy, M. E., & Parrella, M. P. (2015). Crickets Are Not a Free Lunch: Protein Capture from Scalable Organic Side-Streams via High-Density Populations of Acheta domesticus. Available at http://journals.plos.org/plosone/article?id=10.1371/journal.pone.0118785 [Accessed August 23, 2016].
Momagri. (n.d.). Available at http://www.momagri.org/UK/agriculture-s-key-figures/With-close-to-40-%25-of-the-global-workforce agriculture-is-the-world-s-largest-provider-of-jobs-_1066.html [Accessed June 3, 2017].
NPD. (2013). NPD Group reports dieting is at an all time low. Available at https://www.npd.com/wps/portal/npd/us/news/press-releases/the-npd-group-reports-dieting-is-at-an-all-time-low-dieting-season-has-begun-but-its-not-what-it-used-to-be/ [Accessed June 3, 2017].
"Puls (food)." (n.d.). Available at https://en.wikipedia.org/wiki/Puls_(food) [Accessed June 3, 2017].
"Refrigerator." (n.d.). Available at https://en.wikipedia.org/wiki/Refrigerator [Accessed June 3, 2017].

"The food of the Renaissance." (n.d.). Available at http://www.academiabarilla.com/the-italian-food-academy/centuries-dining/food-renaissance.aspx [Accessed June 3, 2017].

UNHCR. (n.d.). Climate change and disasters. Available at http://www.unhcr.org/climate-change-and-disasters.html [Accessed June 3, 2017].

"Urban agriculture." (n.d.). Available at https://en.wikipedia.org/wiki/Urban_agriculture [Accessed June 3, 2017].

US Food and Drug Administration. (2000). CPG Sec. 515.700. Available at http://www.fda.gov/ICECI/ComplianceManuals/CompliancePolicyGuidanceManual/ucm074443.htm [Accessed June 3, 2017].

US Food and Drug Administration. (2016). Code of Federal Regulations Title 21. Available at https://www.accessdata.fda.gov/scripts/cdrh/cfdocs/cfcfr/CFRSearch.cfm?fr=172.862 [Accessed June 3, 2017].

van der Spiegel, M., Noordam, M. Y., & van der Fels-Klerx, H. J. (2013). Safety of novel protein sources (insects, microalgae, seaweed, duckweed, and rapeseed) and legislative aspects for their application in food and feed production. *Comprehensive Reviews.* 12 (6), 662–78.

van Huis, A., Itterbeeck, J. V., Klunder, H., Mertens, E., Halloran, A., Muir, G., & Vantomme, P. (2013). Edible insects: Future prospects for food and feed security. FAO Forestry Paper 171. Rome: FAO.

6

The New Food Industry Business Model: From B2C to B2B, from Product Manufacture to Selling Know-How, and from Now to Then

Since we cannot change reality, let us change the eyes which see reality.
—Nikos Kazantzakis

6.1 The Old: Develop, Manufacture, and Sell ("Demase")

It's an old story, really a very old story. I have written elsewhere that the business model of develop, manufacture, and sell is as old as the food and beverage industry, the oldest written records of a beverage (beer) manufacturing company in Germany being around 1000 years old. The monks in the Weihenstephan beer brewery developed the beverage, which by the way in Germany runs under the definition of "Lebensmittel" (literally translates to "means of life"), even earlier than the first written records of manufacture and sale of beer. But that's what they did, and what today's company—not the monks though—still does: develop, manufacture, and sell. They "demase" if you allow me to use this as a new verb, which describes these actions (develop, manufacture and sell) in short.

By the way, demase sounds similar to demise. Demise, or passing away, pretty nicely describes the situation, at least as I see it and interpret it. Those who simply continue in the old, traditional ways, even if they camouflage them under heaps of consumer research and fancy business plans will experience demise, or in other words, there is a one-way street from demase to demise.

Enough of these puns, although they nicely illustrate the point I am making here: transition from the old, worn-out ways to a promising and more importantly sustainable future by changing the business model substantially. Unfortunately, because most large to very-large food and beverage companies are doing fairly well to very-well, they don't see this need and their preferred pastime is demase-ing. They do it because they can.

Megatrends in Food and Agriculture: Technology, Water Use and Nutrition, First Edition.
Helmut Traitler, Michel Dubois, Keith Heikes, Vincent Pétiard and David Zilberman.

Sometimes on rare occasions, I was involved in discussions that this model would require a good overhaul, and people—employees—should generate ideas as to how this could be done and where to this should and could lead the company. However, as already spelled out, this always was a kind of alibi action and ended in reports, which in turn, ended in a drawer (best case) or in the shredder (worst case). Why am I honing in on this? Simply because consumers have begun to distrust the food industry as a whole. Let me be a bit more specific, without wanting to point fingers but just pointing out a few events that have happened in and around the food industry in the past. Additionally I shall mention a few events that relate to my former company, Nestlé, because I have better insight.

6.1.1 The Fall of the Righteous

The first one is seen by some as the "mother of all events." The 1970s were marked by the well-known baby formula event. It so happened that eager and ambitious marketing representatives pushed a baby milk powder product, which was supposed to be dissolved in water in areas of Africa where availability of safe water was far from a reality and instructions as to how make unsafe water potable were sketchy at best. Rightly so, this became a big thing and tainted the reputation of the company big time. Yes, important strides were undertaken to rectify and improve the situation, but harm was done and it was done in important and lasting ways.

There are more examples, which paint the picture with a broad brush, and I list them in chronological order.

The year 1981 was another big year of a big food scandal, the so-called Spanish cooking oil disaster. It so happened that unscrupulous people in the industry wanted to make a buck, a rather big buck that was, and a really smelly one. Second-grade rapeseed oil was denatured to render it unfit for human consumption. Pretty much what happens with ethanol for industrial applications. Whereas ethanol is typically denatured with methanol or pyridine or a few others, the rapeseed oil in question was denatured with aniline.

This was my first year in my company and I happened to be a lipid scientist and was therefore closely looking into this matter. In those days, we believed that oil tanker trucks were used for both food (oil) as well as industrial (aniline) transports and were not properly cleaned in between loads. This hypothesis was rather incredible because typically, at least in more recent times, trucks would either be used for food or non-food in exclusive ways. So, the fraudulent usage of aniline denatured rapeseed oil and attempts to render it safe for human consumption by an additional refining (cleaning) step looks like the more credible and likely hypothesis. The fact is that this disaster killed more than 600 people in Spain, as the so refined oil was sold to the consumer public typically by street vendors, and cheaper than the real rapeseed oil in supermarkets. Nothing to be proud of!

Another, rather shameful event was Beechnut's fake apple juice that it sold in the early 1980s. The juice was marketed as "100% apple juice," although it did not contain any apple juice and had mostly artificial ingredients. To my knowledge, nobody was killed; however, such behavior can hardly be characterized as trust building with the consumers.

Mad cow disease was another event that shook the food industry, especially in the United Kingdom in the early 1980s and again late in 1990s and early 2000s. Cows were fed bone meal infected with bovine spongiform encephalopathy, a form of misfolded proteins called *prions*. I personally remember the days when hundreds of thousands of cows were killed and burned in the United Kingdom. I sat on a train to Manchester airport and read the big headlines in the local newspaper. What a debacle! British beef was banned for export to the European Union, a ban that lasted 10 years and was only lifted in May 2006. And here again is an event that didn't help build trust with consumers.

I do recall another event that hit the Nestlé Company, although not exclusively, especially in Italy. It concerned Tetra Pak containers that were printed with a fast-curing, UV sensitive ink which contained isopropyl thioxanthone (ITX), a chemical that was permitted for use in food packaging. As the preprinted multilayer carton was rolled up after printing, the future interior of the container came in contact with the printed exterior and small amounts of ITX could eventually be traced inside the infant formula milk that was filled into the container. The problem was recognized in 2005 and the immediate action was to discontinue the use of ITX.

The year 2008 saw the detected appearance of melamine in milk powders in China in several private label products. Melamine is used for instance as binder in industrial applications and is definitely not safe for human consumption. Its chemical formula is $C_3H_6N_6$. As one can see, it contains a lot of nitrogen, and in simple nitrogen analysis of milk and milk powers shows a false-positive, artificially simulating higher protein content. This was clearly fraudulent action that may have been ongoing for many years before its "official" detection.

The last example on my list dates to 2013 and is the discovery of varying amounts of horsemeat in beef, especially burgers in the United Kingdom and in other European countries. Although not typically a case of life and death (many people and cultures eat horse meat; not being a fan myself, I have eaten it a few times), it is yet another case of fraudulent behavior by the industry not helping to build up a trustworthy image.

There are many more cases that one could recount and that fit in this section, however, I believe I have made my point: the food industry and its affiliates (logistics, raw material suppliers, packaging suppliers) are in this negative spiral together, and the one, probably the only, driver is to maximize margin and earn more money. I do not pretend that fraudulent behavior is an intentional part of the business model of any food or beverage company; however, there appears to be ample space in these companies for those who want to follow this pathway of cheating and tainting the food and beverage companies' image.

I do accept that there is much space to hide for such people in large companies and sometimes, wrongdoing can go on unnoticed for longer periods of time than anyone would wish for or fear. One thing is for certain: at the end, all or at least most of the cheaters will be discovered and punished.

I have listed these food related scandals also to help understand that all this has happened during the time when all food and beverage companies followed the business model of demase. Maybe it's time to change this, and maybe it's time to change this in a stepwise approach. Let me discuss and analyze the suggested individual steps in some detail.

6.2 The New: The Customer Is King, the Consumer Is an Enabler, and from B2C to B2B

In all the years I worked in the food industry, I always heard the mantra: the consumer is king, or queen, for that matter. Everybody in the company, and as I learned everybody in basically all food and beverage companies, was indoctrinated with this slogan. And everybody, or almost everybody ultimately believed it. And yes, it makes a lot of sense for a consumer goods' company, especially a fast-moving consumer goods' company such as a food company, to put the consumer on a pedestal and pretend that everything is done for the benefit of the consumer. Yes, the benefit of the consumer is nicely transformed into what is typically called "consumer benefit" and became, and still is, the guiding principle in any activity that happens in a food or beverage company. "We need to listen to our consumers," we "have to define and deliver the consumer benefit," we "cannot afford to develop anything that does not fit the defined consumer benefit," and so on.

It's all about consumer benefit and what is good for the consumers, and in this, the customer, the retailer, often gets forgotten or at least becomes second priority. There were even times when food and beverage companies considered retailers to be their enemies, or at least someone who worked against the company. I have experienced such times and this animosity can still be felt. Retailers were not only seen as enemies but were almost considered evil and agitating against the manufacturing food or beverage company. This was, of course, never entirely true, but the fight for bigger and bigger margins and profits was, and is, the driving force. Manufactures have to "rent" shelf space at the retailers, and such rent, under the name "slotting fee" (euphemistically also called "slotting allowance") could eat a substantial portion of the profit. Although highly undesirable from the point of view of the manufacturing company, it often happened, and still happens, that retailers capture a higher margin from any given product than the manufacturer. This is a rather unsatisfactory situation to the food or beverage company.

6.2.1 Slotting Allowance

It's the retailer who ultimately meets the consumer and speaks to them directly, while the manufacturer, albeit the fact that consumers are the outspoken target, rarely or almost never engages with consumers directly. It's a sad fact, but fact it is. A company such as Nestlé or any other large or smaller food company almost entirely relies on the retailers to be the conduit of their products and thereby becomes totally dependent. There have been steps toward improvement of the situation through closer collaboration between manufacturer and retailer; however, the costs for slotting fees are still disproportionally high.

These fees are mostly a one-off initial payment to the retailer so that new grocery products or new beverages are placed on the retailer's shelf. Here's a typical definition for slotting allowance.

> Slotting allowances are one-time payments a supplier makes to a retailer as a condition for the initial placement of the supplier's product on the retailer's store shelves or for initial access to the retailer's warehouse space. (Federal Trade Commission Staff Study, 2003)

To some degree it is understandable that retailers require a fee from the manufacturer for the introduction of new products, which in the United States alone may amount to between $10,000 and $15,000 per year and per product.

The same report however has surprising statement: "Several retailers and suppliers reported that if a slotting allowance is paid, it does not guarantee any particular shelf placement" (Federal Trade Commission Staff Study, 2003). Or in other words, there is no guarantee that the new product introduction is actually appropriately, let alone, prominently placed.

Moreover, not all retailers charge slotting fees for all new product introductions, and it is estimated that such slotting fees are charged for 50% to 90% of introductions. There is, however some positive to these slotting fees because they typically include services such as shelf maintenance and restocking.

And finally, some numbers.

> For those products with slotting allowances, the average amount of slotting allowances (per item, per retailer, per metropolitan area) for all studied categories combined ranged from $2,313 to $21,768, depending on the particular retailer and metropolitan area. (Federal Trade Commission Staff Study, 2003)

From the retailers' perspective, the additional income from slotting fees is seen as a reduction in the cost of goods sold; in turn, it represents an increase in costs of goods sold for the manufacturer. The winner seems to be the retailer. It is difficult to estimate the total value of such cost-of-goods-sold modification (up or down, depending on which side one stands), but let me assume the following numbers, for the US market alone. Average slotting fee per new product introduction is $10,000 and typical number of new product introductions per year 15,000. This would amount to $150 million per year that the consumers ultimately pay more when purchasing their food or beverage products. Certain large retailers have a policy to require a new product introduction every year, otherwise the retail price for any product older than one year goes down (e.g., by 3%). So, for the manufacturer this means that they either accept a reduction of the retail price or introduce a "new" product, for which they have to pay a slotting fee/allowance. Some choice!

And still, the food and beverage industry worships their consumers and considers the retailer as a necessary evil or something not far from this definition. This needs to change, and this is the first step of a new approach, which is seen by some maybe as too radical, but in my eyes, normal and deeply necessary.

6.2.2 Retailers Become the Most Important Partners for Food and Beverage Companies

As I just wrote, not everyone in the food industry, and most likely also on the retailers' side will like the idea of partnering of the two sides. I truly believe that this is not a matter of "like it" or "don't like it," but a clear necessity that the industry has to adopt and ultimately live by this new standard. Let me describe and discuss in a bit more detail what I mean. When a food manufacturer develops a new product, representatives from selected retailers, one or several, should become active members of the development project, and this at every step from inception to hand over to the manufacturing side. Consumer insight data, objective and subjective consumer benefit assumptions, choice of suppliers, budgeting, marketing spend, supply chain and procurement, and probably a few more should all be elements jointly discussed, jointly decided on and, as appropriate, jointly executed.

Yes, there may be open questions as to how to best split costs, intellectual property and its ownership, and a few more legal questions such as potential market dominance when the "big guns" work together. These questions need to be sorted out, especially with lawmakers; however, I suggest a new way of working that may or may not require a new set of boundary conditions from a legal perspective. As long as results of such collaborations are eventually shared with all, after a lead time for the originators of 2 years, everyone should be satisfied. It would cut down costs, time, and uncertainty, whether my new product is actually liked and accepted by the retailer to whom I intend to offer it.

Slotting fees would become obsolete, savings would be achieved on many fronts, and frustrations of unsuccessful new product introductions would largely be reduced. Sounds to be too good to be true? Sounds too naïve to be realistic? Sounds too easy for not being some hidden catch somewhere? I don't know for sure but much of the rejectionist voices that I seem to already hear might simply be based partly on fear of change, and partly on fear of loss of influence and power. And yes, we do it, at least in part, already anyway, so no need to think along those rather radical lines. My answer to this would be one word: bollocks! Please forgive my short answer but it comes from the heart, and the brain, in whichever order you prefer. It reminds me of the three standard answers that you get, when change is proposed.

We have never done it in this way, everybody could come and dare suggesting change, and finally the ultimate answer, "whatever." But let me not despair and let me continue to develop these ideas of close collaboration between manufacturer and retailer and illustrate this at a few examples and some numbers (money that is)—all based on my personal experience in open innovation and innovation partnerships.

Although these two siblings are based on the inbound innovation and collaboration space—new ideas and innovation coming from know-how and solution providers—I have had some preliminary and personal experience on the outbound side, namely working with large retailers on specific projects very early on.

6.2.3 How This Could Work: A Possible Path and Examples

No innovation and development collaboration that was done in the past together with a retailer was as large in its scope as suggested previously but a few elements were definitely present. The main element was early involvement of retailers in new product development. It's probably the best starting point for this transformation that goes from consumer-centric to customer (retailer)-embraced.

Let me sketch a possible pathway: Consumer-centric status quo (antagonistic, in competition with retailers) → early involvement of retailers in new product conceptualization (jointly discover the opportunities) → joint definition and resourcing (people and money) of new product development programs and ultimately individual projects → in parallel, joint discovery of common target consumer groups and joint definition of consumer benefits → alignment of development programs, including definition of to be shared or out-licensed intellectual property (who keeps what and who pays for continued intellectual property maintenance?) → join forces in definition of product (and brand) architecture and packaging design → joint efforts in marketing → discuss and propose new legal boundaries for this new collaborative approach → exchange and physically put together expert resources needed for this joint new product

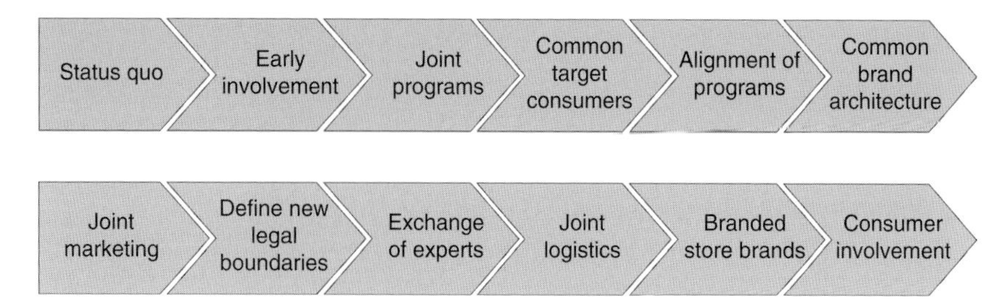

Figure 6.1 Evolution of collaboration between food manufacturer and retailer.

development, profit from in-kind contribution through existing expert resources thereby reducing other resource budgets → join forces in distribution and logistics much more closely than already done to some degree today → in the more distant future, once many or most of the preceding is achieved, create "branded store brands" that combine history of well-known (by consumers) existing brands and somehow limited private label store brands → involve local and regional consumers and clients of the retailer actively in such new product and brand development.

Figure 6.1 depicts the evolution of the collaboration pathway between manufacturer and retailer.

This looks like a rather lengthy and potentially complex and complicated process, but at the end of the day it is not, however, in my opinion it is necessary to embrace this approach or at least parts of it.

Let me discuss and critically analyze two examples that I have personally experienced and was involved in.

6.2.3.1 The Example of KitKat® Chunky®

The first one dates back quite a few years already when I worked in the United Kingdom in the area of chocolate and confectionery. We are looking at the last years before the new millennium and it's the story of KitKat Chunky, the rather heavy and rather large single finger KitKat.

After almost historically lengthy periods between the first KitKat developed even before World War II in 1935 and introduction of the two-finger KitKat in that same year, it took until 1999 when KitKat Chunky was launched after a relatively short development time, which probably started around 1997. So it took more than 60 years to come up with a new product idea and it took approximately 2 or so years to develop the Chunky product. And with its 50 grams in the original version, chunky it was.

There were of course continuous improvements in KitKat over all those years, but they mainly concerned the wrapper design and process and ingredients optimizations. The wrapper was always red, with one short exception in the years between 1945 and 1947. A milk shortage in the United Kingdom forced the parent company Rowntree to use dark chocolate for its KitKat. To indicate this change, the wrapper became blue.

When the cogitation time of 60-plus years seemed to be enough, Nestlé, which had acquired Rowntree in 1988, took on the challenge to "desecrate" a holy cow, KitKat that is, and suggest that a new format, other than two- and four-finger, should give not only a great new product but also new target consumers and consumption opportunities. Questions of "sharability" and into how many parts Chunky should be broken up along predetermined breaking lines were discussed and rediscussed.

Most surprisingly, a large United Kingdom retailer was early on involved in the development, so as to obtain the best consumer insight and feedback not from a theoretical consumer research company perspective but really from the horse's mouth. The retailer gave incredibly valuable feedback, especially when it came to help the decision to ultimately jump and launch the new product in 1999. And it was not only a big success in the marketplace by quickly becoming the most successful new product launch of that year (and for several years to come), but it also ploughed the ground for many more KitKat variants such as minis, flavor variants, initially very much pushed by the Japanese consumers, and white chocolate and peanut, let alone the smaller versions (40 gram) of KitKat Chunky.

It looked like the dam was broken, and an important reason for that was the supporting and backing assurance by the retailer to go ahead and share their generic consumer information with the Nestlé Company. It was an excellent example of how retailers and food and beverage manufacturers could work together in a no-frills and participative approach. Moreover I am confident that because of the role the retailer played in this example the project was pursued and executed in the first place. I am not sure, whether the Nestlé Company and the R&D arm would have been as daring as they were on their own.

6.2.3.2 A German Retailer and Wholesaler Example

The other example is more recent and dates back to approximately 2009–2010. Through some personal and rather serendipitous connections I had the good fortune to meet the then-CEO of a large German retailer and wholesaler. The informal dinner at his house quickly turned into some semi-professional chit chat and to no one's surprise we talked about what ifs, and could we and how if, and shouldn't we. We discussed and potentially formalized a closer collaboration between "his" company and "mine" (Nestlé). Unlike the other example, this one has no happy end, at least at first sight. From this first and informal get-together, a number of actions and further more formal meetings resulted. Many of the elements that I have mentioned in the collaboration pathway in Figure 6.1 were also elements that were initially suggested and discussed between the retailer and the manufacturer. It was seen as important to define the field of potential joint activities in terms of defined program platforms and resulting projects. It was also understood that there had to be consensus as far as the most appropriate target consumer groups and, to some degree, alignment of joint activities. It was not an easy undertaking, and we had several meetings and discussions to carve out a practical as well as doable work approach.

Things became a bit tough when elements closer to logistics, finances, and intellectual property became relevant parts of the discussion and, as far as I can recall, we did not find much common ground there. That probably should be good fodder for those who might not like the approach of retailer and manufacturer closeness at all; however, I have to say that we were at the beginning of something that in its entirety and as proposed here was no industry practice yet. It is hard to say at this point in time as to how we could have cut the Gordian knot, especially as we didn't have real historic industry examples.

The major learning was, however, that we did come together and discussed possible pathways. This ultimately led to a heightened level of trust and improved mutual respect. Even without formalizing a potential collaborative set-up between the two parties, they became partners on many levels.

As mentioned, we stopped short of formalizing a defined collaboration, because there were too many open questions with regard to financial benefits and how to share them as well as questions of ownership, especially of intellectual property. But learn we did! If I were to do it again, I would put much emphasis on the elements of joint brand architecture, defining the legal boundaries, the best approach to joint logistics, and ultimately a joint plan as to how to share the prize.

All this calls for a new business model that clearly involves retailers and their know-how at any level of the new product development activities. However, this is not the only element that calls for a new business model, and I see this just as the beginning of a "brave new world" of the food and food retail industry. This will be the topic of the following section.

6.3 From Selling Products to Selling Know-How

In my previous book, *Food Industry Research and Development: A New Approach* (2016), I initiated a discussion and analysis of such a shift from selling products to selling know-how from an R&D perspective. Before moving to the target of selling know-how, I made the following critical suggestion and called it the new business model 2.0.

- The R&D organization drives the company (similar to what happens in the pharmaceutical industry already strongly)
- Open innovation and innovation partnerships are part of the new DNA of the new food industry, and to top it off…
- Retailers become real partners for the food and beverage industry.

Table 6.1 The R&D-centric company model 2.0.

Functions inside the company	Functions outside the company
→ **Office of R&D Vision**: Develops the major future R&D directions and focus areas, thereby directing company's product, process, and services development → the company's sustainable future	Finances
→ **Office of R&D-Driven Product and Packaging Development**: Drives and manages packaging development including new materials and technologies in this area.	Legal
→ **Office of Design Vision and Execution**: Ensures a solid design basis and understanding for product, process, services and packaging development	Procurement, Logistics, Supply Chain
→ **Office of Manufacturing, Q&S, Compliance and Regulatory**: Manufacturing driven by the initially decided R&D and design vision and principles, especially influencing and driving best manufacturing paractices	Advertising
→ **Office of Marketing and Sales**: R&D and design driven marketing and sales; extensive and close collaboration with retailers	***Retailers: targets and partners*** ***Know-how providers: Suppliers, Universities, others (Open Innovation and Partnerships)***

I had suggested a detailed distribution of functions and responsibilities to be done inside as well as outside the company. Table 6.1 depicts this R&D-centric model 2.0 and the suggested distribution of tasks and roles.

What is called the "Office of R&D Vision" takes the lead and is at the heart not only of the direction that R&D must take but also directs and steers the new product development process and assures that the company has a sustainable future. The strategy of the company is directly linked to this process and is devised rather by R&D than marketing and finances. This is a big step that many will hate, some will belittle and ridicule, and a few will outright fight. But that's not really new because fighting goes on anyway, even in situations when everyone seems to be of the same opinion—making one's point and one's marks is more important, not necessarily for the company but for the ego of the involved. Bet let me not get carried away in ranting but rather remain on a constructive pathway of proposing new approaches for the food and beverage industry of the future. And the approach of an R&D organization that is driving the company, additionally embracing open innovation and partnerships and ultimately recognizing the important and crucial role of retailers, is the basis and are the first steps toward this future.

The outlined R&D-centric company structure depicted in Table 6.1 was first described in *Food Industry Research and Development: A New Approach.*

Other elements of this suggested R&D-centric company model 2.0, which I strongly believe should remain inside the company are new product and packaging development, mainly as the basis to acquire and maintain relevant and valuable knowledge and maintain mastership in outsourcing the actual manufacturing of products. This means that there has to remain the offices of manufacturing, quality and safety, and compliance and regulatory. The office of design vision and execution as well as marketing and sales should definitely remain inside the company however; they will have to adapt to the new approach of not only selling products but also increasingly services and knowledge. All other branches such as finances, legal, procurement, logistics, supply chain, and advertising should all be considered back-office activities that can be outsourced to experts in these fields. I have added retailers as the real targets of all consumer-related activities, and they should become the partners in this important transformation to a knowledge-centric company.

Why do I think that this is so important for preparing the ground for the next step, the one that transforms the food industry from a consumer- and brand-centric industry to a knowledge- and service-centric one? Well, simply because accepting that the knowledge branch of the company (i.e., the R&D group) "takes over," paired with the acceptance that others from outside the company can actually help, and the recognition of the real closeness of retailers and consumers, will almost automatically lead to the realization that the old model of demase might slowly but steadily come to an end and might actually be harmful if not outright hampering further existence of the traditional food and beverage companies.

6.3.1 The Knowledge-Centric Company

I do not advocate a dramatic transformation from one day to the other or not even from one year to the next. I propose to first embrace the described business model 2.0 and then take it step by step to the next level, namely business model 2.1. Let me explain and discuss and analyze this model 2.1 in some more detail here in this chapter and section and again under the challenging aspect of the food company that almost radically transforms itself in chapter 9.

Again, in *Food Industry Research and Development: A New Approach*, I have already briefly initiated this topic and the discussion of its validity.

> One of the prerequisites for 2.1 to work is the acceptance of 2.0, at least in part:
>
> Recognize that distributors and retailers are the real targets and partners of all the efforts of any food company.

This is the real first step that is necessary to successfully progressing towards business model 2.1. So, what's in 2.1? Here's my first list:

- Preparing and eventually executing the gradual transition to selling proprietary know-how through different channels such as for instance and by no means complete, experts who's work place is off-site as well as internet platforms and professional internet based apps.
- Based on the strengthened relationships with retailers use both, their brick-and-mortar locations as well as Internet platforms jointly for activities of selling nutritional and health and wellbeing related know-how.
- Grow and promote understanding and acceptance of consumers for the new "your company name inside" approach, similar to what the computer industry has perfected since many years and what can increasingly be seen in areas such as digital photography emphasizing the usage of CMOS chips.
- Adapt the hiring approach of your company and increasingly hire personnel with health and nutrition expertise combined with excellent communication skills, which will become the new marketing.
- Train your existing personnel accordingly and in preparation for the new business model 2.1
- Most importantly make personal and personalized expert advice the core of your business both, in person as well as on-line.
- Include "chat-with-the-experts" platforms, ideally based on loyalty programs similar to the ones in the airline industry and through membership and long-term subscription."

And I suggested that because the approach calls for the company know-how to become the most valuable asset, more so than the agro-based and agro-transformed branded product, not only the know-how carriers become important drivers in the company but the company in its entirety embraces the idea of selling know-how is (or will eventually be) more important than selling products. (Traitler et al., 2016, p. 214)

Let me take each of the points one by one and analyze and discuss them individually and in their totality. The first point of this list proposes to prepare the ground for a step change in what a food company should really be doing in the future, at least, in my opinion, the one, which has a promising future. The food company of the future needs to gradually transition away from the demase paradigm to the new paradigm of selling valuable proprietary know-how to consumers in need and help them understand the meaning of good food and appropriate nutrition that supports both their health and their wallet. This is not easy to grasp because the reality today looks different. Food companies try to square the circle by pretending that they are highly concerned about good nutrition and the consumers' health and still make the traditional industrial food products such as chocolate, coffee, dairy, culinary, ice cream and a few others. Yes, a company such as Nestlé also sells products in the area of infant nutrition as well as appropriately healthy and health-supporting products for the elderly, but they are still in the minority, even if they seem to be growing in importance in the overall portfolio.

Funny to mention that the very same company does not have a defined group of products under the range of "adult nutrition," yet adults are the major group of people living on our planet. I don't want to go into any detail as to why this is so, but it just proves my point I have made. On the one hand the majority of products that are sold don't necessarily fall under a healthy and nutritious definition, and on the other hand, the company emphasizes its focus on exactly these attributes. Somehow this doesn't go together, and it is my firm belief that not only will it not go together in the future and one of the major ways out of this consists of the transformation from demase to selling know-how.

How to achieve this is briefly sketched in the following points from the preceding list. The second point refers to a strengthened and improved relationship between food companies and retailers. It emphasizes the importance of reaching out to consumers together, and "uploading" and selling nutrition- and health-related information on food and food products to consumers together. This brings me to the next point on the list, namely the gradual transformation from branded product to the branded know-how company. The intrinsic value of any food product for the consumers really lies in the know-how that was and is generated by the food company and not so much in the product itself. Ultimately, the product manufacture could be outsourced to third party co-manufacturers who specialize on the manufacturing part. These companies, still a part of the food industry, might become less important in the future, if trends toward individualization and democratization of food and food products become more prominent and important.

What I mean to say by this is that, in my opinion, there might come a time when people increasingly will prepare their own foods from scratch and so that the only "leftover" products for the food industry might be specialties such as coffee and chocolate and a few other, highly specialized industrial food products in the area of medical foods and foods for infants and possibly the elderly. I have mentioned the specialization away from "adult nutrition" but have my doubts whether the Nestlé Company did this because they have the same foresight that I suggest here, namely away from demase and toward selling know-how.

6.3.2 Engaging, Interacting, and Selling: The New Etiquette

The remaining points of the said list are all people- and talent-related ones. They are a call for reeducation of existing personnel, especially but not exclusively on the R&D side as well as hiring new talents that have the ambition and the ability to sell know-how in all areas such as nutrition, health, home economics, and possibly even manufacturing and marketing. The new food company becomes a knowledge company at its very core. These new people, the carriers and givers of know-how and expertise continuously trained and go out—literally or via communication networks and social media—to meet consumers wherever they can get hold of them. Retailers and their stores or digital presence are the new meeting places, and shopping for the consumer takes on a whole new meaning: involvement and learning about good food, its nutritional value, and its health implications, ultimately what's real "good for me."

This new generation of employees in the food company of the future applying business model 2.1 is a generation of sellers; they mainly sell the company's collective know-how. This ultimately becomes the real value proposition for all parties: food company, retailer, and consumer, and it carries the product and makes the product almost a side aspect, just the carrier. Out of this, many new habits can and will be borne, some of which are elements such as expert chats and loyalty programs that go beyond receiving coupons for merchandise purchased and that lead to a meaningful and rewarding membership in a "club of the wise" with real added value.

The bond between food company and consumers will less be based on a product name—the brand—and more on trust and the realization that I, the consumer, get something really valuable in the format of detailed nutritional and health information. This is a value for which I, the consumer, am prepared to spend my money on. It's not processing of ingredients into a product and subsequent packaging into a fancy and colorful wrapper, sachet, or box that I pay for, but the nutrition and health implication that this packaged good stands for. This is simply a new definition of "brand" and shifts the consumer benefit away from the static element of a product to the dynamic one of ever-new and added valuable information and learning.

The really cool thing about this new model 2.1 is that finally the oft-cited service that a food company gives or rather wants to give to consumers is totally integrated in this model. Service is not everything....it's the only thing! Service and serve are connected, and that's what ultimately happens here: the food company finally not only serves a purpose beyond demase but it also really serves consumers and thereby makes them not only happy but also healthy and loyal.

6.4 The Community of Consumers: It's What They Want that Counts!

It cannot be said often enough: the consumer is the ultimate decider-in-chief when it comes to purchasing a specific product or another one or none; this, despite what I had elaborated on previously in this chapter—namely that the new target for food and beverage companies is rather the retailer and not so much the consumer any more. Sounds like a contradiction? Maybe, but let me explain. It's a tricky and rather complicated situation, almost like a catch-22.

Here's a short definition what a catch-22 situation means. "A catch-22 is a paradoxical situation from which an individual cannot escape because of contradictory rules. An example would be: To get a job, you need to have a few years of experience. But in order to gain experience, you need to get a job first ("Catch-22," n.d.).

One can imagine how this translates to our situation between retailer and consumer: To be able to reach out to consumers you need retailers. But to get the attention of retailers, you need brands and products that are liked by consumers. This is not an easy undertaking at all and requires a lot of "dancing between the lines" and trying to make everyone happy and aligned to the goal to sell products, services, and ultimately know-how to consumers. It is just one more example of why the suggested new model 2.0, aligning with and targeting retailers first and foremost, is so crucial to the future success, and at the same time not forgetting the consumer. Ultimately, this is not only a catch-22 situation but also an attempt to squaring the proverbial circle—impossible or at least almost.

There is an "almost" in this equation because there are ways out from this unique dependence on retailers by increasingly becoming a know-how and knowledge-selling company. This is of course not the only way out, and this section as well as sections in Chapter 9 attempt a discussion and analysis of the situation and promising possibilities. All of these, however, involve change, often important and not-so-easy change. This brings me back to the title of this section: the community of consumers and that it's all about what they want and don't want. From personal experience, I can say that the food industry is not a good listener when it comes to the themes that consumers dislike. The reason is rather obvious; why would an industry—any industry—like to hear that consumers, users of their products and services show a disliking for industrial products in general. Or in other words, one of the scariest consumer dislikes (scary for the food industry) is the fact that consumers in general reject industrial food products. Of course not all of them and not every product or service but especially those that could be done otherwise or in other ways such as most prominently all products that one could prepare and cook at home, which are rather the majority of what we eat.

This is a rather dangerous situation, at least long term, being rejected by those we need for our sustainable success and survival. I mentioned that people in the food industry are not really good listeners when it comes to consumer dislikes, and maybe some readers will criticize what I write here as alarmist or even false. I would hope that the majority would, however, see and understand this clear writing on the wall for what it is, namely a wakeup call and getting prepared for substantial change.

So what is an "industrial" food product? In the most general definition, it is food that was processed; ingredients were mixed and processed to obtain an edible end product, a meal in the largest sense. But hold on, wouldn't that also count for my home-cooked meal as much as for any meal that was prepared and packaged on a larger, industrial scale? In theory yes, because I will use similar techniques—*unit operations* as they are called in the industry—and I will mix, chop, puree, and mince ingredients very much like the food industry does it at a larger scale. And yes, I will apply temperature and sometimes pressure to finish off—cook—my meal, again very much the same as in the food industry. So, you might ask, where is the difference? The industry very much or almost exclusively plays the card of "convenience", or in other words it is easier to open a pack of a prepared meal and heat it up than cooking the entire meal from scratch. And yes, that's true, but it doesn't emotionally connect the consumer, on the contrary, it alienates the consumer.

On the other side, the consumers' side, the one key word that emerges is *processed*, and the one feeling that is mostly repeated is "mistrust." I do not pretend that either of these words are either totally justified or totally false. All I am saying is that the food industry had ample time to come up with emotional answers to these feelings but has miserably failed. And it had many, many years to work on this, and the same answer is given again and again: it's convenient! And yes, more recently most food companies have added the terminology of nutrition and health and some even "wellness"—for whatever it's worth. This is not good enough because consumers still hold the same opinion, namely: processed and mistrust. And it doesn't look like this is going to go away any time soon, or ever. So, what's the way out? I truly believe that the most promising way out of this dilemma is to go the route of gradually becoming the know-how–selling company and increasingly turning to give service and assurance to consumers, and all this with retailers as partners.

I do admit that for an industry, which is as old as the food industry, this requires quite a change in the thought process of its people and especially its leaders. It's an almost Orwellian "thought crime," something that is not only unthinkable but mustn't even been thought. Looking at the various consumer cohorts and groups they have become so diversified that it's virtually impossible to propose appropriate products for just about every one of these. The diversity is principally based on age groups, gender, geographies and cultures, religion, affluence levels, and increasingly is dominated by those who have or pretend to have all kinds of food intolerances such as lactose, gluten, peanuts, nuts in general, all sorts of proteins beginning with milk and ending with insects. And there are probably more that we don't know of yet. And yes, there is more: vegetarians, vegans, fruitarians, all sorts of health diets that require the appropriate ingredients, and yet a few more which I am unaware of today but that might become popular in the near future.

So, how does a globally, or even regionally, operating food company cope with all of this? It's basically impossible and I do know, from personal experience, that a large food company such as the Nestlé Company does not respond to this at all, or only in specific cases. Onc of these cases is hypoallergenic formula for infants (baby milk) because it has been well established that a not negligible amount of babies, once they start to drink formula milk suffer from a cow milk protein intolerance and a hypoallergenic formula almost responds to a mainstream need although it might only affect 2% to 3% of the breast-fed babies. So, a rather small number, however, if not properly recognized can lead to complications (Kids Health, n.d.).

In my many years in the food industry I have not really seen any positive, let alone, enthusiastic response to any of the aforementioned groups when it came to develop a range of appropriate and consumer-accepted products. One stays with the mainstream and rather repeats the umpteenth iteration (called "renovation") of the existing range of products and traditional portfolio of a large food company. The big ones leave this to the small ones, the ones that can respond more quickly and in more agile ways and for which it appears to be a lesser risk should the custom made product range not work as it was hoped for. So, why not abdicate this process of product development, manufacture, and sale to a large number of small to medium-sized local food enterprises, who, on behalf of a large, globally operating food company serve as agile and flexible co-manufacturers and leave the selling of service and knowledge to the big ones (the new types of food companies). Those food companies who are the best in their field of nutrition, health, and food-related issues and who possibly can expand in areas that a traditional food company has never wanted to enter until today, such as supporting the large number of recreational athletes, people who go to the gym or health club, and actually create their own health clubs with professional nutritional advice.

The new food company could team up with the cruise ship industry, with hotel spas, with health clinics, and similar institutions and make good on the promise of nutrition and health and be *the* experts when it comes to food and its health impact in all these and many more areas of our daily lives. There are so many opportunities to earn revenue and appropriate margins that lie beyond the traditional approach of demase that it's just about time to enter into this new space age of nutritional knowledge and its worthwhile and rewarding delivery to the vast amount of diversified consumer groups.

6.4.1 The Consumers Become Involved

There have been rather timid attempts in the past by the traditional food industry to involve consumers in the new-product development process. I use the word *timid* because that's what it really was: a shy attempt to bring consumer in almost real time to the development process and include their opinions and ideas, as much as it is really possible into the new (innovated or rather renovated) product. Although this seems to be frequently used in other industries with more or less success and acceptance by the industry itself—crowd innovating comes to mind—I have only seen this happening a few times in my many years in the food industry. I want to be clear: consumers involved, side by side with the technical and marketing developers of the company in the process of creating something new and not the often used process in which consumers are presented with newly developed prototypes and can give their input and feedback even on rather short notice, almost as if they were fully involved. It's not the same thing: one is participative, and the other one is reactive. There are many innovation and creativity companies that involve consumers in reactive ways but hardly anyone does it in participative mode.

I personally was involved in two such events, which were unrelated but both in the area of confectionery. One was with kids and the other one with adults, both, however, consumers of candy. Let me give you the more interesting insight, namely the one we gained with kids. Quite a few years ago, we invited approximately 10 or so kids, age around 10 years and they were all children of company employees. If you would think they were positively biased, think again. They were all but mostly critical, and some even embarrassed with what their parents did professionally. Not that there was anything to be embarrassed about, but that's what kids are at this age. Their peers become more important than their parents and parents sometimes are a bother. Anyhow, just to say that we felt that the kids were rather unbiased and open to every crazy twist they could think of.

We equipped them with fitting lab coats, notepads, writing material, cardboard, moldable plastic, scissors, glue, scotch tape and gave them just one target: develop a candy that is just right for you and where existing candies miss out. And then we let them run. We didn't tell them how to organize or how, if at all, to form groups and collaborate, neither did we give any hints as to how to proceed when getting stuck. Yes, we answered questions and yes, of course we helped them to get traction again if they ever slowed down. But overall, we left them alone for a good 2 hours or so. Any much longer than that would have been a challenge because the kids would eventually lose interest and turn this into a play session of the open kind, which would have not much to do with the original target.

The results were intriguing and can be summarized in two points.

- First, all kids, without exception as far as I can recall, developed candies that they could share with friends.
- Second, the most striking results were candies that would deliver different flavors and textures in one product sequentially discovered as the candy is eaten.

These findings were so different from what the candy manufacturers typically do, at least as far as the sequential flavor delivery is concerned. But even the sharing aspect is very rarely found in a mainstream candy bar. Even a product such as a two-finger KitKat (or four-finger for that matter) focuses on the breaking and suggests a break in one's activity but not a sharing with a friend. So here we were with these results and looked at them and thought that the kids not only did a great job but also gave us new and potentially important insight, and we probably even applauded them and must have given them some real candy as reward. If my recollection doesn't fail me that was about all we did: looking, taking notes, nodding, and congratulating. We pretended to be in awe but we didn't follow up....we just didn't follow up.

And I believe that there were and still are several reasons for this striking non-reaction to this outcome, and they have a lot to do with the industry's hesitation to accept the unusual, not yet tested, and moreover with the industry's overall suspicion when it comes to full consumer involvement. Although there always is much talk about sharing and "good for us" as opposed to "good for me," but the reality looks different. Continue the old ways; you break for a pause but not to share with a friend!

The other element, the sequential flavor and texture discovery is, of course, a more "tricky animal" from a purely technical and manufacturing point of view. But the prize is for the courageous and not for the timid. But the industry, at least the one that I knew and know best, didn't move on this but decided to continue in the traditional and well-trodden ways and see their market share slowly but steadily erode, especially in the area of confectionery products.

The other example, again in the area of confectionery products happened with a group of selected consumers, adults this time; however, they were not present at the same place and were in contact with the development group by videoconferencing or e-mail. The developer had an idea, prototyped it, and immediately shared the visual outcome with the consumer at the other end. When new flavor variants or textures and shapes were developed, prototypes were over-nighted to the group of consumers and they could give immediate feedback and suggestions. Needless to say, the whole exercise was not really successful and was rather used as an example why consumer involvement in new-product development is not desirable.

Don't get me wrong, I am definitely not of this opinion and truly believe that it is about high time to involve consumer, the crowd, in new-product development in any angle: product shape, taste, ingredients, nutrition, health impact, and whatever is of importance when it comes to create and make a new, healthy and nutritious food or beverage product. And it is totally in sync with my call for the new food company to become a knowledge-amassing and know-how–selling company. This crowd developing of products, and services for that matter, is the best learning and training ground for the "new" food company employee, the one who knows, and the one who sells this know-how.

6.5 Food-Related Trends and Hypes in Today's Societies: An Outlook to the Future

It is extremely difficult to know beforehand what represents a sustainable and meaningful trend and what, on the other, hand might just be a passing hype, a fad, or fashion. Let me however try this for the area of food. Please forgive me if you feel that I become too personal or too opinionated. And yes, this is going to be personal, on purpose, so it is going to be, after all my own opinion based on many years of observations and lived experience in the food and beverage industry, which should serve as a good basis to form opinions. If you don't agree, no problem; form your own opinion.

Probably the oldest trend that has befallen the food industry, right after the 20 tough years after World War II is the one of "low calories." What I call the tough years was especially the time until 1965 and especially in Europe, less so in the United States. It was probably never a trend in most countries in Africa and many parts of Asia and India, but it was a given daily reality, simply because there was not enough food available to voluntarily or involuntarily overindulge. But the 1970s and onward were and definitely are the years of too many calories in affluent societies, and therefore an entire industry emerged from the recognition that something should actually be done to cut down on calories or enhance and stimulate the metabolism so that excess calories would be burned more efficiently. The "gym" was born and workout and especially running became the new lifestyle. Running however is not always done outside but more often inside the workout room, typically on a treadmill. The air quality, especially in big cities might actually be counterproductive to the healthy part of running, and it's safer to read a book or listen to music when running inside and in a closed environment.

Although nutritionists and the health industry have since found out that working out is probably one of the best remedies against being overweight, and there is a felt record number of people who work out, there is at the same time a real record number of overweight and obese persons in affluent Western societies and increasingly also in less affluent societies. Obesity is especially high for children; in Brazil around 35% of girls are considered to be obese and most of the numbers found for adults in Europe, North America, and parts of South America are between 15% and 30%. A "world map of obesity" can be found on the site of "worldobesity.org (World Obesity, n.d.).

Some of the reasons why obesity and being overweight is an increasing problem in less affluent societies, too, are that carbohydrates are cheaper and easier to obtain than proteins and at the same time are a lot easier to overindulge than proteins. And they are a good and fast source of energy and can easily be consumed in ways that the fast-food industry has chosen as their business model. It doesn't help to deplore this and accuse any one single player, in the end it's the consumer who decides and makes choices for or sometimes against their personal health and well-being.

So, reducing calories by counting them conscientiously was and is one trend that exists at least for the better part of 40 or so years. And during the same time, obesity has not declined. So, something is not working and it's maybe the simple undifferentiated reduction of calories (in German there is a popular saying: "*Friss die Hälfte*," simply, eat half). This was an important recognition and gave rise to many specialized diets of all kinds, advocating exclusion of carbohydrates, reducing fat, dissociation of meal components, smart combination of meal components, solely liquid diets containing vitamins and minerals and some proteins, and many, many more. If you want to check, you can find more than 100 diets on the Internet, which all claim to make you slim and healthy again.

So, the two major food-related trends can be summarized as.

- Reduce calories
- Redefine composition of your meal

All accompanied by physical activity, typically in formalized fashion. Thus far there is nothing new. This has led to the creation of not only new and specialized food and beverage products but also to the creation of a new industry, the workout industry with all its machinery and programs.

So what is new, or fairly new, in this space? Well, you can observe it yourself when you are invited to a large family gathering. The poor parents (or grandparents) who organize this have to cook or cater for potentially the following groups of people (and I have briefly mentioned this already in the previous chapter): vegetarians (for ethical or taste reasons), vegans (for personal or ethical reasons), lactose intolerant (for real or imagined reasons), gluten intolerant (real celiac or slightly imagined), protein allergies (real or infrequent history), peanut or other nut intolerances (can really be very serious), kosher or halal (for religious reasons), carbohydrate free (for instance Atkins diet), and maybe a few more. What a nightmare to cook for such a group of people. Gluten-free has especially become very popular and is increasingly requested in restaurants, more often with no other reason than its just being popular.

I am not advocating to disrespect real needs and the problems that peanut allergies or celiac disease can arise for the individual that suffers from these, but the majority of people, especially in affluent societies, does not really require specialty foods that are based on the exclusion of certain ingredients.

The question that can be asked is simple: what's next? There is no solid answer to this, just a few speculations, however, based on observations of facts. One trend seems to become more and more visible, namely the consumers' quest for food that is grown according to certain agricultural principles such as organically and is grown in a place not too far away from where it is consumed. Whether food ingredients, vegetable, cereals, or fruits are then industrially processed or home cooked is, although not unimportant, not necessarily the most critical differentiation criterion anymore. Increasingly consumers will decide what to buy or not to buy on these criteria: sustainably grown and harvested and grown as closely as possible.

Let me summarize this in the simplest format.

- Personalization of food based on perceived or real food intolerances and other criteria will continue to be an important differentiator.
- Sustainable and nature-respecting agriculture will become an increasingly important differentiator for consumers.
- The distance between growing and consuming food will also become an increasingly important differentiator for food choices.

To end this chapter and to lead into the next one, let me just mention here that the role of personal electronics, communication devices, and apps will further grow and become an important part of food and beverage trends and how they might play out across almost every society in today's world.

6.6 Summary and Major Learning

This chapter mainly discussed and analyzed topics of change in the food industry necessary to transition to the food industry of the future. The following topics were discussed and analyzed in detail.

- The old ways of doing business in the food industry are really old and date back almost 1,000 years to the oldest written traces of food and beverage companies. The model is simple: develop, make, and sell or what I called de-ma-se. The food and beverage industry has never changed this model, and it is suggested that it's about time to do so.
- One of the most important reasons for change to avoid demise is a decreasing apparent trust in the food industry by consumers. This decreasing trust has much to do with many bad events that happened in many food companies over the last 40 or so years, beginning with the "mother of all bad events," the so-called baby food scandal that happened to the Nestlé Company in the 1970s. It's just one of a long list of such events in several food companies, often local or regional ones.
- The first step toward a new, only slightly modified business model is the recognition that the real "king or queen," the first target for the industry, is not necessarily the consumer but rather the customer, the retailer. After many years of painting the retailers almost as enemies of the food company, the situation has much improved, however not to the point of total alignment and collaboration yet. Closing the gap to retailers by meaningful collaboration should be one of the first major steps toward the new food company.

- Slotting allowances, which I rather called *fees* to rent shelf space at the retailer's brick-and-mortar store, are a long-standing tradition and basically represent a situation where the retailer does not trust the potential success of a food product, let alone a newly developed one.
- The clear goal for the new food or beverage company is to recognize that retailers become their most important partners, more important even than the consumers who ultimately will eat or drink the product. Close up-front collaboration especially at the level of new-product development is key to this.
- The evolution of such collaboration is composed of the following steps: from status quo to early involvement, joint programs, common target consumers, alignment of programs, common brand architecture, joint marketing, definition of new legal boundaries, exchange of experts, joint logistics, "branded store brands," and ultimately consumer involvement in new-product development.
- I discussed and analyzed two examples of close collaboration between food company and retailer that I had personally experienced, one of which included the development of KitKat Chunky.
- Close collaboration with retailers is only the first step for the food and beverage industry to renew itself and move away from the old ways of demase. The next critical element consists of the transformation away from selling products to ultimately selling expert know-how in the area of nutrition and health-related to food and beverages.
- To do this successfully, the new food company needs to recognize that their experts in the R&D organization have to play a more outgoing and more responsible role in selling such know-how. I suggested an R&D-centric company model 2.0, previously described and discussed in my book *Food Industry R&D: A New Approach*. A detailed discussion and analysis can be found in that book as well as this chapter.
- A further crucial element on the way toward the new food industry is the recognition of the important role of the community of consumers, not only as the ultimate deciders-in-chief but also the ones who have a word to say when it comes to how "their" food should look like, taste, and what it should be composed of.
- The food company in transition to the new company is in a clear dilemma torn between retailer and consumer and has to satisfy the two parties in the most efficient ways. It was discussed that consumers become increasingly critical of industrial, processed food. This has to be a wake-up call for the industry and the voice of the consumer has to be listened to well. The listening has to go beyond what consumers like and has to include consumers in the new-product development process.
- It was discussed that it is extremely difficult, especially for larger food companies, to include the large variety of so many different consumer views and translate them into food products that satisfy consumers and do not appear to be industrial or processed.

- One important way forward is to involve consumers in active ways in new-product development activities. In the past, food companies were rather hesitant to go beyond traditional consumer research that only collects consumers' opinions and stops short of personal involvement.
- Two examples involving kids and adults in new-product development were briefly discussed.
- Finally the topic of food-related trends and hypes was discussed. In the more recent past, a strong increase in food-related trends and health-related restrictions are being observed, making the task to cater to all or most consumer groups even more difficult for the new food company.
- Major trends, which have not changed substantially over many years are reduced calories, redefined meal composition to respect perceived or real food intolerances such as lactose, certain proteins, gluten, and others. It is difficult to judge whether some of these trends are passing hypes or if they are here to stay.

References

"Catch-22." (n.d.). Available from: https://en.wikipedia.org/wiki/Catch-22_(logic) [Accessed September 2016].

Federal Trade Commission Staff Study. (2003). Available from: https://www.ftc.gov/sites/default/files/documents/reports/use-slotting-allowances-retail-grocery-industry/slottingallowancerpt031114.pdf [Accessed June 3, 2017].

Kids Health. (n.d.). About milk allergy. Available from: http://kidshealth.org/en/parents/milk-allergy.html [Accessed June 3, 2017].

Traitler, H., Coleman, B., & Burbidge, A. (2016). *Food Industry R&D: A New Approach*. Chichester: Wiley Blackwell.

World Obesity. (n.d.). Available from: http://www.worldobesity.org/resources/world-map-obesity/?map=overview-boys [Accessed September 2016].

7

The Internet of Just about Everything: Impact on Agriculture and Food Industry

Second to agriculture, humbug is the biggest industry of our age.
—Alfred Nobel

7.1 Modern Cooking: Forward to the Past

The gradual introduction of microwave technologies and appropriate kitchen versions approximately half a century ago was seen as an important step forward to a new way of cooking food: quickly and conveniently. Since the discovery that fire and subsequently heat and smoke could be used for cooking purposes of any kind, this microwave technology looked like the first major innovation in cooking since millennia, if not longer. After all, cooking can simply be described as a combination of heat transfer and mass transfer, although with a good portion of passion and tender loving care largely based on personal preferences. Cooking is personal and, more recently is increasingly supported by technologies.

Cooking has three dimensions: first the process of combining various food ingredients in smart and personal ways to achieve a beloved and preferred taste experience and second, the use of certain techniques and supporting machinery to prepare something that looks good, tastes great, and is hopefully healthy, nutritious, and reenergizing. And yes, there is a third, an emotional and social dimension to cooking as well, without which cooking could be done by a robot. It is, by the way rather likely and can be expected that the core processes of cooking are being taken over by robots, and not the "kitchen robots," such as good old mixers and choppers, but real ones that can be programmed and do many tasks without any or very little human interaction or supervision.

Let me take the two major dimensions, ingredients and technologies, separately and discuss and analyze some relevant elements in this context. Much has already been written about the evolution of recipes over many centuries,

Megatrends in Food and Agriculture: Technology, Water Use and Nutrition, First Edition.
Helmut Traitler, Michel Dubois, Keith Heikes, Vincent Pétiard and David Zilberman.

so I will limit this discussion here to the more recent times, maybe the last 50 years or even the last 25 years. While during these years, the trend toward fast foods has steadily increased, it is also true that more and more emphasis has been given to foods and recipes that fulfilled the quest for healthy and nutritious diets. This seems to be a contradiction, but it is not and just reflects today's complex and diverse societal situations.

I have my personal, and maybe overly simplistic theory on this, which is not only based on society, but the different fractions in the religion of Christianity. While predominantly Protestant, especially Puritan (Calvinistic) groups see food as fuel to reenergize and quickly go back to work (fast-food territory), Catholics see food as more of a feast and social gathering during times of celebrations and keep scarcity and restriction to times of fasting and lent (indulgent and rich/overly rich, nutritious food). I have no other proof of that theory than my own experience and observations, having lived my entire life in this field of tensions (a very pleasant one, by the way) between these two main Christian groups.

The Spanish Saint Teresa of Avila, who lived in the 16th century, supposedly formulated this "dualism" regarding food from a Catholic perspective very nicely: "When it is fasting time, fast. When it is partridge time, partridge." In other words, be happy and eat well, rich, and probably much during the regular periods of the week or the year, be strict with fasting when required (in this case by the church). One should probably apply this every time, even if you fast for other reasons than religious ones. In other words, more Protestant societies are more likely to tend toward a fast-food approach, and predominantly Catholic societies are more on the side of food for indulgence.

I do realize that this might not be seen the same way by some or many of the readers, that's why I called it my personal theory of food, fast versus indulgent. Fact is that nutrition and health, together with high-quality and well-tasting food have become important elements in modern diets in the more recent past and to this very day. The difficulty clearly lies in the difficult situation of nutrition (not the biochemistry side of food components!) not being an exact science, hence not being able to propose clear recommendations of one kind, without another nutritional finding suggesting another recommendation, possibly even the opposite of the first one. This statement might not please many or most nutritionists but it is a fact that there exist more than 100 diets that supposedly are good for everyone's health. There are major differences as to what is healthy and nutritious in many or most of these diets, and all pretend that they are efficient, unique, and great to support one's health.

It is also true that some of these diverging recommendations are possibly justified by the fact that not every human being is equal in their personal dietary needs and requirements, and hence the great diversity. Let me not delve too much on nutrition not being the role model of exact science but rather go back to the task at hand, namely to further elaborate on today's cooking that is possibly as old as humankind. It is clear that modern nutrition findings have had an impact on mainstream dietary knowledge and have found their way into recipes of today. Good examples are for instance the role of carbohydrates, proteins, and fats in the metabolism, digestive pathways, passage times, metabolic degradation and end products, energy consumption, energy storage, the role of minerals and other trace elements, and many more. And yes, there is a fine line between the biochemistry side of this, which with all due respect to nutritionists is the clearer, more precise side of the medal, and the nutritional findings based on the biochemistry.

That is the real difficulty here: how to exactly value and acknowledge the role of modern nutrition and its impact, on the one hand on food and recipe development, including clear-cut indications to agriculture as the precursor, and on the other hand on sustained health and well-being of the individual consumer of food and beverage products. Because this is an almost impossible task, there is so much confusion out there with consumers that, at the end of the day, everyone seems to find his or her individual solution. There is, however one important and critical common element in all of this: consumers, or at least the vast majority of them, reject "processed food." As a big downturn to the industry, such processed food is most often equaled to industrial food, whereas home-cooked food, although it undergoes the same "heat and mass transfer" actions is seen as the good food.

So, there is kind of a trend that would suggest a back-to-nature direction, yet, not necessarily wanting to give up on the positives of industrial food, especially convenience and safety that comes with it. There is a trend to more "organic" food, food that is not grown with the help of artificial fertilizers but rather with natural ones (animal feces), which could become a mandatory label for organic produce, one probably not very much appreciated by consumers. I am not ranting against organically grown food but simply stating a truism, namely that consumers rather hide behind "all-natural, all-organic" than admit reality and spell it out for what it is. The really modern kitchen and modern cooking has at its heart the combination of back-to-nature type of food raw materials, mainly produce, and the ready availability and usage of the most modern and technologically advanced kitchen gadgetry.

7.1.1 The Role of Robotics and Connectivity

As just discussed, there appears to be this convergence or rather rapprochement of back to nature (if not to say back to basics), and the most modern and technologically advanced kitchen utensils. Actually, one can hardly call robots utensils, at least not for the time being given their relative novelty in the kitchen. Yes, so-called kitchen robots of the first generation such as mixers, choppers, juice makers, and a few other siblings, have existed for some time, but the real robot, the one that possibly cooks for you an entire meal, is not ready to go yet; at least as far as I am aware. Maybe from the time I write this in October 2016 until you read this, things might have evolved and such a real kitchen machine, the "independent" robot, will have come into existence.

So what do I mean by this intelligent and almost independent machine? Well, a priori a device that does it all. Based on the information that I provide to the device such as what meal I would like, how many people will eat, dietary requirements and possible restrictions, nutritional goals, diets, and so on, the device will do almost everything else. I still may have to do the shopping for produce and other ingredients, although also this might be done differently in the future, and I shall discuss and analyze this in the next section. Back to what the advanced kitchen robot, the AKR as I might want to call it, will be doing. In principle, it's everything else, including preparing all ingredients, cutting, cleaning, mashing, mixing, cooking in the largest sense including operations such as boiling, steaming, broiling, searing, frying, and some as well as automatically and independently preparing, organizing, and decorating the dish, as well as serving it, and at the end of the meal, making everything disappear in the robot's kingdom, formerly known as kitchen.

To the best of my knowledge, there is no such thing out there in today's world. The closest—and without wanting to be sound demeaning—is the proverbial "Italian Mamma" who is the master, or rather the mistress of all gestures and actions listed. And yes, there are many more mammas, housewives, housemen, chefs, amateur chefs, even kids, and probably especially grandmas and even grandpas who fall into this category of perfect kitchen "device." I am fully aware that none of them is a device, but I just wanted to make a point about complexity and diversity of anything that happens before, in, and after a kitchen, which is not an easy feat to master for a true device, the AKR. As I write this, Moley's kitchen arms seem to be the closest thing to an AKR, although it is still in the crowd-funding and development phase (Moley Robotics, 2017; IFLScience, 2017).

Although Moley does not come close to any human touch and perfection when it comes to cooking, the vision goes right into the direction of an AKR, especially when it comes to connecting ingredients know-how, recipes, and technology in one smart way. It may well be that when you read this, Moley is either already successfully in the market space of early adopters or has been superseded by some even more advanced and successful technology. When looking at what is happening today at the "brain-machine-interface" and the technologies and algorithms that have been developed over the last few years and are still advancing further, it might even be the case that the person who wants to cook a meal only has to think about some meal details and the AKR will do everything on its own—I believe we can safely assume that it is an "it" ("Brain–computer interface," n.d.).

And there is more. A group of students at MIT developed a robotic kitchen concept that they named "Spyce," which in supposed to replace human interaction if fast food kitchens (Business Insider, 2016). And here's yet another example. The *Los Angeles Times* (2017) wrote a story about a "barista robot" in a San Francisco coffee shop.

As you can see there is a lot going on in the area of robots and food and beverage preparation. But let's not look that far out into some almost fiction devices but look at what's happening in mainstream today on the ingredients and recipe side. I have a few friends who cook on Youtube® or exchange recipes on Facebook® (which is not any longer really fancy) or via WhatsApp® or other social media. Although Twitter® is not really a suitable medium to share recipes, I would call this whole thing the "twitterization" of cooking. Connectivity is everything, and it's the underlying principle that enables all this to happen. It's like the rail tracks for the railways—the necessary platform to operate.

7.2 Everything Is Online and Everyone Is Online—All the Time

Although I hear everyone I know complain about this "online-itis," everyone really seems to be online all the time or at least most of the time they are awake. It's a curse says one, it's a necessity say many others, and it just happens says the majority. So everything and everyone are online, so what's new? Let's not forget, however that we were able to exist even before Berners-Lee created the World Wide Web and the Internet found a platform. How did we do it then? As I have discussed, good food has always existed and has always been consumed by those who could afford it. In the past they were called "nobility" and the more affluent parts of society, today they are called the rich and the more or less affluent middle class in most societies.

What I try to say is that despite the democratization and the cannibalization of information through constantly open communication channels, food has not yet really been democratized and is not yet at the reach of everyone. Every person seems to have a cell phone, mostly a smartphone, and when you see images of refugees, for instance in refugee camps on the island of Lesbos in Greece, almost every person has such a device. They don't really have much to eat and are heavily reliant on the goodwill and hospitality of those who have; communication and connectedness appears to be more important than eating and drinking or should I better say has gained the same level of importance. This is an important observation in regard to Maslow's pyramid of the hierarchy of needs, communicating and being connected is now right at the same basic level as surviving through food, water, and physical shelter.

So, being connected has almost—or entirely?—become a basic human need and is as important as eating and drinking, so we believe, or are at least, made to believe. Be it as it may, the reality is simple: everyone is connected and online and so are things, more and more things, seem to be. Just a few days ago, remember, I write these lines end of October 2016, there was a large-scale hacking into the functioning of major social network providers such as Twitter® and a few others by simply using mundane devices as baby phones or DVR players as "individuals" that in overly large quantities ask questions or request information from the Internet and thereby totally overwhelmed the system. This attack was a so-called DDoS or distributed denial of service. Nothing fancy, just a glimpse into the "opportunities" that the internet-of-everything (IOE) might represent in the future. The IOE is only growing, and increasingly takes over not only our homes but also our kitchens.

Not only is it about to take over the kitchen but also how we get food, how we consume it, and possibly also how we share it. Your fridge is online or maybe will be in the not-so-distant future. The precursor to this is simply taking pictures of the contents of the fridge with one's smartphone and creating the shopping list that I have always on me. It's not this piece of paper that I have studiously planned and written and that I do vividly recall lying on the kitchen counter while I am in the supermarket frantically trying to recall what I should purchase. My always-connected smartphone has visual proof of what I still have in my fridge and my pantry (should I have one) or my kitchen cabinets, and the Youtube recipe that I can immediately consult on my still-connected smartphone give me all the necessary information to shop everything I need for this wonderful night, save of course, the stereotypical one item that I have still forgotten.

The next step is the smart fridge that knows exactly what it still has in its belly and with which I can communicate any time I would like to. It brings me out of my potential solitude if I have a chance to converse with my fridge, which has just been promoted by me to have a name and go from an "it" to a "she" or even a "he." Joke aside, it can be really useful, provided that my fridge isn't busy hacking in DDoS-style to an important social network and knocking it out. I shall have a serious talk with her (or him) when I come home, believe me!

And there will be more kitchen- and cooking-related devices online, remotely controllable through apps on your smartphone or tablet or any other linked mobile gadget. I do remember when my in-laws owned a cabin in the Swiss mountains, about an hour or so away from where they lived. They didn't want to leave the heat on in the cabin during times they were not there but they wanted it nice and cozy and warm on their arrival Friday afternoon or whichever day they had planned to go there. So, in the late 1970s they had a system installed that replied or rather reacted to a phone call (analog phone that was).

They had to dial the cabin's phone number and, like an answering machine, the heating system answered. To switch the heat on they had to whistle in the phone, yes, whistle. This was quite a funny undertaking when they were not sure, whether they had already whistled to the heater; funny, because the next call and whistle would switch the heater off again. I seem to recall that more than once they had come to a cold and un-cozy cabin. Luckily they also had a wood-burning fireplace and could correct the situation rather rapidly.

The 1970s whistle through a phone line is today's pushing a button of an app on your smart and connected mobile device. Times have changed and technology has changed, but the desire to control and start things from afar is still there and is probably stronger than ever before. The other side of the coin seems to be the almost total availability of food just about everywhere, and this not only in the so-called developed world, but in many other parts of the world, too. Numerous are the food carts, food stands, hawker stalls, and small neighborhood stores that offer food and beverages almost during the entire day and parts of the night. And I am not forgetting ever-more sophisticated vending machines that serve food, however limited in choice, and beverages 24 hours a day. In the United States and many European countries, out of home (OOH) consumption has reached an almost 45% of the population and appears to hover around this number.

The US Department of Agriculture (2016) reports the following trend and numbers.

> Consumption of food prepared away from home plays an increasingly large role in the American diet. In 1970, 25.9 percent of all food spending was on food away from home; by 2012, that share rose to its highest level of 43.1 percent.

It further lists the following major reasons for this trend toward increasing out of home food consumption.

> A number of factors contributed to the trend of increased dining out since the 1970s, including a larger share of women employed outside the home, more two-earner households, higher incomes, more affordable and convenient fast food outlets, increased advertising and promotion by large foodservice chains, and the smaller size of U.S. households.

So, there appear to coexist two opposing trends: The first one that encourages in-home food and beverage consumption supported through an increase in usable and useful kitchen and cooking gadgetry in combination with online capabilities, and the second one that takes all food preparation out of the hands of the individual consumer and gives him or her the opportunity to eat and drink, to consume without having to do anything except three things: first order, second consume, and third pay (not necessarily in that order). So, what's it going to be? Maybe one answer can be found in the model that the mobility company Uber® had tried out for a short period of time back in March 2016. They have announced that they want to compete in the business of food delivery, especially to deliver your pizza or any other gourmet meal that you have ordered from your neighborhood take-away place or restaurant. They have launched "UberEats" and Instant Delivery in spring 2016, only to pull it from the market 5 weeks later (Kosoff, 2016). There are and were many more start-ups, mainly in New York City such as Sprig, Munchery, DoorDash, and a few more trying to do business in the same area of food delivery.

As for technology, in the future the delivery vehicle might be self-driving, autonomously delivering your food and beverages. It might even be a drone that lands on your patio and literally drops off what you have ordered.

We might just be at the dawn of the "battle of gadgets," the one, which do something (e.g., cooking) and the ones that bring something already cooked to you (the drones and self-driving vehicles). I don't know. None of us can by the way know who will win, and it might be ending like it always ends: both will have their space with us, the consumers, the users, and benefactors of all this technology. There are still a few important questions around these scenarios, one of the most important ones may be: where is all the fun of cooking and socializing going? Will it still be the same if a robotic cooking arm does all your cooking and you just watch and smell, or if a drone whirrs in to your place every evening to deliver you the meal that you had ordered through an app on your smart device already around noon and you just have to take it in and eat? I am not sure, maybe again the middle way will be the winning one: sometimes I shall order food to be delivered to my home, some evenings I will switch on my robotic kitchen helper to prepare some delicious meal for me, and yet there will be other occasions, when I prefer to cook myself.

The one, probably most important element around food, at least in my eyes, will never go away: it's the emotional element, the sharing with friends and loved ones and even the sharing with strangers. Food does that to you; it normally brings the best out in you and helps you to connect, emotionally that is, not through the Internet or your smartphone provider.

7.3 Food and Agriculture: The New Hardware and Software

Big data seems to be the buzzword these days. When companies like Google® began their operations many years ago, they started to collect an awful lot of data from their users from the very start. They didn't call them big data, but that's what they principally are. Over many years so many data on just about everything were collected, stored, and brought to good use; sometimes maybe to not so good, and most of the times are immediately turned back to target consumers, individually, or in groups for better business purposes. Companies of any kind, which deal with customers or consumers, collect data today, one way or another. Mostly it is pretended that such data are depersonalized and are therefore anonymous. However such data are definitely used to predict trends, habits, consumer preferences, consumers' shopping behaviors, flow of persons to shops, and other places of commercial transactions. Over time, over many years now, the data collection has really become big and someone coined the term *big data* for these. A definition looks like this:

> Big data is a term for data sets that are so large or complex that traditional data processing applications are inadequate to deal with them. Challenges include analysis, capture, data curation, search, sharing, storage, transfer, visualization, querying, updating and information privacy. The term "big data" often refers simply to the use of predictive analysis, user behavior analytics or certain other advanced data analytics methods that extract value from data, and seldom to a particular size of data set. ("Big data," n.d.)

The definition is rather complex and "big," just like big data themselves. There are really two underlying factors to big data: first, they are a large collection of data coming from basically all areas of life, and secondly, they can be, and are, used for financial gain (in other words to make a profit). There is nothing basically wrong with this, Google and other large firms do this all the time and use our data for profit. We profit when our search is successful, and "they" profit from collecting data related to our search and search profiles. Since the time that my cell phone provider and its affiliates know exactly where I am at any time my phone is switched on (and possibly even if it is switched off?), my credit-card company knows exactly what I have shopped for and where and has analyzed my spending patterns on all levels such as goods shopped, prices paid, locations where acquired, and every time I search something on Amazon, it's only a matter of hours until I receive a mail suggesting merchandise I also might be interested—since all this takes place, I am no longer too surprised that even more sinister searches for big and more sensitive data take place in the background.

Despite all assurances by all companies that use our data for profit that our privacy is perfectly protected, I am not naïve enough to believe that this is the case. There are of course possibilities to escape such intrusions in my privacy by giving my credit-cards back, cancelling my cell phone subscription and no longer use the phone, donate my computer to Goodwill, and give my tablet away to a distant relative. Next thing to do is to close my bank account, shred my checkbook, and commence to pay everything in cash. The question is of course open where I get my cash from unless I can convince my employer to pay me only in cash, ideally every Friday or so. And wait, don't forget to give my modern car back and drive an old-fashioned cruiser with no on-board electronics. Such modern cars know an awful lot about their drivers and even "collaborate" with the police against you in cases of accidents and possible disputes. And then there are utility companies, which seem to know everything about you as well, and there is probably no way that you receive electricity anonymously, unless you live off-grid and make your own.

7.3.1 Big Data are Here to Stay

You probably feel by now, pretty much like I do, that to do all this is not only rather outdated and outlandish but impossible to achieve, and to live in such a way, totally off the grid with few exceptions is not feasible in the vast majority of societies and cultures around the world. So, data continue to be collected for all kind of purposes, and we can only hope that most of the times it happens for rather good and not sinister causes. Big data, whether we like or acknowledge their existence, are a simple reality and can and should be used for many applications that will eventually help society as well as industry to improve their ways of living and doing business.

In the world of agriculture, such big data have increasingly gained importance and can help in areas such as generating climate prediction models in different geographies, crop forecast, improved irrigation patterns, and optimized use of water, fertilizers, and pesticides. They can especially help in the increasingly popular field of organic farming by fine-tuning and refining growth patterns, crop rotations, and efficiency of organic fertilizers and predator insects or other animals replacing pesticides. Agriculture is probably the largest single user of the application of relevant big data collections to improve the agricultural output.

The agricultural industry is much encouraged by food companies, especially the large ones, to increasingly use the algorithms and findings of big data for one major reason that is probably dominating all other possible reasons, which is food security. Food security is to be understood as the assured availability of agricultural products that serve as raw materials to manufacture a vast array of industrial food products. Food companies are also keen on other elements in this equation such as quality, nutritional value, price and stability but again, most of all they need to be assured that there is an ample and steady—secure—flow of raw materials that arrives at the delivery docks of their factories. From my personal experience that is the requirement that stands above all others and because the smart use of big data can help in this endeavor, the food industry encourages agriculture to go after such big data, big time!

7.3.2 Agriculture and Space Science: The New Connection

It's been approximately 60 years or so that humankind put its first satellites out in space and some 50 years that NASA landed its first mission on the moon. There are conspiracy theorists who insist that the moon landing was a hoax and secretly filmed somewhere in the desert of Nevada or Utah. I think that this is a pretty crazy belief but then one will always find some people who believe in just about anything as long as it is different and sounds outlandish. Well, this is not our topic, and I apologize for having digressed. What I want you to understand that while looking down from satellites, mostly for reasons of "intelligence"—commonly known as spying—the cameras of the first satellites could also see and observe things other than troop movements, military construction sites, nuclear sites, and such but still of interest for the military such as large crop fields and their actual status of growth.

Although not of the same level of strategic importance, it is in my personal opinion more important to know how much food is grown and where than to know which bombs could or should explode here or there. But then I have worked in the food industry my entire professional life and not in the military and that may explain why I am of that opinion. Be that as it may, satellite observation of crop fields has become an important additional element in the drive toward collecting meaningful and usable data that can help improve the agricultural output, especially food security. At present I am personally involved in establishing several climate forecast model-related projects jointly between the Jet Propulsion Laboratory (JPL and NASA) in Pasadena, California, and a large food company. The projects will go beyond climate models; this is just one more sign that space agencies such as NASA and others increasingly focus on the planet to make good use of their knowledge acquired in space exploration.

It is clear that information acquired by satellites is only one part of the equation and is complemented by such gathered through low-flying specialized and appropriately equipped aircraft or drones. All these satellites, aircraft, and drones give complementary information, and even a measuring device and camera fitted at the tip of a long stick can survey a crop field and give meaningful and useful results to optimize the output of the crop field monitored in this way.

I personally believe that such projects between space agencies and agriculture- or food-related companies is just the beginning of a much larger collaboration in this area, and many more important results can be expected from this. Just to name a few, here is a list of important expected outcomes.

- Optimization of crop health by monitoring chlorophyll levels and its distribution across crop field.
- Monitor humidity levels and optimize irrigation across crop field.
- Detect and measure water table under crop field (today limited to a few feet depth).
- Monitor potential migration and approach of dangerous insects (dangerous for the crop).
- Optimize fertilizer usage by monitoring growth and yield in different sectors of crop field.

And there are certainly other parameters that can ideally be checked and observed from above, either close by or from afar or a combination of both.

As always there is a cost involved—a price to be paid for such rather novel approaches. However, it is an important investment in the future of the agriculture and food industry and optimizing both will help improve food security in important ways.

The topic of the chapter is fittingly about the Internet of just about everything and this "everything" includes space and space-related science and technology. It actually lends itself ideally to the endeavor of feeding an increasingly growing world population. This is not a simple thing to do, and if we want, like some of us avoid too much fiddling around with plant and eventually animal genes, optimizing agriculture through technologies just described is most likely the way to go. However, such optimization requires a combined approach of modern breeding and the approaches described here. I would call this a combination of new "hardware," the crops with new and improved traits achieved mostly through conventional-yet-accelerated breeding, and the new "software" as described in this section. The major question that we have to ask ourselves here is: how will this impact the industry as a whole: agriculture and food industry?

7.3.3 Impact on the Food Industry and the Consumer in the Middle

Having stated several times in this book and in my previous books, it can safely be said that both agriculture and the food industry are conservative and resistant to change. And this is probably a euphemism; they are downright opposed to change—any change. Yes, as discussed, they will follow certain trends in case there is no other way out and renovate their ways of doing things and some of their product lines and then call it innovation. And now they should accept space science and related technologies to help them improve their business? No way, that's not for us, far too complicated, and too distant from our own science and technologies that we—the industry—are so proud of. Let's wait another 1000 years. Let's wait until some competitor has tried it out first and hopefully found out that it didn't work, and it didn't result in anything new.

Luckily, not all of the food companies act like that, as evidenced by one company that actually works with the JPL noted a few paragraphs previously. But trust me, it was a tough and stony way to get all parties involved to the point that there was even a small field test trial. Anyhow, we got it finally going, and I do hope that this is just the beginning of many more projects and that many more agriculture and food companies will follow this example. My prediction is that not only will agriculture and food profit from such collaborations vastly, and ultimately it will be really good for the consumers. And this is not only because of improved security but also because of improved quality, both from a taste and nutrition point of view. So, there is a lot in for all members, close or far, of this new approach of collaboration: agriculture and farmers, food industry, consumers, customers (the trade), logistics and all such space agencies that open up and make their technologies and findings available to the outside world. It is understood that such opening up has its price, but from personal experience, I can say it's less expensive than you may think.

The question that I haven't really asked yet is: does the consumer appreciate all such efforts? I have kind of answered the question whether the food industry in general wants to go down that road, and I have concluded for myself that some are willing. However, it takes an enormous effort to get it going. The consumer is a different "animal" altogether. The famous last words of the genetically modified organisms (GMOs) supporters were: "it's good for the consumer"; however, nobody believed it because, at least in the beginning, it was simply not true. It was not good for the consumer, but it was good for the industry that had developed certain GMO plants and pushed these into the agricultural pipeline.

So, let's not make the same mistake here; let's be honest and not only tell the consumers what good can be achieved but actually make sure that all or most visible achievements are actually positive for the consumer. This is not an easy feat and great and honest communication is of the utmost importance. If we cannot achieve this, it is really not worthwhile to continue. However, continue we must because so much is at stake—improved and increased food security.

7.4 An Attempt at Peaking Ahead: Will There Still Be an Agriculture or Food Industry?

This may sound like a rather provocative question,but it has to be asked: will there still be an industry that covers agriculture and food? Don't get me wrong; it is certain that there will always be agriculture and there will always be food, but will there be real industries, as we know them or even new forms linked to them? If I take up the theme of demase again, which is ongoing as a model for a good 1,000 years in the food and beverage industry, then we are in for a long lead time until something substantial is likely to change. But that is an argument that industries and societies have brought forward in the past, namely that new ways are not really necessary or beneficial for the individual or the community. And almost all the times, such assumptions were profoundly wrong and change, whether willingly accepted and embraced, there is no going back any longer. There are many examples in the history of humankind such as the creation and introduction of the automobile that put an end to horse-drawn carriages, electricity that put an end to gas lights, telephone and telegraph that put an end to couriers on horse, diesel and electric trains that put an end to shoveling coal, and probably quite a few more from the past, and many more to come.

And when looking backward we can always observe struggles: struggles between those who want to hold on to the old and proven ways, and those who want to jump ship and fully embrace the new ways. The latter we would call the early adopters, and the former, the conservative crowd who believes that everything is in order the way it is. Those are typically the ones who have most to lose through change; the others belong to a group that has nothing to lose.

The reality, however is that change always happens and most of the time it's for the good. The industries that this book covers and discusses will inevitably go through changes, although the agriculture industry will increasingly seek to go back to the old ways, back to the roots, back to nature, using animal manure as fertilizers and predator insects as pesticides just to name a few. Even if this sounds cynical, it's not, not at all. The challenge, however in this is to make it work for so many more people and make it work in just and sustainable ways.

So the first, probably obvious and trivial answer is, yes, there will be an agriculture industry, and yes, there will be a food industry because both cater to and satisfy our most basic needs of supporting life and our survival. Sounds pretty pompous, but it's the simple truth. Whether agriculture and food will continue to exist in their actual industrial setup is an entirely different question, and my assumption and answer to this question is simple and I have kind of insinuated it already. No, the structures will be different and most likely after the present-day wave of agriculture companies growing by mergers and acquisitions, there will be a different trend again toward smaller and more agile units.

7.4.1 Bigger Is Not Always Better

One can see a harbinger of this in an industry that has migrated to larger and larger production units and company structures, namely the energy and electricity industries. After decades of units and companies growing bigger, be it in the nuclear or the coal sector, the introduction of alternative, renewable energy sources has created a lot of nervous commotions in this sector. The fear of the established energy giants is not really whether alternative energy sources, decentralized and combined in a kind of crowd-sourcing way can and will deliver the energy needs of tomorrow. They know the answer and they know the inevitable and the simple answer, yes. What they fear is to lose the well-established platforms of rather gigantic units that can be organized and managed from an office desk in a large corporate headquarter. It's the fear of loss, and therefore, fear-mongering has become the name of the game. But the break-up will come, and a new management infrastructure will be in place to manage thousands of smaller units and link them in smart and secure ways. Energy security here means the same as food security, namely the assurance of its constant availability.

Table 7.1 Growth trends in different industries.

Agriculture Industry	Food Industry	Electricity Industry
Increasing	Stagnating	Declining

Agriculture is still in the build-up to larger units phase and has similar arguments to the energy industry: only big corporations applying the most advanced technologies, especially genetic modification of plants, can solve the problems of feeding the increasing world population and assuring food security. The argumentation is right out of the playbook of any other industry that heavily relies on large conglomerates centrally organized and controlled.

Interestingly enough, at least in my eyes, the food industry is somewhere in the middle of this development. The various companies that comprise the food industry have grown to a certain size, partly by merging and partly by internal growth, and there is not much left for large companies to pick up and become even larger.

Table 7.1 is a simple depiction of this trend of growth versus stagnation and decline in the three industry examples chosen for the discussion here. Note, *decline* means from few large units to many smaller ones.

Of course, surprises are always possible and we might see mergers. However, I rather anticipate the opposite. Let me use the business of ice cream as an example. Making ice cream in a factory way is inherently expensive and the price one can command for one cone or stick of ice cream is relatively modest. It's expensive because ice cream has to be manufactured cold, very cold that is, and has to be kept cold throughout the entire supply chain until such time that the consumer has decided to consume it. That costs a lot of money and is not necessarily done in the most environmentally friendly ways by using, for example, evaporating ammonia as a coolant in a closed system.

The point here is that to make this worthwhile, cold temperatures generated in such a fashion have to be used to make as much ice cream in this environment as humanly, or rather "manufacturingly" possibly. So, this means that ice cream factories typically are large monsters, even XXXL monsters in some cases such as the largest ice cream factory in the world that is located in Bakersfield, California. The factory is there because dairy agriculture is close, labor is close and affordable, and consumers are rather close, given that California is rather populous and rather hot. Speaking of hot, Bakersfield is probably one of the hotter places in the state, so the cold rooms require the proverbial extra layer of insulation. So, big is the motto.

And despite that the factory is really large, and the economy of scale should play into the hands of the business, it actually doesn't perform the way it should and is expected. My strong belief is that wishful expectations and industrial realities do not really match when it comes to ice cream. Yes, one can earn money with ice cream, but on average, profit margins are typically only 40% to 50% of profit margins of other product segments in the food industry.

Economy of scale does not play an important enough role because generating and maintaining cold is so expensive. It is a typical case where small is really beautiful. Just imagine a cup of shaved ice (just water ice crystals with syrup) for which you pay approximately US$7 in a small shop on Front Street in Lahaina, Hawaii. And people line up in front of the shop and business is doing pretty well, to say the least. I am always surprised that the ice cream industry has still not picked up on this and found a way to make shaved ice industrially, but maybe one day, when it's probably too late they will attempt to do so. But that's not my point here; my point is that either ice cream manufacturing units and related businesses become even bigger and earn money through the shear volumes of sales or alternatively they are completely cut up into smaller units and are much closer to the retailers, maybe even producing in the retailers' warehouse and thereby substantially shortening distribution pathways. I can rather see the latter, but only the future knows.

7.4.2 Elements that Will Stay and Others that Might Disappear

I have said this before but let me say it again: "Hindsight is the only exact science." Predicting the future is a tricky undertaking, especially because most of the time predictions and future realities are two different pairs of shoes altogether. Do you recall Faith Popcorn's 1991 *The Popcorn Report*? One of her predictions was "cocooning" would become an important trend and that people would stay home more often again. Well, pretty much the opposite has happened in the years after the book was published, and more and more people were going out and eating out. This is not to say that she was wrong with all her predictions and ideas. but it reminds me a bit of the old Soviet weather forecasting, probably a joke but with underlying wisdom: there is a 50% chance that tomorrow's weather is like today's.

That's exactly what predicting the future is all about, taking a chance and being either right or wrong. One can improve the quota of the chance in one's favor by reading the signs correctly and interpreting past and current trends in ways that help predictions to become future reality. The food industry relies heavily on consumer and market research to predict the immediate future, and the success rate might be just slightly better than the Soviet weather forecasting. The agriculture industry increasingly relies on a combination of many observations, local knowledge, and reading the big data right. So the question here is what will remain rather stable in these two industries and what will change or even completely disappear?

That's a really tough question but let me try to find a few personal answers, mainly based on observation and trying to read the writing on the wall from an external position and not tainted by wishful thinking. Let me start with a list of likely future scenarios in the industry of agriculture, including farming. It's not so much about what might disappear but what is likely going to happen.

- The large agriculture conglomerates that control seeds, chemistry and biogenetics of plants, and also the feed side of animals for meat will again transform to smaller units.
- Industry and farmers will collaborate even more closely.
- R&D in agriculture will be intensified and will become a cornerstone of the industry.
- Necessary yield improvements of crops will be increasingly based on modern breeding, non-GMO–based.
- Urban agriculture will play an increasingly important role.
- Genetically modified plants will find growing resistance in all parts of the world.
- The approach to pesticides will be revolutionized and they will eventually become "all natural."
- Bees will be bred and developed to fulfill specialized tasks in the optimization of crops both for yield and nutritional value of plants.
- As robots and drones will increasingly become part of the agriculture industry, they will take on many roles....why not potentially replace bees?
- Animal farming will be under increasing public scrutiny and consumers will become more demanding with regard to how especially cattle is bred, fed, grown and slaughtered.

There are probably a few more that you the reader could think of, and I encourage you to extend this list of future scenarios in the agriculture industry based on your own professional experience.

The important message derived from this list and possibly a few more is simple: democratization of agriculture can and will take place, mainly thanks to the support of the Web and the connectivity that derives from that. Agriculture, like many other industries too, will be increasingly operating like a social network and "crowd-farming" may become the new way of growing food, vegetable as well as animal, for the future.

Figure 7.1 illustrates these possible changes and new approaches.

So, how do I believe the list would look like on the food industry side? Here's my take on it.

- Food companies will sell mainly know-how together with products.
- Manufacture of food will be closer to consumers and consumers will have a greater impact.
- Large factories might still exist, but they will be filled by many different product lines producing regionally rather than continentally or even beyond. This trend can already be observed.

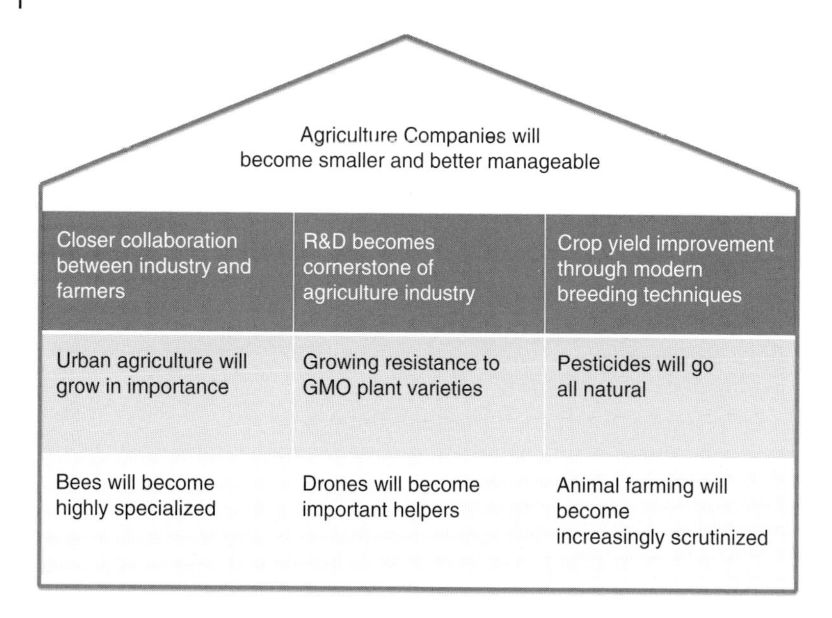

Figure 7.1 Expected changes in the agriculture industry.

- Consumers will be more closely involved in the development of great new food products.
- R&D will play a pivotal role in the food industry being simultaneously involved with consumers, selling their know-how and gaining new knowledge, especially regarding tangible nutritional value of food and food products.
- Processed food, as we know it today, will increasingly disappear.
- All natural and organic will become the new normal.
- Food companies will become more specialized with regard to the type of products they make.
- Food companies of the future will be smaller than they are today.
- Robots will replace more and more manufacturing personnel and why not the CEO of the company? After all, emotions are not required for that job.

Figure 7.2 illustrates these possible changes and new approaches, some of them maybe surprising.

Again, you might think of even more important elements that would need to be listed here, and I encourage you to find out for yourself and discuss this topic in more detail. The same holds true here as for agriculture; the food industry will increasingly rely on Web connectivity and individuals and crowds of individuals will become much closer to the transformation of agricultural raw materials to healthy and nutritious food products. And that's a problem for the food industry as we know it today. That's by the way, at least

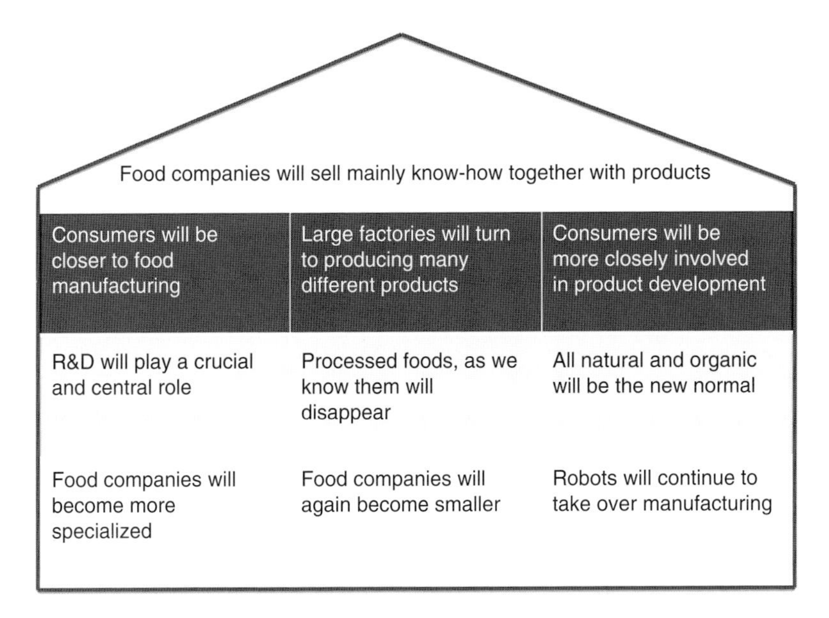

Figure 7.2 Expected changes in the food industry.

in my eyes, one of the reasons why food companies increasingly search to expand into areas that are presently occupied by the pharmaceutical industry: everyone (or almost everyone) can cook and make a good tasting and healthy meal, but hardly anyone without highly specialized expertise can make a pill. In my opinion it's the wrong strategy, even if it looks like an attractive one. I have worked my entire professional life in the food industry and have seen the onset of this venturing out toward the area of pharmaceuticals more than 15 years ago.

The plan was to deliver health-supporting functional ingredients in food carriers, the proverbial wolf in sheep's skin. There was and still is, however, a big question mark surrounding this approach: is the food industry willing to invest into R&D in amounts that come even halfway close to those done by the pharmaceutical industry? We are talking about R&D investments of approximately 1% to maximum 2% of annual revenue for the food industry and between 10% and 20% for the pharmaceutical industry. The simple answer was (and is) always no, because the food industry would only use approved functional ingredients and would therefore not have to pay for their development, hence less investment. All I can say is that this has never really worked, and I strongly believe that food companies' strategies that define their future as a hybrid, "half-baked" pharmaceutical company are clearly not going to succeed. Feel free to judge me on this statement in a few years' time, and if I'm wrong, so be it. Today I feel that there is no future for this approach.

Don't get me wrong; I am not saying that the future of the food industry is not in the area of healthy and nutritious food because it most certainly is and will be. However, I can hardly see a Nestlé or Unilever Company turn into a pseudo-Novartis or Roche. I actually believe that due to the connectivity enabled through the Web and Internet, it will lead to a different, almost opposing trend. People will make their own health-supporting, even healing foods and beverages in personal approaches by using platforms such as YouTube and will lose the lust for pills and will not accept "pizza with cardiovascular benefits" from a large food maker either. Consumers will search the Internet for just about everything and turn to the next wave of super foods, health foods, novel "secret" ingredients from a hitherto unknown exotic plant, and prepare their personal and personalized smoothie and other healthy product.

It is my strong believe that this is the real impact that the IOE will have on the industries of agriculture and food. And that's the great fear of both of these industries, namely that the individual consumers and stakeholders will take possession of agriculture and food in much more direct ways and, at least to some degree, sidestep these industries and the large companies that operate in these areas. Watch the space, there will still be much commotion, and it might pretty much look like scenarios happening today in the industry of energy and electricity: small will become beautiful again and connectivity is and will be everything.

7.5 Summary and Major Learning

This was a difficult, if not outright tricky, chapter to write because it combined old and traditional with a fresh view from a modern and almost futuristic perspective. The following topics were discussed and analyzed in some detail.

- The chapter introduced a discussion and analysis of some aspects of modern cooking paired with the mechanistic approach based on heat and mass transfer. It was postulated that cooking has two major dimensions: first the smart combination of ingredients to cook, to compose something great from a health and taste point of view and secondly the smart use of appropriate kitchen tools and techniques.
- It was mentioned that there is a third dimension to cooking: the social and emotional one.
- Cultural and religious differences when it comes to food preparation and consumption were discussed and analyzed in some detail.
- The nature and role of modern nutrition was discussed, and it was suggested that despite the important findings and recommendations that emerge from work related to nutrition, there is no such thing as nutritional science, albeit the claims and assurances from nutritionists. The methods may be scientific, but the results are not or only on the surface. Often nutrition is confounded

with the underlying biochemistry of metabolism and organ functions with regard to food and food intake.

- The increasingly important role of real kitchen robots was discussed and analyzed, and it was concluded that it is still a long way to an artificial chef.
- The role and importance of being connected most or all of the time was discussed and analyzed, and it was suggested that perpetual connectivity, either in the foreground or in the background, is at the centerpiece of most of what we are doing, increasingly also in the world of food. Fridges become smart, shopping lists are in the cloud, my stove knows how I want my steak or fish cooked, and my wine cooler (should I have one) will recognize the wines I store and will switch to the optimum temperature all on its own.
- Contrary to predictions from the 1980s, out-of-home food consumption has steadily increased and new consumption locations and opportunities have constantly been added.
- There is an apparent dilemma: modern kitchens have become bigger, more integrated into the home ("open plan"), and have been equipped with more and more kitchen gadgets and robots, and at the same time, the very same people who own these kitchens go out to eat more often. There is no end to these two opposing trends in site yet.
- Big data (i.e., the collection of just about any relevant set of data that one can access through the Web) has become the new treasure and the treasure hunters of today are called Google and the like.
- Whether we like it or not, big data are here to stay, and except for emigrating to a tiny and remote island somewhere in the middle of nowhere, there is no way to avoid to be one tiny speck in the vast flock of potential data providers to the collectors' industry that transforms these data to what they call big data.
- Big data are not necessarily a bad thing and can be useful especially in the agriculture industry.
- Another more or less surprising rapprochement takes place between agriculture and space science and technology, especially in the areas of monitoring crops and crop health as well as predicting regional and local climate through appropriate climate models.
- The important question as to how the consumer who is in the middle of all this reacts to this and especially how she or he can profit from it. My personal assumption is that at the end of the day, the positive elements of big data, space collaborations, climate models, and perpetual connectivity will outweigh the negative ones, such as having the feeling of being spied on or living in an environment in which no one speaks to anyone else "live" anymore.
- The future of GMOs is questionable, at least in its original approach, namely good for the industry and no real benefit for the consumers. Modern plant breeding technologies are likely to gain in momentum and importance when it comes to the gigantic task of "feeding the world" with healthy and nutritious food and assure food security.

- An attempt to peaking ahead to the future of the two industries in this totally connected world was made. It was suggested that the industries of agriculture and food, respectively, are not at the same evolutionary stages, especially in comparison with other industries such as the industry of energy and electricity. Although agriculture companies are still growing in size mainly through agglomerations, food companies seem to have come to a kind of standstill and size growth rates have stagnated to some degree. The major growth drivers for my former company is still "real internal growth" combined with "organic growth," both of which heavily rely on factors other than just acquiring yet another still small food or nutrition company, although it still does happen. In contrast, times are just about right for the electricity industry to become cut up in smaller industrial units again. Something that is suggested will happen to agriculture and food in the not-too-distant future.
- In the future, agricultural companies will become smaller and better manageable, companies and farmer will work even more closely together, R&D will become a cornerstone for success, crop yield will be improved through Web connectivity and modern breeding techniques, urban agriculture will grow in importance, pesticides will become all natural, bees will become highly specialized, drones will come everyday farm helpers, and animal farming will continue to be critically monitored.
- In the future, food companies will mainly sell know-how together with food products, consumers will be closer to manufacturing, large factories may still exist but they will cater regionally many different types of products and thereby avoid long distribution distances, consumers will also become more closely involved in development of food products with emphasis on health and nutrition, R&D as in agriculture will play the crucial role, processed foods as we know them today may well disappear from the marketplace, all natural and organic will be the new normal, food companies will become more specialized with regard to what type of foods the will make and sell, food companies will again become smaller units, and robots will continue to take over manufacturing, and why not management, for example, the position of the CEO of the company.
- Contrary to strategies and announcements of some large food companies it is my firm belief that the field of pharma and hospitals is not the future of food companies, especially for the simple reason that investments into R&D are worlds apart and the future of agriculture and food is not in healing the sick but helping to avoid that people become sick in the first place through healthy and nutritious food.
- Personalization and democratization of food and its manufacture is the most important trend and the IOE and perpetual connectivity place the right tools into the hands of individual consumers to achieve exactly this.

References

"Big data." (n.d.). Available from: https://en.wikipedia.org/wiki/Big_data [Accessed October 2016].

"Brain–computer interface." (n.d.). Available from: https://en.wikipedia.org/wiki/Brain–computer_interface [Accessed October 24, 2016].

Business Insider. (2016). MIT students invented a robotic kitchen that could revolutionize fast food. Available from: http://www.businessinsider.com/mit-students-invented-a-robotic-kitchen-2016-4 [Accessed February 18, 2017].

IFL Science. (2017). Robot chef that can cook 2,000 meals set to go on sale in 2017. Available from: http://www.iflscience.com/technology/robot-chef-home-could-arrive-2017/ [Accessed June 4, 2017].

Kosoff, M. (2016). Uber shuts down experimental "instant" food delivery service. Available from: http://www.vanityfair.com/news/2016/04/uber-shuts-down-its-experimental-instant-food-delivery-service [Accessed June 4, 2017].

Los Angeles Times. (2017). San Francisco loves tech and fancy coffee. So of course it has barista robots. Available from: http://www.latimes.com/business/technology/la-fi-tn-cafe-x-automation-20170217-story.html [Accessed February 18, 2017].

Moley Robotics. (2017). The world's first robotic kitchen. Available from: www.moley.com [Accessed June 4, 2017].

US Department of Agriculture. (2016). Food-Away-From-Home. Available from: http://www.ers.usda.gov/topics/food-choices-health/food-consumption-demand/food-away-from-home.aspx [Accessed October 2016].

8

Nutrition: The Old Mantra ... the New Un-Word

There is no bad food, there is only bad diet.

—Ed Fern

8.1 Nutrition: What's All the Fuss about?

I may not have made many friends by having insisted a few times that nutrition is not really a science. And I truly believe this, although armies of nutritionists may argue against me on this. My main argument here is that nutrition never can say anything, draw any conclusion with authority and absolute or even relative assurance. It's all about might, could, should, would, maybe, and most prominent of all, we need to do yet another study. I agree that this sounds a bit negative and demeaning, but based on many years having worked side by side with nutritionists, it's the truth. Often nutritionists were their worst own enemies, immediately casting doubt on an exciting result from their most recent study.

Nutritionist are always careful of what they say and how they say it when it comes to conveying a nutritional message. Almost like lawyers, they traipse around the message and put up more questions than they give answers. I also agree that this might sound similar in other areas of scientific discovery, however at least I can safely say as a chemist that two molecules of hydrogen and one molecule of oxygen once they are excited above a certain energy barrier will give two molecules of water. And I could cite examples from physics like the first law of thermodynamics is the law of conservation of energy that states that the total energy of an isolated (closed) system is constant and nothing can get lost or be gained. And there is more in chemical engineering such as Arrhenius's equation that describes and calculates the temperature dependence of reaction rates. I searched long and hard, and I couldn't find similar postulates, laws, or equations for the field of nutrition. And again, I use the

Megatrends in Food and Agriculture: Technology, Water Use and Nutrition, First Edition.
Helmut Traitler, Michel Dubois, Keith Heikes, Vincent Pétiard and David Zilberman.

term *nutrition* to describe the field of chemical, biochemical, and physical interactions between food and its ingredients and the human or animal body. Almost everything seems to be possible in this area, but nothing seems to be clear-cut from a scientific point of view.

Does it matter? Not really, unless nutritional findings of any kind would be taken as the absolute truth and ultimate wisdom, which they always seem to be yet they never really are. The trouble is that so many people take every new discovery, every new finding, and every new message coming from nutritionists or those who work in this field at face value and almost never have the knowledge or even the courage to critically question such messages. It is so much easier and certainly hopeful to learn about yet another miracle food, ingredient, or combination of ingredients that are supposed to have these fantastic virtues and help support our health and well-being. And yet it would be wise to look at every such new promise from experts or pseudo-experts in the field of nutrition with a critical eye and be careful when starting a new diet or taking a new set of pills just because they have been discussed in a TV ad or came up in an article in the press.

I remember a good friend who had followed nutrition and health news almost religiously and believed every single new finding that Dr. X or Dr. Y had talked about on any of the popular food- and health-related TV channels. When I visited her and her husband and had breakfast together, she typically had 8 or 10 pills lying on the table next to her plate and proudly reported that the blue one was just last week newly recommended by Dr. Z and was supposedly the new miracle supplement for imbalance "abc" and has helped many of his followers already. And that's why she took all these pills, because they were all supported by anecdotic evidence and therefore were really efficient. I am not saying that some of these pills with all kinds of new and exotic ingredients wouldn't be good for something other than the bank account of those who make and sell them, but probably more often than not, they do nothing for you at best. They might still have some placebo effect, which they would have in common with many serious pharmaceutical over the counter drugs. Again, no harm done apart from your wallet, and if it makes people happy, so be it. The trouble, however, is that this approach of advertising new miracle ingredients—the more exotic and unheard of the better—does not really help giving credit to serious nutrition research, which is already trying to escape this grey zone of might, could, should, and "more studies are needed."

8.1.1 The Hottest New Food Trends

The Wall Street Journal (2016) published an entire supplement on "the next hot trends in food" and focused on this topic in several articles over eight pages. The major topics they discussed dealt with topics of new superfoods such as moringa, the importance of grazing cattle for better meat quality, consumer friendly products with reduced sugar, the trend of plant water consumption beyond coconut such as cactus (unfermented that is!), jackfruit as meat substitute, and algae like spirulina, the latter as a natural food color source. I must say that in my native Austria during my student years—and these were quite some time ago—spirulina was already popular, and cattle didn't know anything else than grass or hay full of dried flower buds cut with the grass. Also coconut water was well known and most sweet products contained far less sugar than today's. I am not saying that "those were the good old days," but it just proves that when you wait long enough, everything comes back because the "old ways" were forgotten in the meantime. Even kale was known and consumed then but was rather considered as cheap and not too sophisticated food. Eat the Seasons (n.d.) writes the following:

> Kale has been cultivated for over 2,000 years. In much of Europe it was the most widely eaten green vegetable until the Middle Ages when cabbages became more popular. Historically it has been particularly important in colder regions due to its resistance to frost. In nineteenth century Scotland kail was used as a generic term for "dinner" and all kitchens featured a kail-pot for cooking.

In other words, kale was pretty un-fancy and was even replaced by the slightly more fancy cabbage, which is rich in minerals and vitamin C. Cabbage is pale; kale is nicely green so the latter makes for really green smoothies, homemade and most likely with the most sophisticated kitchen machine. Again, there is nothing wrong with the fact that people rediscover the old/new foods, and, if they like to, give them new names, for instance "superfood" and claim this to be the most important discovery and invention since the first egg was broken up and sunny-side-up was created. Do I sound cynical? Maybe, because the run toward such new superfoods with wonderful virtues doesn't really help the serious work that nutritionists put behind potentially important and credible discoveries still to come. Everything gets drowned in the ocean of the most fantastic health news and newly discovered super potent foods and food ingredients.

After all, it's rather simple: a balanced mix of most food and food ingredient is always a good thing. There is nothing basically wrong with well-prepared French fries, just not every day. It's ok to eat meat of any kind, just in moderation. Well, the latter is not good for those who are vegetarians or vegans; but be it as it may, balanced means a bit of everything and in moderation simply means not too much. There are so many different nutritional recommendations as to how the macronutrients (i.e. proteins, carbohydrates, and fats) should be distributed in your diet on a caloric basis. To repeat what has been said and written time and again, we need to keep in mind that fat contains more than double the caloric amount of either carbohydrates or proteins; in

Table 8.1 Macronutrient compositions: Weight versus calories.

Macronutrient	Weight (g)	Calories (%)
Fat	20	36
Protein	30	24
Carbohydrates	50	40

other words, 9 cal per gram versus 4 cal per gram energy. That's old stuff you might say, and it is. However, a caloric (energy) proportion of fat of for instance approximately 35% in your diet means that you actually must not exceed something like 20 grams of fat in 100 grams of food. Proteins could for instance account for 30 grams or 24% of calories, and 50 grams of carbohydrates for 40, (always calorie percentages and not weight percentages).

Table 8.1 depicts this correlation in the most simplistic ways.

It goes without saying that to simplify the calculations and show the correlations clearly I have not taken any micronutrients into account. Anyhow, this would not change the calculations in any substantial ways. The question was still open as to what macronutrient mix accounts for a balanced diet? Please note that I am speaking about balanced diet, not necessarily balanced food. Recommendations by different authorities have changed over time, especially with the increasing popularity of diets based on reduced carbohydrate intake.

The Australian Ministry of Health (2014) has the following recommendations regarding macronutrient energy distribution.

- Proteins from 15% to 25% energy
- Fat from 20% to 35%, and
- Carbohydrates from 45% to 65%

The US Food and Drug Administration (n.d.) publishes similar numbers.

- Proteins from 10% to 35%
- Fat from 20% to 35%
- Carbohydrates from 45% to 65%

This is a high level view, and I do not want to enter into a more specialized and detailed discussion as to how much soluble versus insoluble fiber, how many polyunsaturated fatty acids of the Ω3 versus Ω6, and how many monounsaturated fatty acids of the Ω9 type should be in your balanced diet. It is safe to say that with the average home-cooked meal (and I don't count putting a frozen pizza in the oven as home cooked) you pretty much get a distribution as suggested. And it's probably not such a bad thing to count a balanced diet over a period of 1 to 2 weeks rather than 24 hours, although the nutrition purists would not agree with my former colleague's statement that there is no such thing as bad food but only bad diet. They might find this statement even dangerous and counterproductive to the efforts to fight food-related maladies such as obesity and diabetes.

8.1.2 The Debate Continues: What's Good and What's Not Good for You?

Food Navigator (2013) published the following statement:

> Is the "there is no such thing as bad foods, only bad diets" argument helpful?
>
> By Elaine Watson, 11-Feb-2013
>
> A new position statement from the Academy of Nutrition and Dietetics (AND) which can be paraphrased as 'there is no such thing as good and bad foods, only good and bad diets' is eminently sensible, but will play into the hands of 'junk' food companies opposed to any government intervention in their industry, claims one academic."

I am not sure whether this claim has merit or whether it's too hardline a position given the actual obesity trends in Western Europe and North and Latin America, or even to a lesser degree in Asia. Stringent dietary recommendations have been given since the early 1980s and were apparently not really followed by the individual consumers. The food and beverage industry, however, came up with a whole range of light and reduced-sugar individual products and failed to create food and beverage combination, "lunch boxes" or whatever one would want to call it, and thereby did not help the consumers make the right, complementary choices.

The State of Obesity (n.d.) project recently published an article on childhood obesity based on data between 1980 and 2016 with these major findings.

> Since 1980, the childhood obesity rates (ages 2 to 19) have tripled—with the rates of obese 6- to 11-year-olds more than doubling (from 7 percent to 17.5 percent) and rates of obese teens (ages 12 to 19) quadrupling from 5 percent to 20.5 percent.

Harvard T. H. Chan School of Public Health (2011) had the following headline to a publication related to this topic: "Obesity has doubled since 1980, major global analysis of risk factors reveals." The article goes on and mentions that in 2008 10% of all adults were obese with a higher rate in women than in men, which amounted to the staggering number of half a billion obese adults worldwide in that year.

The question of how much dietary recommendations and guidelines have actually really helped to achieve the opposite of what was intended—very well intended that is to say—remains unanswered. Simply looking at the facts one could come to the conclusion that building a dietary, health-related architecture around individual products, be it food or beverages, combined with the constant availability of such products for the consumer does not lead to the overall desired public health goal. But then we are getting older, aren't we? So, something must be good—or not too bad—in our daily food and beverages?

The situation is of course more complex because there are many other factors beyond food and beverages that influence our lives and especially life expectancies. These are factors such as personal hygiene and although they can be going too far, exercising, positive stress, genetic factors, security of food without overindulgence, and societal affluence.

I have a personal favorite, which might come as a surprise to you, traveling and moving around physically, especially mentally. Travel not only means physical displacement but even more so mental displacement and the need to rearrange with and adapt to new, often unexpected circumstances. I think that being on the move and especially being and remaining alert plays a big role in our well-being, and ultimately our health. I do not suggest that everyone should get up now, jump in the car and drive to the airport or directly to a far-away destination and see what's happening elsewhere, but some of this for all of us, from time to time, is a healthy and happy undertaking and definitely plays a critical part in our increasing longevity.

8.1.3 And Here We Go Again: Fasting Can Do You an Awful Lot of Good

So, where is the truth? What dietary recommendation is the right one and which of the more than 100 diets, more serious ones as well as fads and fashions, is the one for you that not only gives you the desired physical outcome such as for instance losing weight and keeping it off but also gives you pleasure and happiness at the same time? I don't have the answer, but simply looking back in history, most cultures and religions had one important recommendation: regular and recurring fasting, for whichever cultural or religious reason, typically for a day (the Catholic church) or a month (Islam) or other similar frequencies. Fasting, even for one day or not eating for say 24 hours, gets you into a ketogenic state called *ketosis*, and can, among other factors, help stimulate detoxification and purification of your body and your metabolic system. Such ketosis can also be reached by only ingesting proteins and fats in your diet and may replace total fasting, thus eliminating the negative effects of fasting such as increased weakness and becoming increasingly tired.

It has, however been observed that with prolonged total fasting, which I certainly do not recommend, mental acuity and sharpness can to some degree again increase. This article is a good source of information regarding ketosis ("Ketosis," n.d.).

> Longer-term ketosis may result from fasting or staying on a low-carbohydrate diet, and deliberately induced ketosis serves as a medical intervention for various conditions, such as intractable epilepsy, and the various types of diabetes. In ketosis, fat reserves are readily released and consumed. For this reason, ketosis is sometimes referred to as the body's "fat burning" mode.

One could consider fasting as a "ketogenic holiday," if it wasn't for the hardship of not eating anything at all for a more or less prolonged period of time. So, to make this really a pleasant holiday, it would be advisable, for the period of fasting to only consume protein- and fat-containing foods. This is, by the way, the underlying science of any low-carbohydrate diet, especially the Atkins diet. However, I am a bit skeptical about prolonged periods of ketosis. As suggested, it would be best to go on a 2-day "ketogenic holiday" accompanied with the right food and to stay happy and alert, yet profit from all the positive effects of fasting (Ketogenic Holiday, Ellen Mitchell, personal communication, August 2014).

8.1.4 A Few Simple Tips When It Comes to Healthy and Happy Eating

So, all this leaves you probably confused and still not sure which trend to follow, if trends to follow is your desire. I truly believe that there are a few, simple nutritional guidelines, dietary guidelines, if you will, which are probably worth to follow for every healthy person.

- Try to avoid eating the same food all the time unless you are a 6-year old and all you ask your mother to prepare for you is franks with mashed potatoes or spaghetti with tomato sauce. You will get over this phase and will ultimately add variety to your diet.
- Do eat just enough and stop eating before you feel full.
- Have a glass of wine from time to time; it's the alcohol that is positively correlated with improved cardiovascular health and is not dependent on wine color.
- Ideally eat and drink in company because eating is also a social event bringing people together.
- Find your personal eating schedule or in other words listen to your body and to your stomach. My good friend Heribert Watzke (2010) gave a much-lauded TED talk about the stomach being the second, complementary brain: start thinking with your stomach.
- Do not use pills to complement your food; it's not necessary unless you are an 18th-century sailor on a ship exploring the vastness of the Pacific Ocean and you might run into a problem with scurvy; lemons may help, but again no pills!

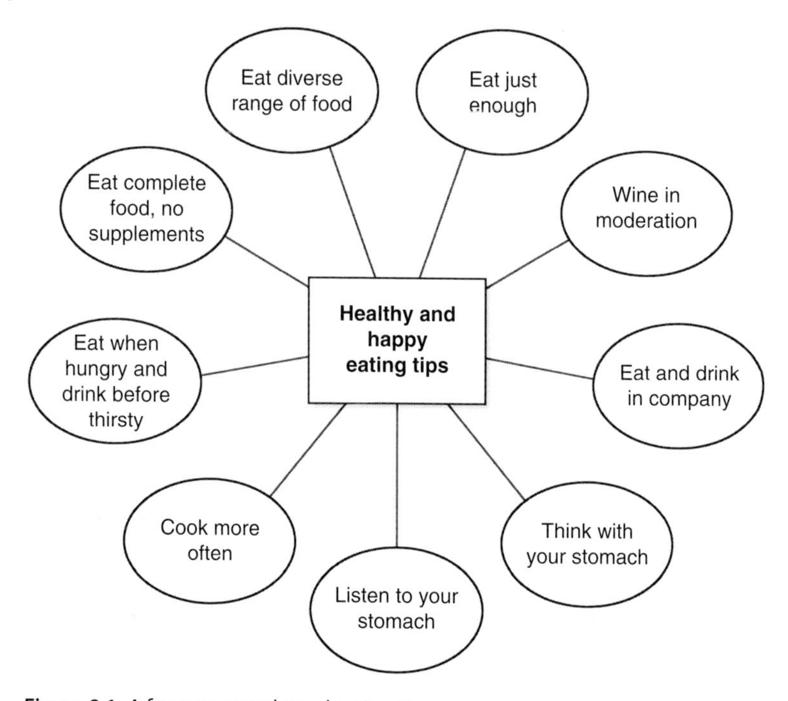

Figure 8.1 A few suggested good eating tips.

- Listening to your stomach and your cravings will tell you more about what food you should eat than any dietary recommendation, even the most advanced one.
- Eat when you feel hungry, drink (water) before you become thirsty.
- Cook yourself more often or at least take a greater interest in cooking; you will get a better idea about the beauty and complexity of food. This includes taking a greater interest in shopping for food as well; take the time because it's well spent.

You may miss the one or the other of your personal favorites in this list, so please feel free to extend it with your items and make it your very personal list of what good food and healthy eating and drinking do mean for you.

Figure 8.1 illustrates these simple good eating tips.

8.2 A Bit of Nutrition History

I do realize that history of anything, except for historians, might always be a bit of a hardship for any reader, that's why I didn't begin this chapter with history but rather hide it in this section. On the other hand, history is important and tells us what was and what led us to today, with all the wrongs that we did and redid or hopefully avoided. The history of nutrition is not much different, and I always use the example of the discovery of essential fatty acids by the Burrs as an important milestone in the history of nutrition; however, the story begins much earlier. Let me use a short overview that was concisely and elegantly put together by Nutrition Breakthroughs of Glendale, California, and who kindly gave me the permission to use their take on the history of nutrition.

> Nutritional discoveries from the earliest days of history have had a positive effect on our health and wellbeing. The word nutrition itself means: "The process of nourishing or being nourished, especially the process by which a living organism assimilates food and uses it for growth and replacement of tissues." Nutrients are substances that are essential to life, which must be supplied by food.
>
> Today more than ever, obtaining nutritional knowledge can make a big difference in our lives. Air, soil, and water pollution in addition to modern farming techniques, have depleted our soils of vital minerals. The widespread use of food additives, chemicals, sugar and unhealthy fats in our diets contributes to many of the degenerative diseases of our day such as cancer, heart disease, arthritis and osteoporosis. Here is a brief history of the science that offers the hope of improving our health naturally.
>
> 400 B.C.—Hippocrates, the "Father of Medicine," said to his students, "Let thy food be thy medicine and thy medicine be thy food." He also said "A wise man should consider that health is the greatest of human blessings."
>
> 400 B.C.—Foods were often used as cosmetics or as medicines in the treatment of wounds. In some of the early Far-Eastern biblical writings, there were references to food and health. One story describes the treatment of eye disease, now known to be due to a vitamin A deficiency, by squeezing the juice of liver onto the eye. Vitamin A is stored in large amounts in the liver.
>
> 1500s—Scientist and artist Leonardo da Vinci compared the process of metabolism in the body to the burning of a candle.
>
> 1747—Dr. James Lind, a physician in the British Navy, performed the first scientific experiment in nutrition. At that time, sailors were sent on long voyages for years and they developed scurvy (a painful, deadly, bleeding disorder). Only nonperishable foods such as dried meat and breads were taken on the voyages, as fresh foods wouldn't last. In his experiment, Lind gave some of the sailors sea water, others vinegar, and the rest limes. Those given the limes were saved from scurvy. As Vitamin C wasn't discovered until the 1930's, Lind didn't know it was the vital nutrient. As a note, British sailors became known as "Limeys."

1770—Antoine Lavoisier, the "Father of Nutrition and Chemistry" discovered the actual process by which food is metabolized. He also demonstrated where animal heat comes from. In his equation, he describes the combination of food and oxygen in the body, and the resulting giving off of heat and water.

Early 1800s—It was discovered that foods are composed primarily of four elements: carbon, nitrogen, hydrogen and oxygen, and methods were developed for determining the amounts of these elements.

1840—Justus Liebig of Germany, a pioneer in early plant growth studies, was the first to point out the chemical makeup of carbohydrates, fats and proteins. Carbohydrates were made of sugars, fats were fatty acids, and proteins were made up of amino acids.

1897—Christiaan Eijkman, a Dutchman working with natives in Java, observed that some of the natives developed a disease called Beriberi, which caused heart problems and paralysis. He observed that when chickens were fed the native diet of white rice, they developed the symptoms of Beriberi. When he fed chickens unprocessed brown rice (with the outer bran intact), they did not develop the disease. Eijkman then fed brown rice to his patients and they were cured. He discovered that food could cure disease. Nutritionists later learned that the outer rice bran contains vitamin B1, also known as thiamine.

1912—E. V. McCollum, while working for the U.S. Department of Agriculture at the University of Wisconsin, developed an approach that opened the way to the widespread discovery of nutrients. He decided to work with rats rather than large farm animals like cows and sheep. Using this procedure, he discovered the first fat-soluble vitamin, Vitamin A. He found that rats fed butter were healthier than those fed lard, as butter contains more Vitamin A.

1912—Dr. Casmir Funk was the first to coin the term "vitamins" as vital factors in the diet. He wrote about these unidentified substances present in food, which could prevent the diseases of scurvy, beriberi and pellagra (a disease caused by a deficiency of niacin, vitamin B-3). The term vitamin is derived from the words vital and amine, because vitamins are required for life and they were originally thought to be amines → compounds derived from ammonia.

1930s—William Rose discovered the essential amino acids, the building blocks of protein.

1940s—The water-soluble B and C vitamins were identified.

1940s—Russell Marker perfected a method of synthesizing the female hormone progesterone from a component of wild yams called diosgenin.

1950s to the Present—The roles of essential nutrients as part of bodily processes have been brought to light. For example, more became known about the role of vitamins and minerals as components of enzymes and hormones that work within the body.

1968—Linus Pauling, a Nobel Prize winner in chemistry, created the term Orthomolecular Nutrition. Orthomolecular is, literally, "pertaining to the right molecule." Pauling proposed that by giving the body the right molecules in the right concentration (optimum nutrition), nutrients could be used by people to achieve better health and prolong life. Studies in the 1970s and 1980s conducted by Pauling and colleagues suggested that very large doses of vitamin C given intravenously could be helpful in increasing the survival time and improving the quality of life of terminal cancer patients.

1994–2000: Have you ever wondered why vitamin bottle labels and nutritional web sites include a phrase saying that their products and information are not intended to diagnose, cure or prevent any disease? These also usually state that the health claims have not been evaluated by the Food & Drug Administration (FDA). Here's why: The Dietary and Supplement Health and Education Act was approved by the US Congress in October of 1994 and updated in January 2000. It sets forth what can and cannot be said about nutritional supplements without prior FDA review.

While this law limits what vitamin manufacturers can claim about preventing or curing diseases, its passage has been a major milestone in the natural health field. It acknowledges the millions of people who believe dietary supplements can improve their diets and bestow good health. It opens the way for people to obtain the information they need to make the best nutritional choices for themselves.

In January of 2000, the FDA clarified that supplement makers can state their products can improve the structure or function of the body or improve common, minor symptoms. Allowable statements include things such as: "maintains a healthy heart," "helps you relax," "is good for symptoms of PMS," "strengthens joint structure," etc. Overall, due to this law, vitamin, herb and nutrient manufacturers have greater freedom to say what their products can do to improve our health.

Although this story focuses more on the United States at the end, it still provides a good overview on the history of nutrition. As you also may have seen, Jobee Knight from Nutritionbreakthroughs.com, and the author of this "history of nutrition" has not mentioned my personal favorite, the discovery of essential fatty acids.

Back in 1929 and 1930, George and Mildred Burr were curious and patient enough to discover that certain types of fatty acids were essential to our health and had to be present in our daily food because our organisms are not able to synthesize such fatty acids. They are used as starting point for further desaturation and elongation to long chain, polyunsaturated fatty acids, all-important in different ways for our health and well-being. You can find more information in Mukohpadhyay (2012).

I am probably not the only one who might want to add a few more lines to the history of nutrition, and I encourage you, again, to do your part and add what for you is missing or expand on the bullet points already written down in this story. For my part, I would like to add a few more, more recent elements but will pick up the topic on admissible claims, and especially on nonadmissible claims when it comes to food and beverages in a later section of this chapter.

8.2.1 Low and Reduced, Lower and "Reduced-Er": Low or Reduced Fat

So let us begin with the "low" and the "reduced" trend that is ravaging the food and beverage industry probably since the 1990s. I have personally lived through quite a few cycles of low and reduced, and the only thing that was not low and reduced during all these years was the effort that was put behind the R&D activities in trying to find not only the right answer but also the right solution with the right taste and especially at the right costs. The latter was, and to the best of my knowledge still is today, the biggest hurdle to having successfully introduced low and reduced products.

Yes, you will find a lot of low-calorie products, especially in the beverage aisle, but in the food aisle they may be there but are not very successful, especially when it comes to low sugar or reduced salt. Low fat seems to be the easier undertaking. Although all the historic efforts of the 1980s to design and tailor low-calorie fats, for instance Proctor & Gamble's (P&G's) sugar esters were pretty much unsuccessful, at least with regard to the industry's expectations. Olestra®, P&G's sucrose polyester, was such a fat. At the height of its popularity in the mid 1990s, it received partial approval from the US FDA for use in certain snack products.

> Olestra is a fat substitute that adds no fat, calories, or cholesterol to products. It has been used in the preparation of otherwise high-fat foods such as potato chips, thereby lowering or eliminating their fat content. The Food and Drug Administration (FDA) originally approved olestra for use as a replacement for fats and oils in prepackaged ready-to-eat snacks in 1996. In the late 1990s, Olestra lost its popularity due to side effects, but products containing the ingredient can still be purchased at grocery stores in some countries. ("Olestra," n.d.)

I worked a few years in this area, and with my team we developed fatty compounds based on esterification of fatty acids on the backbone of 2,3-butanediol.

The molecule contained two fatty acids of any given mix, saturated, mono- or poly-unsaturated. The idea behind this was to keep the external regions of the backbone, the ones that would typically be attacked by the enzyme, lipase free from fatty acids. In this particular case we dealt with two methyl groups. The resulting products had basically no taste or at best the same taste as any other fat, and we did quite a number of trials with obese rats. Yes, those were the days in the late years of the 1980s and the early 1990s. We could show that there was no adverse effect to be observed such as anal leakage of undigested fatty matter or greasy hair or anything else. Two groups of animals were eating at will the two different diets, one with regular fats the other one substituted for 2,3-butanediol esters.

After several weeks of trials, the differences were striking and we estimated the caloric value of the butanediol esters to be approximately 1 cal or approximately only 10% of the energy density of a regular fat. We also checked for possible liver toxicity but couldn't observe any increase for instance in liver transaminases. My team developed several possible applications, one of which resulted in a patent for cosmetic applications. The underlying rationale was the inability of lipase in the skin (dermis and especially epidermis) to attack the fatty phase in a cosmetic cream, thereby potentially prolonging any film-forming and protective effect of such a cream (Cosmetic compositions containing 2,3-butanediol fatty acid di-esters US 5474775 A, H. Traitler and J.L. Viret, 1995). My company never really pursued the "making of ingredients" route; therefore we quickly lost all interest in this and, like many others in the industry, followed other strategies and directions to achieve low-fat and low-calorie solutions.

I tell this personal story, which reflects a work period of approximately 1 ½ years because it's yet another example of one of the many efforts that went into finding the "magic bullet" of low- or no-calories fats, good for you, good-tasting, and not too costly to make—ideally at zero extra costs—and that would, most importantly be approved by the regulatory authorities. This dream fell quickly apart for us at Nestlé and must have been much more hurtful for a company such as P&G, which had put years and years of valuable and professional efforts behind the development of their Olestra, and which ultimately was more or less abandoned except for only a few application examples such as in the area of snacks.

8.2.2 Low or Reduced Salt

Salt reduction is another only partly fulfilled dream of the food industry. The underlying rationale is simple: there is medical evidence that in many people, especially with increasing age, high levels of salt, sodium chloride that is, can contribute to elevated blood pressure.

De Wardener and MacGregor (2002) concluded that based a genetic disposition we should not consume salt at too high amounts. Typically our diets contain a multiple of the salt needed by our bodies. Such important overconsumption can lead to all kinds of blood pressure–related diseases and reduction of salt intake can help alleviating these.

There is, however some debate about such a straightforward relationship and some scholars suggest that this is not valid for all people. In an article in the *Scientific American*, Moyer (2011) indicates that a large meta study that combined results from seven epidemiologic studies involving more than 6,000 subjects was not able to find strong evidence for a correlation between reduced salt intake and reduced cardiovascular and similar health problems.

I do not intend to add to the controversy, but old habits die hard (or never), and so, despite certain more or less founded doubts, the public, and in turn, the food industry still hangs on to the simple belief: less salt equals lower blood pressure. It is likely that this concept works for many and they have personal proof for this, and others struggle to reduce their increased blood pressure by reduced sodium intake and don't see the desired results. Be that as it may, the food industry follows this simple, however, valid recommendation and strives to replace salt in formulations that contain salt and have come up with many strategies and acceptable solutions. From ammonium chloride to potassium chloride—the latter rather bitter than salty and the former rather sour—many approaches have been tried.

I was personally involved, again you might say, in some of this work in my former company and can say based on personal experience that there were quite a few good solutions, but most of them were too expensive or at least not cost-efficient enough. Despite this there are many low-sodium product solutions on the market, and consumers have become acquainted to the new taste quirks, sometimes surprising and unexpected. I do, however find it almost comical to think of exchanges between marketing and sales colleagues that they now ask the consumers more money for giving them something less. This, by the way, was one of the main reasons to look at such "reduced or low" products from a benefit point of view—benefit for the consumer that is.

8.2.3 Low or Reduced Sugar and No Sugar

Low sugar or no sugar in sweet products, food, or beverages seems to be the "holy grail" of the food industry. It is true that an important contributor to increased obesity in many populations is sugar in all its forms. Hence, the strong drive in the food and beverage industry toward sugar replacers, artificial sugars (e.g., aspartame®), or natural high-intensity sweeteners such as Stevia. The latter was approved by the FDA as food additive in 2008, and later by the European Union in 2011.

Personally I can see an important role for sugar replacers, and there are many more than just the two mentioned, for example, in dental hygiene with a positive long-term effect on personal health and well-being. It is still difficult to judge how healthy it would be in the long run to stimulate an insulin response by the organism when sensing sweet and then not finding the culprit to attack; it's almost like a false fire alarm. Joseph Nordqvist (2013) in *Medical News Today* concludes that consumption of the artificial sweetener sucralose may trigger an insulin response in our bodies.

I do not intend to enter any further into this debate other than by saying that everyone who has doubts regarding this topic may find many scholarly articles discussing this topic in much detail but even better, find out the response in your own body under the guidance of your physician.

There is yet another approach to reduced or very-low sugar in beverage and food products, and that is the one that works on modification of sweetness taste receptors in mouth and tongue by using almost subliminal amounts of specific taste receptor modulators mixed into a sweet beverage or sweet food product. There is still sugar but most likely much less, maybe only 50% or even only 25% as compared to the original formula. The company Senomyx of San Diego, California, is one of the leaders in this approach to taste modulation, or more accurately, perception enhancement. The following statement comes from their Web site.

> Senomyx discovers novel flavor ingredients that boost taste sensations, such as sweet. Our flavor ingredients help food and beverage companies create or reformulate better-for-you products, for example, by allowing them to significantly reduce the use of traditional, higher-calorie sweeteners, while maintaining the same taste. We also develop savory flavors, bitter blockers and cooling agents. (Senomyx, n.d.)

There seems to be some dispute about Senomyx's development practices of some of the molecules that serve as such taste enhancers. A few years ago I was personally involved in a few taste testings of beverage formulations that one of our larger flavor supplier partners had created. The effect of the enhancer on sweetness perception was quite striking, and it was hoped that the approach could eventually be introduced into some of the company's sweet snack and cereal products. The big draw back in all of these formulations was the artificial origin of the molecule because my former company did not want to use any artificial molecules with regard to flavor enhancement. Although the approach was seen as smart and elegant, the fact that it involved artificial components was a no-go. I do recall discussions of finding possible naturally occurring candidates with similar molecular structures and enhancement efficiencies.

I do know for a fact that Nestlé has tested different approaches to reduced sugar at equal sweetness perception in relevant products, and I am not aware, at least based on present public knowledge, that the all-natural

"Senomyx-approach" is still being pursued by Nestlé. This just proves, once again, one of the present strongest trends in the entire food and beverage industry, the desire to go all natural.

8.2.4 Low Saturated Fats, Good Monounsaturated Fats, More Polyunsaturated Fats, and Lots of Ω3 Fats

I discuss this topic separately from the one on low or reduced fats in general, because it nicely leads over to the next section on nutrition controversies. It also shows the kind of confused approach by many, including the nutritionists, when it comes to fat in general. On the one hand, the recommendations generally hint toward reduced fat intake. This is especially pointed out on a pure energy related basis, as depicted in Table 8.1. What I mean to say is: almost everything should be reduced when it comes to the macronutrients, especially fats such as saturated fats, not to speak of other food components such as salt or cholesterol. On the other hand, one should increase consumption of proteins and specific fats, rather oils, which are rich in certain types of fatty acids, such as monounsaturated fatty acids of the Ω9 family, double unsaturated fatty acids such as linoleic acid, C18:2 Ω6, or triple unsaturated acids such as Gamma-linolenic acid, C18:3 Ω6 or Alpha-linolenic acid, C18:3 Ω3.

Moreover, there is a strong push to consume increasing amounts of even polyunsaturated fatty acids (PUFAs) such as C20:5 Ω3 (eicosapentanoic acid) or C22:6 Ω3 (docosahexanoic acid). The latter are predominant fat components in fish and therefore, next to other virtues, it is recommended to eat more fish. On the other hand, you find more and more people critical of eating fish, because doing so will dramatically reduce fish populations in our oceans and lakes and thereby destroy the fine balance of nature in such habitats. Personally I am all for eating fish, in moderation though, and probably with a careful look where the particular fish comes from. When I go to eat in my preferred taverna on my preferred Greek island that I have visited for many years, I always go to the kitchen and ask the owner and chef to show me what fish he has tonight. He shows me his "collection," which would elsewhere be called "catch of the day," and I point to a fish, which I like because the size is right and his (her?) eyes are pretty clear and tell me that they were really caught last night. I digress, so let me come back to the confusion when it comes to recommend less of this, more of that, or was it the opposite? Let me try to summarize the different messages and recommendations that I have discussed in this chapter thus far.

- Reduce the overall food intake to a level that is at par with your energy expenditure.
- You can do this in different ways, ideally based on observation of your own body and personal experience.

- Temporary fasting, either by total elimination of any food intake or with the help of an exclusive fat- and protein-based diet is your personal choice.
- Look up Figure 8.1 and try to follow the major tips listed in there.
- Follow the trend to a targeted reduction of individual food components, macro- or micronutrients alike, with a grain of salt (no pun intended), and be smart and selective.
- It is generally accepted that high intake of saturated fats is not recommended; in simple words, it's not a good thing to do to your body.
- A balanced portfolio of different fatty acids can be achieved through a smart and happy selection of different food raw materials, including especially fish.
- Be careful with the consumption of sugar; don't fool yourself that unrefined sugar is so much better. Yes, it's less bad but still bad consumed in too large amounts.
- Best thing to do is to reeducate your pallet when it comes to sweetness perception. Take for instance unsweetened ice tea and reduce the amount of sugar that you add ever so often by a tiny fraction. If you are used to two teaspoons of sugar (8 grams) reduce it by 5% (40 mg) and see what happens. If you drink this new tea for a week or two and are eventually happy with the sweetness, take another 5% off, and so on, and so on. This way you could reach a 50% sugar reduction after 10 to 12 steps and have trained your pallet back to become happy with less sugar. It's all in your mind anyway. You will need a small scale and some discipline to follow this procedure, and you may want to write a little protocol for your own personal sake to believe that you eventually have come this far.

Figure 8.2 attempts to depict suggested frequency (how often) and importance (how seriously) you should follow a select number of supposedly good eating habits.

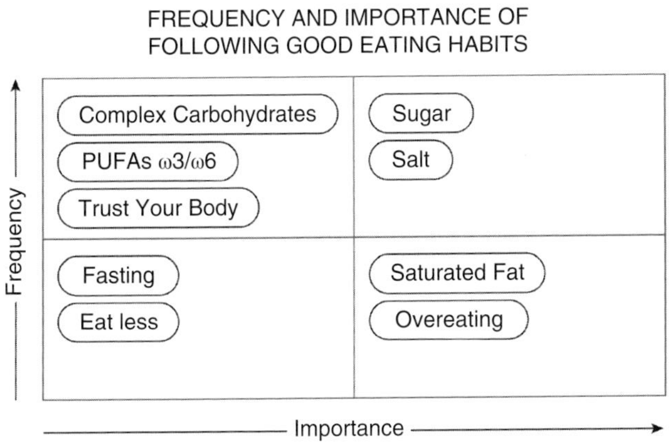

FREQUENCY AND IMPORTANCE OF
FOLLOWING GOOD EATING HABITS

Complex Carbohydrates
PUFAs ω3/ω6
Trust Your Body
Sugar
Salt
Fasting
Eat less
Saturated Fat
Overeating
Frequency
Importance

Figure 8.2 Good eating habits prioritized.

To come back to the topic of sweetness perception, I truly believe that there could be a wonderful new product opportunity for food or beverage companies to sell you a multipack of sweet snacks or beverages that are already formulated in this fashion: from level 10 to level 5 in 10 weeks. And you can buy level-5 products after that. I can hear the voices of new product developers who might say: great, we can do this, and on the other hand the engineers in the factories who might say, bollocks, we can't do it, it's too complex. Anyhow, you can always do it yourself, similar to the way I just described.

8.3 Typical Nutrition Controversies

As you may recall, I argued that nutrition is not an exact sciences, and therefore, that many findings that have been unveiled in past and present nutrition research are most of the time ambiguous to say the least. Some would say they are contradictory and others would say they are outright useless. Personally I would never go that far but would put my bet on "ambiguous." Let me explain this with a few examples.

Let me start with the vast family of grains as well as fruits and vegetables or carbohydrates in general. Looking at the typical recommendations of the nutritional or food pyramid, you will see carbohydrates as the most abundant component in the diet, both on an energy as well as weight basis, and this by far. Table 8.2 shows a rather detailed view of such dietary recommendations as they would typically be found in a more fancy pyramid shape in nutrition- and diet-related publications. By far the most important candidate is carbohydrate with and average consumption recommendation of up to 75%!

8.3.1 So Many Recommendations...Too Many?

The controversy comes from the many other dietary recommendations that propose low-carbohydrate intake, and some even temporary zero-carbohydrate consumption, as the better alternative to sustained health and well-being. So, it's complicated out there, and I for once have no clear answer as to where the truth may be. It's really not easy to decide because if I would slavishly follow the low-carbohydrate route I still have the dilemma to overcome what to replace carbohydrates in my diet with.

Atkins followers would suggest high-fat intake instead of carbohydrates. Ok, but haven't we all heard that fat in large quantities and especially saturated fats are actually outright bad for you and not only may lead to obesity but all kinds of cardiovascular health problems on top of it? Is that all true? Here's the next controversy. There is some doubt about this as expressed in Siri-Tarino and colleagues' study (2010) that summarizes the evidence of a number of

Table 8.2 Dietary recommendations.

Dietary factor	1989 WHO Study Group recommendations	2002 Joint WHO/FAO Expert Consultation recommendations
Total fat	15–30%	15–30%
Saturated fatty acids (SFAs)	0–10%	<10%
Polyunsaturated fatty acids (PUFAs)	3–7%	6–10%
n-6 PUFAs		5–8%
n-3 PUFAs		1–2%
Trans fatty acids		<1%
Monounsaturated fatty acids (MUFAs)		By difference
Total carbohydrate	55–75%	55–75%
Free sugars	0–10%	<10%
Complex carbohydrate	50–70%	No recommendation
Protein	10–15%	10–15%
Cholesterol	0–300 mg/day	<300 mg/day
Sodium chloride (Sodium)	<6 g/day	<5 g/day (<2 g/day)
Fruits and vegetables	≥400 g/day	≥400 g/day
Pulses, nuts and seeds	≥30 g/day (as part of the 400 g of fruit and vegetables)	
Total dietary fiber	27–40 g/day	From foods
NSP	16–24 g/day	From foods

Source: https://en.wikipedia.org/wiki/Food_pyramid_(nutrition) [Accessed on November 20, 2016].

prospective epidemiologic studies on a possible positive correlation between high saturated fat consumption and increased risk in coronary heart disease (CHD), stroke and cardiovascular disease (CVD). The conclusion of this particular study is plain and simple, namely that there is no significant evidence for a correlation between saturated fat consumption and increased cardiovascular health risk. And it goes on to say, as expected, that more studies are required to find possible other culprits that may cause higher incidences of cardiovascular health problems. Wow, that not only would suggest that saturated fat is actually not so bad when it comes to cardiovascular health risks, and I don't go into the obesity side and risk of diabetes here, but the authors of this study indirectly suggest that there might be a greater hidden risk in food components that we eat instead of saturated fat!

I would not be too astonished if I have lost you and that you are, by now, asking questions about not only my possible insanity but the sanity of nutritional recommendations at all. Maybe as a consolation, epidemiologic studies and meta-analysis are heavily reliant on statistical evaluations, and it has happened that some statistics were actually wrong. I am not suggesting that this is the case here, but you and I should take this information with some caution and not throw out the baby with the bathwater. Epidemiology is showing us most likely trends by making assumptions about correlations between different factors. A simple, and totally misleading example for this would be the following: every living being drinks water in one form or another. At the same time every living being eventually dies. Therefore, it might be concluded that there is a strong positive correlation between drinking water and mortality. I do admit, it's not only simplistic, but it's also obvious that I have not taken other important factors into account that correlate much more closely with intake of such and such food or food component with disease and ultimately with mortality. I just wanted to briefly illustrate that one can quickly take a wrong term in the domain of epidemiology.

8.3.2 More Controversies

I still owe you a few more examples for dietary recommendations that are sometimes recommending the exact opposite and may add a lot of confusion, although they may claim to represent the truth. Cholesterol is one of the more prominent candidates. Cholesterol is one of the crucial elements in our body to build, fortify, and maintain membranes that are so important to the functioning of our body. I admit that this is a simplistic description, but it's pretty accurate. It's the lipoproteins, which contain cholesterol, and which build these membranes. Fats, as the lipo– would suggest are also involved in this membrane-building, and depending on the type of membrane and location in our body, the fatty composition might differ slightly, but that's beside the point.

Historically, or if you want "back in the day," cholesterol was measured in total and was, and still is in most places, expressed in milligrams per deciliter (mg/dL). It is also measured in millimoles per liter of blood. Let me use mg/dL for my example here. For instance a total value of 250 mg/dL was considered a high cholesterol level in the past and efforts were undertaken through education of the public that diet can help to reduce the cholesterol level in your body. This may have worked for some and was more difficult to achieve for other. To be a bit controversial myself, I see the evolution of the recommended levels as a bit of a catch-up game with the pharmaceutical industry. Once many or most people in many communities have reached a level of say 220, the recommendations changed to 200 and then to 180. And the industry had specific cholesterol-decreasing prescription pills solutions, just in case.

And then there came the sophistication of cholesterol measurements: not all cholesterol is equal, and it depends on the type of lipoprotein in which it is partaking; lipoproteins are classified by their density, which is related to how many fat molecules they can transport and their diameter. Low-density lipoproteins (LDL), for instance, have a diameter range of approximately 21 to 27 nanometers (nm), whereas high-density lipoproteins or HDL have a typical diameter range of 7 to 13 nm. There are more classes of lipoproteins, but these two are the ones most frequently used when it comes to the distinction between "good" (HDL) cholesterol and "bad" (LDL) cholesterol (German, Smilowitz & Zivkovica, 2006).

So, today if your overall cholesterol is a bit high, your physician might be more forgiving if your HDL was 60 mg/dl or even above and your LDL is "only" 140 mg/dl or close. Be cautious, however, because there might be new recommendations coming, which might lead to new uncertainties and new controversies.

On the other hand, it must be said that most of the controversies about cholesterol and food are linked to the opposing opinions by nutrition experts, as well to some degree the medical profession, whether or not eating food that contains higher levels of cholesterol actually can add to your own body's blood cholesterol. Some say it does, yet many others say it does not. I suggest that you read and find out for yourself, which might be pretty much based on who you talk to and which scientific paper you may have read.

I promised you more controversial nutrition-related topics and here we go; let me just list them, and shorten the process.

- Are eggs good or bad for you?
- Is gluten really bad for you?
- Does a high-fat diet really contribute to becoming overweight? Not a high-calorie diet but specifically high fat.
- Is sugar really only bad or doesn't your body actually need sugar?
- Is consumption of fish only beneficial? What are the risks and issues involved?

You may have additional topics and questions you may want to add to not only this list and also the entire section, and I encourage you to do so. It is clear that so many seemingly controversial recommendations are difficult to discuss at once and give it a much-needed legal framework that can bring some order into this highly difficult field. The next section will discuss and analyze this in more detail.

8.4 Food and Claims, Food and Benefits

"Heart healthy California walnuts" is a line from a commercial frequently broadcast on TV while I write this chapter at the end of November 2016. It is, by the way, a claim that the FDA permitted back in 2008. For the first time ever, the FDA permitted a qualified health claim made by a conventional food. Permission was especially given to a series of nuts, including the mentioned walnuts.

From a purely semantic point of view, this could mean that Californian walnuts have a healthy heart, whereas the other possible meaning of this message is that it's good for the health of your heart is more of an interpretation by the viewer. This may sound a bit too much nitpicking, but it nicely illustrates the difficulties that the agriculture, and especially the food industry, are confronted with. Ambiguity is typically the name of the game, and it can ultimately lead to a regulatory "whiteout" when it goes too far and as has happened with the European Food Safety Authority (EFSA) during the second half of the 2000s. But let me back up a bit and discuss and analyze in some detail the topic of health and nutrition claims for foods and food ingredients and the desire of the industry to show and describe special and differentiating product benefits through such claims.

I specifically refer to health claims made by the industry with regard to probiotic bacterial strains and stomach health. Toward the end of 2009, a large number, actually the totality of health claims for probiotics was thrown out by the EFSA because a review of the scientific literature amassed over a period of almost two decades was inconclusive and therefore did not warrant acceptance of any claims in this context. Hickman (2009) published a story on the topic of EFSA's total refusal to accept any of the large number of 180 health claims that food companies have made with regard to probiotics.

Looking at EFSA's Web site (n.d.), one finds the following definition for the topic of nutrition and health claims.

> An increasing number of foods sold in the EU bear nutrition and health claims. A nutrition claim states or suggests that a food has beneficial nutritional properties, such as "low fat", "no added sugar" and "high in fiber". A health claim is any statement on labels, advertising or other marketing products that health benefits can result from consuming a given food, for instance that a food can help reinforce the body's natural defences or enhance learning ability.
>
> EU Framework:
>
> In December 2006 EU decision makers adopted a Regulation on the use of nutrition and health claims for foods, which lays down harmonized EU-wide rules for the use of health or nutritional claims on foodstuffs based on nutrient profiles. Nutrient profiles are nutritional requirements that foods must meet in order to bear nutrition and health claims. One of the key objectives of this Regulation is to ensure that any claim made on a food label in the EU is clear and substantiated by scientific evidence.

It sounds like a really good definition, simple and clear, however was apparently not so clear for many new-product developers and marketing representatives in the industry, otherwise they wouldn't have gotten it that wrong with the claims for probiotics. I was not personally involved in the work that went on in my former company for probably the better part of 15 or so years, and was in awe of the number of studies, including many clinical studies, that were carried out to substantiate something.

And yes, I met many people that had told me that a shot of a probiotic concoction did them an awful lot of good and others who told me about their unpleasant diarrhea experiences after consuming the very same probiotic product. EFSA was probably as confused as I was then and this, together with the missing substantiating evidence for claimed efficacy of probiotic strains, made them decide the way they did.

Although I could add more stories on this topic I intend to limit myself to the one just described. It is a dire warning that the industry, both agriculture in growing new "miracle" produce into, as well as the food industry, attempting to transform such miracle foods into claimable industrial food products, should be careful when going this route. The baseline is, and always has to be: what's in it for the consumer? What's the consumer benefit? If the only answer would be "heart healthy walnuts," it might be a bit weak although this example throws up three really crucial points, with which I want to end this chapter.

- First, is the agricultural raw material, in our example the walnut, a key ingredient to the industry, or at least can it become one? Note that 99% of the commercial US walnut production happens in the San Joaquin Valley of California, and the United States is the third-largest producer of walnuts worldwide.
- Second, can the claim be substantiated with solid science and not statistically embellished nutritional data and expensive yet inconclusive clinical studies?
- Third, is the claim unambiguously understandable and positively understood by the consumers?

My little walnut example, which is not so little given the size of the Californian walnut industry, probably fulfills two of these criteria, which I have formulated as questions. Yes, it's a key ingredient and yes, the claim is most likely understood by the consumers the right way, namely "good for your heart health." It might, however fail on one ground and that is irrefutable evidence that walnuts are good for your heart to remain healthy in the first place. There is, however quite some evidence that the combination of antioxidants present in walnuts may have beneficial effects after all.

However, I have a suspicion that "heart healthy walnuts from Germany" might not fly with the folks at EFSA in the same way as in the United States.

European Union regulations, in addition to the basic health claim "heart healthy," do also require additional information why this is claimed, and which specific ingredient is contributing to the claimed benefit. It appears to be a bit more complicated in Europe compared with the United States when it comes to claiming specific health benefits. My suspicion though is that the size and importance of the industry, especially agricultural industry, has some weight in the discussions around claims and what might or might not be allowed. But then I have of course no proof for this suspicion of mine. So let's end the chapter with this unanswered open question.

8.5 Summary and Major Learning

This chapter was all about nutrition in the context of agricultural raw materials and food and food ingredients. I started this chapter with the question "what's the fuss about nutrition anyway?" I discussed and analyzed the following topics.

- I repeatedly suggested that the field of nutrition is not really a science, although many, especially those who work in the area of nutrition, would suggest otherwise. My argument, however, was that results from any type of nutrition research are never really conclusive and always give answers that involve "would," "could," "might," and similar ones, often followed by the phrase "further investigation is required." It was also mentioned that ultimately it doesn't really matter; however, always let the reader and the user of such "preliminary results" be always in doubt. It's always possible that a specific recommendation simply doesn't lead to the desired result in the health and wellness status of the individual.
- This can lead to ambiguity, and sometimes complete refusal of any, even the most promising and efficient, recommendations.
- This conversely may lead to an overly naïve approach to nutritional recommendations, and individuals just will eventually believe just about any recommendation as dubious it may appear to be and begin to eat increasing amounts of supplements and turn to every new and highly publicized super food.
- An article from *The Wall Street Journal* discusses the hottest new food trends.
- As an example, the vegetable kale was discussed, and it was mentioned that kale was cultivated and consumed for the better part of 2,000 years, and it's not really a new super food. There was a time when it was seen as not noble enough and was eventually replaced by cabbage in many cultures. It's only been recently that it was rediscovered.

- The notion of a balanced diet was discussed, and it was emphasized again that not all food ingredients are alike from an energy point of view. This is of course trivial but merits to be repeated over and over again because people tend to forget.
- The introduction to this chapter stated a saying from my former colleague: "there is no bad food, only bad diets." This is not seen this way by everyone in the field and an article by Watson discussed this topic in some detail and suggested that this may look like a free ride to overeating and could potentially help aggravate the status of obesity in many societies.
- Obesity and its prevalence, especially in affluent societies, was discussed and analyzed, and it was concluded that despite many good and well-meaning dietary recommendations when it comes to healthy food intake, obesity is still on the rise in many places.
- The role and virtue of infrequent or frequent short-term fasting was discussed, and temporary ketosis was described as a vehicle for body detoxification as well as weight control. Contrary to most other restriction diets such as low calories or low fat, reduction of carbohydrates in the food mix or their total short-term exclusion has probably the highest merit when it comes to sustained weight management.
- Several simple tips with regard to a healthy and happy food intake were listed and discussed. These are the tips: Eat just enough, wine in moderation, avoid eating and drinking alone, think with your stomach, listen to your stomach, cook more often, eat when hungry but drink before you are thirsty, eat complete food and avoid supplements, and eat a diverse range of foods.
- A short overview on the history was laid out and it turns out that it is probably as old as written history of humankind. My personal favorite in the history of nutrition is the discovery and first description of essential fatty acids by the Burrs in 1929 and 1930.
- The popular trends of "low" and "reduced" were discussed and analyzed. It was stated that these trends are ravaging the food and beverage industries like a fire going through dense and dry forest: they leave a large trace behind. Every food and beverage company goes to great lengths to develop, manufacture, and sell reduced-salt, reduced- or low-sugar, reduced- or low-fat products and has invested probably hundreds of millions of dollars in such development.
- The development of low-calorie fats and fat substitutes has been enormously long (probably the better part of 20 years) and has cost a few of the large food companies a rather heavy sum, which I don't dare to estimate.
- I briefly mentioned my personal venture into the field of very-low calorie fats—or rather fatlike compounds—via the development of butanediol fatty acid esters, a story that didn't really have a happy ending, like so many others in this field.

- The development of low- or reduced-salt solutions in especially culinary industrial products is an equally long ongoing activity in the food industry and has kept many researchers busy since probably the 1980s. Most of the resulting solutions were either not good tasting (too bitter or too sour) or too expensive or both. Nevertheless, one can find such products in supermarkets these days.
- Low-, reduced- or no-sugar solutions to beverages and sweet snacks (confectionery) are an even greater prize in the portfolio of the beverage and food industry, and many products are offered that rely on artificial high-intensity sweeteners (e.g., aspartame) or more recently natural not so high but still high-intensity sweeteners (e.g. Stevia).
- The approach of molecules that can modulate taste receptors and thereby enhance sweetness perception was briefly discussed.
- The topic of "fake insulin response" to the sweetness perception through high-intensity sweeteners was briefly mentioned, without really conclusive evidence.
- Finally the role of various types of fatty acids and whether one should reduce the ones (e.g., it is commonly suggested to reduce the intake of saturated fatty acids) or increase the others (e.g., it is suggested that increased intake of polyunsaturated fatty acids is good for your health).
- Deriving from all previous topics I suggested a number of tips for healthy eating, such as having a balanced energy level between food intake and expenditure through physical and mental activity (don't forget that the brain is said to use up to 20% of your body's energy), using short-term fasting and ketosis as a cleansing and weight reducing habit, and eating a nutritionally differentiated portfolio of foods.
- Figure 8.2 depicts suggested frequency and importance consumption of the different food components.
- Some space was given to the discussion of nutrition controversies and they were specifically analyzed at the example of the so-called food pyramid. It was concluded that not only are there probably too many well-meaning recommendations, but some, or even many of them, are contradicting each other.
- Cholesterol and food intake was briefly mentioned and a few beliefs in the context of healthy eating were discussed and analyzed, such as the role of eggs in the diet, gluten, high fat intake, role of sugar, and specifically, the consumption of fish and fish oil.
- Finally the chapter ended with a detailed discussion on product health benefits, how they would be claimed, and the role of regulatory organizations such as the European Food Safety Authority or the FDA.
- The case of prominent and disappointing refusal of all health claims in the context of probiotics was discussed and analyzed in much detail.

References

Australian Ministry of Health. (2014). Summary. Available from: https://www.nrv.gov.au/chronic-disease/summary [Accessed November 11, 2016].

De Wardener, H. E., & MacGregor, G. A. (2002). Sodium and blood pressure. *Current Opinion in Cardiology.* 17 (4), 360–67.

Eat the Seasons. (n.d.). "Kale." Available from: http://www.eattheseasons.com/Articles/kale.php [Accessed June 4, 2017].

European Food Safety Authority (EFSA). (n.d.). Nutrition and health claims. Available from: https://www.efsa.europa.eu/en/topics/topic/nutrition [Accessed on November 21, 2016].

German, J. B.,. Smilowitz, J. T., & and Zivkovica, A. M. (2006). Lipoproteins: When size really matters. *Current Opinion in Colloid and Interface Science.* 11 (2–3): 171–83.

Harvard T. H. Chan School of Public Health. (2011). Obesity has doubled since 1980, major global analysis of risk factors reveals. Available from: https://www.hsph.harvard.edu/news/press-releases/worldwide-obesity/ [Accessed November 12, 2016].

Hickman, M. (2009). Health claim of probiotics not accepted. Available from: http://www.independent.co.uk/life-style/health-and-families/health-news/health-claim-of-probiotics-not-accepted-1796375.html [Accessed November 21, 2016].

"Ketosis." (n.d.). Available from: https://en.wikipedia.org/wiki/Ketosis [Accessed June 4, 2017].

Moyer, M. W. (2011). It's time to end the war on salt. Available from: https://www.scientificamerican.com/article/its-time-to-end-the-war-on-salt/ [Accessed November 15, 2016].

Mukohpadhyay, R. (2012). George and Mildred Burr upended the notion that fats only contributed calories in the diet. *Annual Review of Microbiology.* Available from: http://www.asbmb.org/asbmbtoday/asbmbtoday_article.aspx?id=18162 [Accessed November 13, 2016].

Nordqvist, J. (2013). Artificial sweeteners affect metabolism and insulin levels. Available from: http://www.medicalnewstoday.com/articles/261179.php [Accessed on November 17, 2016].

Nutrition Breakthroughs. (n.d.). A history of nutrition. Available from: http://www.nutritionbreakthroughs.com/html/a_history_of_nutrition.html [Accessed November 13, 2016].

"Olestra." (n.d.). Available from: https://en.wikipedia.org/wiki/Olestra [Accessed December 3, 2016].

Senomyx. (n.d.). Available from: http://www.senomyx.com [Accessed on November 17, 2016].

Siri-Tarino, P. W., Sun, Q., Hu, F. B., & Krauss, R. (2010). Meta-analysis of prospective cohort studies evaluating the association of saturated fat with cardiovascular disease. *American Journal of Clinical Nutrition.* 91 (3), 535–46.

State of Obesity. (n.d.). Obesity rates and trends overview. Available from: http://stateofobesity.org/obesity-rates-trends-overview/ [Accessed June 4, 2017].

US Food and Drug Administration. (n.d.). Dietary Reference Intakes: Macronutrients. Available from: https://fnic.nal.usda.gov/sites/fnic.nal.usda.gov/files/uploads/macronutrients.pdf [Accessed November 11, 2016].

The Wall Street Journal. (2016). The next hot trends in foods. Available from: http://www.wsj.com/articles/the-next-hot-trends-in-food-1476670682 [Accessed November 10, 2016].

Watson, E. (2013). Is the "there is no such thing as bad foods, only bad diets" argument helpful? Available from: http://www.foodnavigator-usa.com/R-D/Is-the-there-is-no-such-thing-as-bad-foods-only-bad-diets-argument-helpful [Accessed November 12, 2016].

Watzke, H. (2010). The brain in your gut. TED talk 2010. Available from: https://www.ted.com/speakers/heribert_watzke [Accessed June 4, 2017].

Part 3

The New Food World

9

A Food Company Transforms Itself

Gastronomers of the year 1825, who find satiety in the lap of abundance, and dream of some newly-made dishes, you will not enjoy the discoveries which science has in store for the year 1900, such as foods drawn from the mineral kingdom, liqueurs produced by the pressure of a hundred atmospheres; you will never see the importations which travelers yet unborn will bring to you from that half of the globe which has still to be discovered or explored. How I pity you!

—Jean-Anthelme Brillat-Savarin (1755–1826)

9.1 The Not-So-New Realities

In a previous chapter of this book and in chapter 10 of my previous book, "*Food Industry R&D: A New Approach*, I described possible scenarios for the future of the food industry, as well in part, the agriculture industry, with the main message being that the industry will most likely convert from making and selling food products to increasingly selling know-how. Let me briefly reiterate the major elements.

A new company business model 2.0 comprised of these three cornerstones:

- R&D takes the role of driver of the company.
- Open innovation and innovation partnerships are the new way forward.
- Retailers become real partners of the food industry as well as the agriculture industry.

Figure 9.1 depicts this in a simple overview.

In *Food Industry R&D*. I suggested a company structure that has at its core the "Office of R&D Vision." You can argue about the title of this office; you may even want to argue whether or not this is feasible and whether the industry can afford such rather radical transformation. It is my firm belief that it's not a question of whether this will be possible but that it will be necessary.

Megatrends in Food and Agriculture: Technology, Water Use and Nutrition, First Edition.
Helmut Traitler, Michel Dubois, Keith Heikes, Vincent Pétiard and David Zilberman.

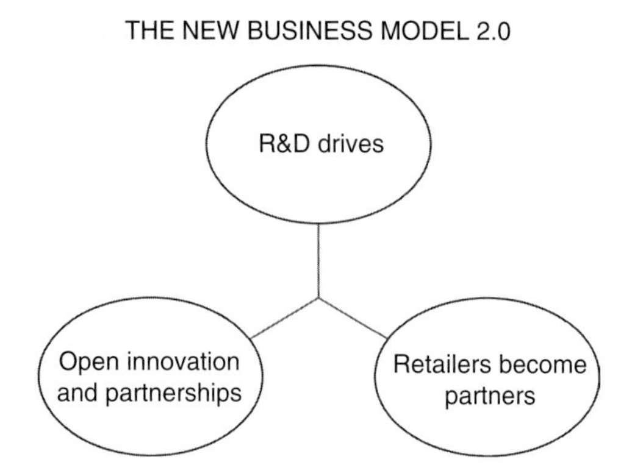

Figure 9.1 The new business model 2.0. Traitler, Coleman, & Burbidge, *Food Industry R&D: A New Approach* (Chichester: Wiley Blackwell, 2016).

Both industries, food and agriculture are still too much encrusted in the old ways of their 1,000-year-old business model of demase, and every change that may be imposed on these industries may be difficult to accept. On the one hand, this is probably a good thing because it ensures continuity and sustainability of the industry as a whole; on the other hand, aversion to change can substantially hinder the representatives of this industry to embrace necessary changes.

I can see that industry, especially in the area of food, are slowly getting ready for change, whether it's the one that I described in my former book, in Chapter 6 of this book, or it's another one that no one can see yet. Change will happen though, and I will discuss this at the end of this chapter.

Let me quote a short paragraph from an interview that the outgoing CEO of Nestlé, Paul Bulcke (2016) gave recently

> We act out of conviction, not out of convenience. We touch billions of lives every day. While everything is constantly changing around us, with consumer expectations evolving rapidly and competition becoming more intense, our purpose and our values provide a reference that defines why the world is a better place with Nestlé: for individuals and families who enjoy our products, for farmers with whom we cooperate, for communities in which we operate and for the environment on which we all depend.

Even if this may sound in part a bit like a commercial for the company, I really think that the more important part of this quote is the one when he talks about how many people and groups a company like Nestlé, similar to all the other big and smaller players in food and agriculture touch upon. So, every change, as subtle as it may be, is potentially of not negligible consequences to many people in this world and to many communities. But he sees the need for change, too. Although at the pace that it might happen, it could take many more years, if not decades, until something noticeably like I suggest in business model 2.0 may happen even to a small degree. Again, averting risks is the nature of these two large industries. The sad reality may however be that this or any other dramatic change is going to be imposed on the industries and that may not be a healthy situation or outcome for anyone. As you can see every day in your own environment, there is a really large revolution happening, which at its core is replacing human manual labor with robots. Some say that it's good because it will free up people to do "more valuable" things, and others see this as the doomsday for humanity. Reality, as always, is probably somewhere in the middle.

If Foxconn announces that over time it will replace up to 60,000 workers in their factories by robots that can do the assembly of parts to the finished products such as smartphones, it's rather frightening news. On the other hand, it's happening for quite some time in the automotive industry and food industries, in principle in every industry that you can think of. We just didn't realize it coming so fast and or so efficiently.

9.1.1 Automation Is Here...For Quite Some Time Actually

I do not want to enter into a debate about the value of human labor, the need for people to have a meaningful occupation, or the right to work as written in many constitutions of many countries around the globe. Neither do I intend to defend or condemn the increasingly fast introduction of robots of all kinds into agriculture and food; however, let me discuss a few of these new, or rather not-so-new realities that we see happening around us every day. Mechanization in both industries, agriculture and food, has been an ever ongoing process for many years, or rather decades and even centuries. The invention of the fridge and thereby the possibility to apply cold temperatures in processing dates back to the early 1800s when Thomas Moore patented his cooling vessel in 1803.

The invention of devices generating temperatures well below freezing allowed for new processes in the industries such as freeze-drying of soluble coffee or "field freezing" of freshly harvested green peas. Ice cream and its safe industrial distribution was and is of course only possible because such technology exists and is applied in the industry.

And there are many more well-known examples that have existed for a very long time and that have paved the way for a new dimension in automation of the food and agriculture industries. Harvesting techniques today are largely automated and have, with few exceptions, replaced manual harvesting. Even the coffee industry, wherever topographically possible, has turned to mechanical harvesting in flat and easy to work on fields of rows of coffee trees. Among other players in this field, the University of Manoa in Honolulu,

Hawaii, has developed coffee-harvesting machines based on blueberry picking machines (large vibrating combs or rods) that, depending on the type of coffee bean, can be very efficient in specifically harvesting ripe (red or yellow for yellow varieties such as "Yellow Caturra" or "Yellow Catuai") coffee beans. Machine harvesting in many instances can even be more efficient than hand picking.

Another branch of the industry that is also in transition from manual to mechanical harvesting is the wine industry. There are regulations such as limitation of output per vine and the subsequently expected quality improvement, which will continue to ensure some manual labor in this industry. However, the switch is happening and will continue to happen. In California, mechanical harvesting of grapes grew to an average of 50% of all grape harvests in 2008 with some areas such as Napa or Paso Robles still well below this number. In 2012 it was reported that wine grown in the state of New York was 90% mechanically harvested and, similar to coffee, it is suggested that mechanically harvested grapes may be of better quality (Goldfarb, 2008; Adams, 2012).

These are just two well-known examples of what happens in the industry already in areas of "orphan crops" or smaller crops as compared to the really large ones such as corn, soya, wheat, or rice. Automation is the name of the game even for crops where the finished product may cost $30 or more for a bottle of wine or $20 or up to double of that price for a kilogram of coffee beans.

There is a strong movement in Europe as well as North America to buy products that are either locally grown (not all can be locally grown, coffee is a good example) or locally made. We began to despise "Made in xxx" as soon as this "xxx" did not represent our country. My family and I lived in several countries over the years and still do and when we lived in the United Kingdom, we had to buy "British beef" because it was the only one you were made believe to be safe; in Switzerland it had to be Swiss meat, "then you know what you get," and so on and so on. It's rather confusing and rather meaningless too and yet it is seemingly imprinted into the heads and minds of consumers.

In my eyes there is one argument and one argument only to buy locally bred or grown food products: avoid long transportation routes and thereby minimize, or at least reduce, energy use. The rest, in my eyes, is just make believe. I am not arguing that there may be and must be quality differences in food products for which the raw materials and ingredients are grown in different places and the consumer can pick those up, at least sometimes. The other side of the coin is locally manufactured (e.g., "Made in USA" or "Made in Germany" and so on). There is an even stronger movement these days to "bring manufacturing jobs back." There was an interesting piece by Kornberg (2016) describing local jobs, local to Los Angeles that is, in the garment industry. Its title was, In L.A.'s garment industry, 'Made in the USA' can mean being paid $3 an hour.

Again, I don't want to get into an argument with readers who may believe that it's better to have any job, even a $3 per hour one (after all there are the infamous €1 jobs in Germany) than being out of a job at all. However, I strongly believe that this can't be it either. Reminds me of some players in the Italian leather industry who brought Chinese workers into the country, paid them Chinese wages, and were then able to put a "Made in Italy" label into their merchandise. To the best of my knowledge that is not happening any longer, but then I might just be wrong.

My whole point here is to say that beware what you wish for, and sometimes it's better to have products that are not locally grown or manufactured and give a real livelihood and sustainable work to farmers and manufacturers elsewhere than the false belief that everything is perfect at home provided there is a "Made in my country" label on the product.

9.1.2 The Novel Directions in Food and Agriculture are Governed by Regulatory Involvement

If you haven't realized it yet, and I don't mean this in any way disrespectful, then let me say it again: regulatory authorities call the shots these days as to which directions the agriculture and food industries will be heading toward in the months and years to come. They are backed up by the fear of insufficient food safety in case they don't do their job properly, and they come out with new rules and regulations every so often. I am not belittling the importance of food safety–related regulations and strongly believe that the food industry has a great responsibility in making sure that their products are safe; much as the agriculture industry has the responsibility to ensure food security, or in other words, making sure that there is always enough agricultural output to feed the world population in all regions of our planet. I admit that this is a tall order, but it needs to be fulfilled. Hence some changes are necessary and will inevitably come.

In this context the regulatory authorities will continue to put pressure on the industry, especially the food industry, to follow to the letter every guideline they may come up and which may be different in different countries. The maximum allowed iodine level in infant formula may not be the same in China and in Germany, and food companies have to become smarter realizing this. And this is just a trivial example; much more complex ones exist and will become more important with the ongoing improvement of analytical techniques, which allow for measuring just about everything at every amount, even the tiniest ones. When consumers read about the presence of a toxic substance in a food product, even if it was just reported at femtogram level, they will not make the distinction and accept that this is really a tiny, certainly negligible amount, they just hear: "toxic substance xyz found in product abc."

The point I am making here that regulatory authorities have an important responsibility to remain reasonable and measured in their directives and statements. This should not be taken as an easy pass for the industry, but the assurance that at such levels it is not necessary to even mention the presence of such ingredients that may be or are toxic at higher levels. When I was a chemistry student in my second year, we were taught about the "omnipresence" of all elements. If I had the right technology, I could probably detect just about every element of the periodic table, for instance, in water, even down to the molecular or atomic level.

If the two industries really want to "feed the world," they need to adopt a few changes. They are both financial as well as technological. On the financial side, there has to be an end to speculating and trading with the large crops such as corn or wheat. I do realize that this may sound a bit radical, but such trading only helps the traders and investors but does not help those who are in need of these agricultural products. Stopping trade will not solve everything, far from it; but it may be an important element in the quest for food security and especially affordability of food products. Regulatory authorities could play an important role in this; they could have an important control and oversight function, which would go well beyond what they do today. This suggestion of course blows right in the face of "free market," and the archaic belief that the market regulates everything. May as well believe in Santa Claus!

The other changes have to be technologically based: new breeding technologies for plants, much of which was already discussed, new growing techniques, novel harvesting and online (in-field) manufacturing techniques, decentralized manufacturing, and probably a few more that we cannot even think of yet. I shall discuss and analyze this topic in more detail in a later section of this chapter.

9.1.3 All-Natural Industrial Food Products: The Way Forward?

For a few years the food industry has undertaken many efforts to develop and manufacture food products with the majority of ingredients being claimed to be "all natural." Although I can confirm, based on many years of personal experience and involvement in such efforts, that a lot has been achieved, much still needs to be done. The new reality here is linked to a few simple questions: Do consumers honor the efforts by the food industry? Do they accept the idea that industrial food can be natural? Isn't it a catch-up game played between the food industry and consumers and in which the industry is always too late?

It could well be that many consumers reject the idea of all-natural industrially manufactured food altogether, and the industry's efforts are actually in vain and "much ado about something" that cannot be resolved in this manner. It's not a trivial dilemma because it goes to the heart of the existence of the food industry as we know it today: can the industry continue to exist and thrive in the eye of consumers' demands for natural and ideally industrially unprocessed food? It is well understood that all cooked or otherwise treated food, for instance, through fermentation, drying, smoking, or salting is ultimately processed food. The idea of some big Moloch doing this instead of myself is at the heart of the increasing rejection of industrial food products.

It's happening in other areas too. At my most recent visit to my hairdresser for a haircut, Kathy tells me about her new approach to using hair and salon products both for sale as well as for direct use in her shop. That's what she wrote on her booking site in November 2016:

> Shop Local / 20% OFF
>
> Dear customer,
>
> This year I pledge to shop local. I will be supporting small, independent local businesses, not the big box retailers.
>
> I am offering 20% off all gift certificates sold between now and December 10th. Gift certificates are easily emailed to you directly. You can call or text me at 808-xxx-xxxx or I can be reached by email, bookwithkathy@xxxxx.xxx.
>
> Happy Holidays!
>
> My love to all...Kathy
>
> (Kathy Perry, November 2016, personal communication)

And she forcefully added to this that many of her colleagues and competitors in this industry are doing the same things: shop local, insist on preservative-free products, and support smaller players. I do admit that this is a small beginning; however, a beginning it is, and it remains to be seen how far this can go and how successful this will eventually become. I argued that even the small ones have one big objective and it's "growth," or if you want, sustainable and profitable growth. So, even the small ones will eventually be bigger and may end up very big ones. Just read the story of L'Oréal or other big players in that industry.

The same holds true for the food industry: all or almost all of the big players started small and almost in insignificant ways. I wrote "almost all" because some of the big players have reached their present size through many acquisitions. So, Kathy's support for the small players is on the one hand an exciting and commendable approach, and on the other hand, she has to realize that she always has to be on the lookout for the "new kid on the block," small enough to fulfill her criteria of local, preservative free, and as natural as possible.

9.2 From Product to Know-How Seller: An Encore

I have discussed this topic several times already, but I bring this up again here because I truly believe if food companies don't complement their offerings (i.e. selling products with sustainably and profitably selling know-how to consumers, with the help and collaboration of retailers, both brick and mortar as well as online), they will not be successful in the future.

The situation is slightly different in agriculture because, for all it's worth, farmers get a lot of support directly from the seed and fertilizer companies. This is part of the product offering and complements their range of products and services. Even a food company like Nestlé has an "agronomic service" and works with more than 1,000 experts around the world who help local farmers grow crops relevant to the company's raw material portfolio in the best possible ways, sustainably and profitably. Nestlé's interest is clearly in being able to purchase the best possible agricultural raw materials from such farmers.

It's part of the deal and has a historic foundation. In the 1970s, the company created and organized a milk collection system and network in rural Ecuador. This was done to ensure that the milk got to the milk factory as fresh and as quickly as possible, thereby avoiding spoilage, bad quality. Yes, of course it was in the best interest of the company, but it was also in the best interest of the farmers and ultimately of the consumers. If a food company can work with farmers and sell them advice and knowledge about best practices and help them to grow crops in sustainable and profitable ways, why don't they approach consumers of their food products in the same ways? If they can work with and ultimately pay more than 1,000 agronomists (1,300 was the number I last heard!), why couldn't they use part of their experts in nutrition and health through food or use all their experts part of their time to meet consumers, to speak to consumers, to consult consumers, and to explain nutrition and good food and the virtues and values of healthy food and beverages to consumers?

It does not make sense to let marketing people, of all, go and sell something to consumers of which they have no clue, or very little! So why not let the experts go and do the job? By experts I mean those who helped create the products, the R&D and technical community, especially the nutritionists and food scientists. Of course they need to learn to speak in simpler terms than the alphabet soups they tend to use a lot. Yes, it's not an easy transition but a dearly needed and necessary one that will help the food industry get into a new era, an era that is coming, whether they participate or not. Some will do, others will follow and ultimately the current will pull everyone in this direction.

9.2.1 Some Assumptions as to How This May Function

In *Food Industry R&D*, I made some assumptions as to possible financial platforms and how such an approach of selling know-how to consumers may help a food company earn additional income. Let me briefly repeat some of this hereunder. Please note that models 2.0 and 2.1 were more extensively described and discussed in Chapter 6. The major criteria are as follows.

- Model 2.0 suggests that food industry and retailers team up in novel ways, R&D drives the food company, and open innovation and partnerships are a complementary basis for innovation.
- Model 2.1, as an extension of model 2.0, suggests the all-important transition of becoming a product and know-how company.

Some Calculations, Just Examples

Estimates of investments versus expected overall returns for the various players in this model 2.1 are the most crucial element in this entire equation and it will not be an easy task. Although it is fairly easy to calculate the investments—or costs, as some may want to call it—that any given food company would have to bear by simply determining how many experts would use which percentage of their productive time to cover a given, desirable size of consumer segment in any given market, maybe one or two to begin with. It goes without saying that one would kick model 2.1 off in carefully crafted first steps and with a limited time investment from required experts. Let me give a simple example: if a globally active food company would start in a place like Singapore it would have a wonderfully compact, rather homogenous and affluent test market with a well-defined population mix as well as mix of retailers. If the same food company would collaborate with two different supermarket chains in for instance overall 50 of their local outlets and would use a total pool of 60 experts at one-third of their time to be present in the different locations to interact with consumers and this on fixed days and as an example three times a week, the total investment would be 20 full-time equivalents (FTEs) at an average of US$ 350,000 per FTE (i.e. $7 million) per year. If the same company would use another third of the experts' time to participate in chats with consumers, we would add another $7 million for a total of $14 million.

The Company Can Earn More with Model 2.1!

If the food company makes an estimated annual revenue of $1 billion and an annual profit of for instance $140 million the additional investment into the experts' time amounts to 10% of profit. Wow, we cannot accept this, many of you would say. My prediction, however is that additional profits that these experts can generate for the company by far outweigh the additional costs, maybe not in year one but, like for most new products in that industry at least at the end of year 3 of the inception of such a model. Out of Singapore's total population of approx. 5.6 million (numbers for 2015) it would only require for instance 3.5% or 200,000 who would become members of such a "chat / interact with the food experts" program, paying for instance $100 per year; this would give them total access to nutritional and food related advice, which potentially can be offset by a tax relief or an insurer contribution (very much like the insurer pays for part of gym membership costs). This would mean a zero additional cost for consumers in such a program based on model 2.1. The expected additional income for the company in this example would amount to $20 million leading to an additional profit of approximately $6 million (based on the estimated costs of $14 million); not bad and simply an improvement of the margin by some 4.3% or 430 basis points. In my book this looks quite a dramatic improvement, achieved by a simple and very careful evolutionary change of the existing business model, without dismantling anything of the old model (yet)! The best part of this is that it might not even be necessary to hire and additional 40 FTEs (remember: the example called for two-thirds of the work hours of 60 people). To begin with, existing R&D resources should be used more efficiently by re-dimensioning ongoing R&D programs and by increasingly using open innovation and innovation partnerships. The company's experts would undergo a tremendously efficient in the field training by speaking directly to consumers—either in person or through Internet chats—and over time would become so much more valuable for the company. There are only advantages to such a business model 2.1. The only difficulty still remains: the "old brains," which know it all, did it all, and are very resistant to change, any change that is. I do hope that these arguments and examples can serve as a solid basis for further in-depth debates in many food companies that see their own future in a different light than the one that is shining today. At least, the arguments should be a basis for critical discussions and refinement of numbers in real-life scenarios. You may come up with different numbers, maybe even better ones, who knows? It is definitely worth a try, and business model 2.1 may become a much bigger reality than today's company executives have ever dared to dream of.

Figure 9.2 illustrates the suggested business model 2.1 in a simplified overview.

Again, depending on which side you are on, you may either praise or ridicule these suggestions and assumptions. If you ridicule you may better think twice, you may just remain on a sinking ship, while the forward-looking companies and their executives will embrace such a model, maybe with slightly different twist and calculations, yet embrace they will.

9.2.2 What are Possible Consequences for Food Ingredient Suppliers?

Food-ingredient suppliers have traditionally played important roles in their collaborations and partnerships with the food companies. It is safe to say that

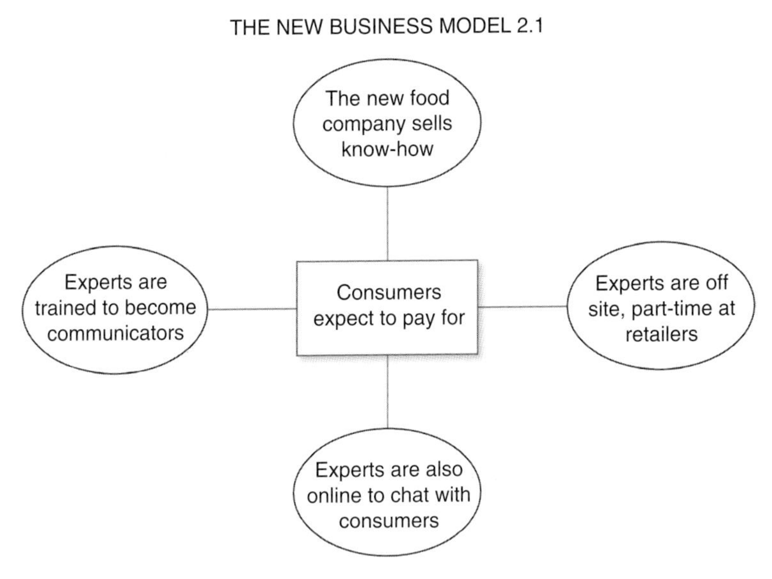

Figure 9.2 Business model 2.1. Traitler, Coleman, & Burbidge, *Food Industry R&D: A New Approach* (Chichester: Wiley Blackwell, 2016).

whatever happens to a food company and in which direction the food company may be headed has important consequences for the food-ingredient supplier industry.

Let me just paint one simple scenario in the area of aromas and flavors. In spring 2013 I wrote in a personal communication about some desirable and expected changes in the flavor industry. Although the food industry still buys flavors in form of liquid or powder concentrates, with the current drive to shorter ingredients lists and the need for all-natural ingredients, the flavor industry might want to add competences around in-process flavor generation based on proprietary knowledge of optimal precursor mixes and process conditions.

I do realize that this might be quite a sea change for the industry and neither easy to accept nor to become this in-process flavor generation expert. Interestingly enough, the Nestlé Company had in their portfolio of subsidiaries exactly such a company in the 1980s and 1990s before they sold it to Givaudan. It was the FIS Company (food ingredients specialties), and their expertise originated in-process flavor generation. Process optimization (i.e., cost cutting), however, got the better of it, and in addition to the fact that FIS products were almost exclusively sold and used within the Nestlé world of companies the trend in food manufacturing headed toward using individual flavors or bespoke flavor mixes simply added to the raw material mix before final manufacturing.

As a consequence, all flavor ingredients had to be listed and ingredients lists on label became longer and longer and almost reached the length of the Chinese Wall, sometimes so long that fitting them on a product label meant miniaturizing them and basically rendering them unreadable to the normal eye. As a further consequence, consumers became more and more confused with such long ingredients lists, and they also became suspicious. Consumers for once don't ask for more but rather for less. They want fewer ingredients in food and beverage products, more natural ones or even better, no ingredients at all, as if this would ever be possible. So, a company like Nestlé, who had access to know-how as to how flavors and taste profiles could be generated in the process of making a specific food product lost this know-how when they divested this activity, in the name of efficiency increase and cost optimization. Hence, they had to buy, like every other competitor, flavor concentrates—mostly liquid, some solid powders, or encapsulated flavors and aromas—from one or more of the large or not-so-large flavor houses. Again, the same Givaudan Company who had purchased the flavor activities from Nestlé (and who has slowly, over time phased it out in favor of developing and selling flavor additives) was and is one of the flavor and aroma suppliers.

This is where you the reader should ask yourself the question: why was this done? I gave two possible answers: efficiency increase and cost optimization. However, I don't think that these were the real drivers. When this happened in the early 2000s, this was just a fashion: companies were supposed to stick to their core competences, and flavor and especially flavor generation did not seem to have made the list, at least in the case of Nestlé. This was and is a shame, at least in the eyes of many, and even some in the flavor industry begin to rethink their traditional position of simply selling flavor concentrates to the food industry. Many flavor houses pride themselves on being able to offer the most authentic and sustainable vanilla flavor concentrate, with full knowledge of the entire product chain from harvesting at the exact known spot in Madagascar down to the filling in smaller or larger bottles to be shipped to the user company. I do agree that vanilla is an example that requires the "original," and a liquid concentrate is probably the only avenue to pursue when producing vanilla ice cream. Or is it not? There might be other, process embedded pathways to achieve vanilla flavor, maybe not in ice cream but in products for which heating is required.

The entire field of savory flavors, in other words flavors that one finds in nonsweet food products such as soups, meat dishes, pizzas, pasta, and the like, could potentially be generated in the cooking process by the smart selection of appropriate precursor ingredients. It is a similar approach to creating flavors through smart selection of enzymes and letting fermentation work its miracles. I do not pretend that by doing so the number of listed ingredients on the label of a food package will ever go to zero, but it would definitely be substantially reduced. And this would help to regain consumers' confidence and make industrial food products look less industrial and processed again. It's closer to home cooking, just on a bigger scale and using grandma's tricks of the trade or other pertinent know-how. It definitely would make food that was considered processed more human and homely again, and that is something that is desperately needed.

Having described the two scenarios, vanilla and savory flavors, nicely illustrates that there is space and need for the two approaches, flavors from plant sources as well as flavors generated in the process of making the food product. In my opinion this clearly means that both ends, the flavor industry and food industry and with them the agriculture industry have to change and adapt to the new needs and new consumer expectations of shorter ingredients lists on food and beverage labels. The consumer understands the need for an added flavor such as vanilla because there is most likely no good and acceptable process pathway that would replace the plant source. In other words when it literally comes natural, it will be understood and accepted.

However, if a flavor is added which could be created in the cooking process such as for instance caramel or grilled or cooked meat flavor, the consumers will have a more critical eye on this and will increasingly revert to products that create such flavors rather than add them.

The flavor industry has to accept this dualism of approach and needs to adapt to this by becoming the go-to address when it comes to know-how concerning in-process flavor creation. The ink industry and its specialists play such an on-site expert role in the packaging conversion and printing industry. The ink know-how resides with the ink company and the printers make good use of their presence in their factories. It's otherwise called a "hole-through-the-wall" (HTTW) situation, and simply means that despite intensive collaboration on site, each of the partners keeps their inherent know-how as a manufacturing secret. I can see the same happening in the food industry by teaming up with the flavor industry in such a HTTW approach: the food industry knows all about their product, the flavor industry knows all about precursors, their smart selection and especially how to "cook" the product so that it achieves its final and most desired and preferred taste profile.

I have a last rather historic example from the chocolate industry, the traditional versus modern crumb making. Chocolate crumb is a premix of some of the chocolate ingredients, such as sugar, some cocoa liquor, and milk, and I am talking about milk chocolate here, to help preserve the milk ingredients during storage of ingredients to be able to better plan the chocolate manufacture. Until the mid- to late 1980s, the Cailler chocolate makers of Broc in Switzerland had a traditional way of making their crumb to give it the distinct and much-loved flavor to the final fine Cailler chocolate. Without wanting to get into too many, even old and outdated manufacturing secrets, I can say that the crumb making was time consuming. So, there was "dead capital" sitting around for far too long, said the financial guys.

I let you guess what happened over time. The Cailler brand, which was perceived premium—and still is to this day—lost its luster and taste profiles, which were specific and inherent of the Cailler chocolate were becoming increasingly "mainstream," just like most chocolates. Most chocolates, by the way, received their base chocolate mass from one global supplier, obviously for cost-improvement reasons, and so most of the chocolate brands you find in the marketplace today have similar, generic taste profiles.

Interestingly enough, for what it's worth and whether you like the taste of it or not, Hershey's chocolate kept its old-fashioned, acquired taste profile from back in the olden days and is doing pretty well in the United States. They are, by a large margin, number one in the United States, and this is quite a sizeable market.

This is, by the way a recurring pattern: many food or beverage products seem to do better if they stick to the more traditional recipes. The original Coca Cola beverage is one prominent example, and so is Red Bull®, even if the latter is still in its infancy compared to the big red one. Yes, these companies continuously try out new formulas or new variants, and some of them are pretty successful such as for instance Diet Coke® or other variations of popular beverage brands such as for instance Rivella® in Switzerland.

None of these companies would dare to meddle with their flagship products in ways that the chocolate industry has meddled with theirs. This is what you get when you let the financial and marketing people in your company decide on the details and delicacies of major brands. Once things don't go so well any more, after the "optimization" has been undertaken, fingers are quickly pointed towards the "technical guys" who have screwed up. This sounds like getting rid of frustrations, and I do admit there is some truth to it. However, facts are not so easy to discuss away, and traditional products such as chocolate, ice cream, dairy products, and even coffee require traditional solutions, yet not traditional approaches and means. This is an important lesson for the food industry: get rid of the cost optimizers and trust those who can make a real difference, those who dare and bring new ideas to the game, those who are creative and can innovate in areas such as product development but also process development and food-related know-how gathering and selling such know-how profitably.

9.3 Anticipating the Inevitable: Possible Scenarios

If we could only understand the writings on the wall and not only read them! More importantly, if we could only act on them on the basis of what we have read and understood! So, repeating what has been said time again in this book and by many more authors, politicians and journalists, these are some of the more prominent and important writings on the wall.

- Barring any major catastrophes of more global nature, the world population will increase to grow, faster and faster.
- Arable land is not increasing at the same pace as world population is growing.
- This will inevitably lead to reduced and disproportionate food security (i.e., the availability of enough nutritious, safe, and affordable food for all).
- Consumers are increasingly becoming skeptical with and almost hostile to industrial food products, "processed" food in their eyes.
- Consequently, more and more consumers will, whenever possibly revert back to home-cooked, some even home grown.
- Diet diversity mania is likely to continue, maybe even grow as more people will find even more outlandish diets and recommendations because more and more consumers will find new discomforting situations when it comes to food and beverage intake.
- Collision points of all these different directions and interests are inevitably increasing, leading to more and more confusion and ultimately leading to the refusal of many of the fashions, fads, and nonsustainable trends.
- It is safe to assume that both industries, given their combined sizes and inertias, will attempt to wipe away these writings on the wall and pretend that everything is just fine if society would only let them do what is best for them (the industries that is).
- Many companies, especially the smaller ones will change more easily and rapidly and will lead the way toward this new reality of simplicity, honesty, naturalness, affordability when it comes to agricultural raw materials as well as industrial food and beverage products.

Table 9.1 represents these topics in a concise overview.

So, what are the ways out, what possible scenarios that together with what was already proposed go even further could we think of?

Table 9.1 Scenarios of relevance.

Scenarios of Relevance to Agriculture and Food Industries
Population growth continues
No increase of size of arable land
Unbalanced food security
Increasing hostility toward industrial (i.e., processed) food
Back to home-cooked foods
Diet mania and personal food differentiation continues
Consumers become increasingly confused with food and nutrition
Large corporations continue to reject necessary change
Smaller agriculture and food companies will lead the way to change

Let me briefly repeat what was proposed, discussed and analyzed thus far.

- Agriculture increasingly applies modern breeding, growing, and plant-protecting technologies that are in sync with nature and respect soils, the environment, animals, and people.
- Agriculture finds more ways to improve growth efficiency and yield per surface, especially but not exclusively including urban agriculture approaches.
- The industry of agriculture really responds to the needs of the consumers and does not continue to operate in the old style of contingents and planned economy as we had seen in the old world of communist countries: make a plan for 5 years and just produce, whether the resulting product or item will be needed or not and not attending to the real needs. *Note: I do realize that this may sound a bit strong but from many years of personal observations and experience in the area, I have seen much of this approach.*
- Food and beverage companies, after some 1,000 years of the same business model of demase have to finally come up with something new, more exciting, fresher, and closer to both the agricultural raw materials as well as the world of retailers (not necessarily brick-and-mortar) and ultimately consumers. The industry has to gain back the trust of the consumers, even of society at large.
- This won't succeed by just continuing the old ways and pretending that the umpteenth flavor variant and the additional color on the package combined with a new graphic font will do the trick. It won't! There are oodles of examples in the industry when relaunches based on these criteria have miserably failed; I know this for a fact as I was involved in some of them.
- The food industry has to embrace the idea that working much closer (and I mean MUCH!) with retailers is a first important step into a new direction. So-called "branded" food companies have to accept this, without having to give up their own brand identities.
- This may lead to completely overhauling the entire value chain, especially the parts of manufacturing and distribution.
- Food companies have to be run by people—not lawyers, finance experts or marketing or sales experts—who understand food. They do function, better or worse, during good weather and when communities and consumers are, rather were, happy with what was proposed by the industry. This time is over, out, finished, The End. These people are food-savvy engineers and scientists with additional business education. R&D takes over! *Note: it can take as little as 2 years to acquire a degree of MBA; it takes at least 8 and more years to obtain a Ph.D. in any of the food-related areas.*
- The world of R&D, having more business responsibility, needs to rub shoulders with the world of retailers and consumers on an almost daily basis. Scientists and engineers have to escape from their labs, pilot halls, and factory floors and see what's going on outside their own perimeters.
- An important consequence of this is that R&D goes out of its own walls and sells know-how to retailers and consumers, in addition to selling products and other services.

- Consequently, selling food- and beverage-related nutrition and health benefits know-how will become an important income segment for the food industry. *Note: the agriculture industry already acts in similar ways, by dispatching experts, agronomists, to farmers in basically all areas of farming, animal or vegetal.*

Table 9.2 illustrates these suggested "softer," less disruptive pathways that may lead out of standstill or even decline

These were the topics of possible future actions and activities that the food industry at large may have to pursue, by anticipation and their free will or forced and coerced into action. Many companies in the world of food and beverages have discovered that there is a whole new world out there, which is based on the increasing digitalization around us. Especially in areas such as supply chain and manufacturing, acceptance of this digitalization, the so-called "Industrialization 4.0" is rather high, if not to say total.

Industrialization 4.0 describes

> the current trend of automation and data exchange in manufacturing technologies. It includes cyber-physical systems, the Internet of things and cloud computing.
>
> Industry 4.0 creates what has been called a "smart factory." Within the modular structured smart factories, cyber-physical systems monitor physical processes, create a virtual copy of the physical world and make decentralized decisions. Over the Internet of Things, cyber-physical systems communicate and cooperate with each other and with humans in real time, and via the Internet of Services, both internal and cross-organizational services are offered and used by participants of the value chain. ("Industry 4.0, n.d.)

Table 9.2 Possible ways out of impasse: Nondisruptive steps.

Agriculture rides the wave of change with consumers in mind
Agriculture becomes more efficient in sustainable ways
Agriculture responds to the real needs of all involved
The food industry overcomes the demase syndrome
Food evolves from renovation to real innovation
Closest collaboration with retailers helps pave the way
Value chain needs reassessment and overhauling
Good companies need to be run by those who understand food
R&D personnel go out into the world and meet real people
Selling food- and nutrition-related know-how is part of the equation

More recently, digital marketing has befallen many companies, including also food and beverage companies. It took a long time until, for instance. the Nestlé Company went this route and created a small group dealing with digital marketing in San Francisco, California. It's a step in the right direction, but seen by many insiders in the industry almost as a not productive undertaking, and they see this rather as a pastime. This attitude is still palpable but will have to change soon and new concepts of logistics, manufacturing, marketing, and especially, sales have to be developed and fully accepted by management.

9.3.1 Possible Future Models and Scenarios

Many readers may believe that the scenarios suggested and described are too far-reaching and almost too utopian and therefore not workable. I respect this opinion; however, I do not accept it. What has been described and discussed is by far not enough to find answers to consumers' questions and doubts, and especially, to find answers as to how society at large, communities and individuals, perceive and continue to accept the food industry. I strongly believe that more radical change has to come. Let me suggest, discuss, and analyze a few topics that come to mind. They are based on two assumptions and they follow their logic.

- First, globalization as we know it will slow down, if not come to a standstill and reverse toward more regionalization and even local levels.
- Second, industrial food (processed food as its called by many consumers, even if not entirely true) will decline in its importance and home-made (however this may be defined) will take a more important part in the food mix of consumers.

The reader may want to challenge both assumptions; however, signs of these happening can be seen all over the world, and the overarching assumption is that both suggested scenarios will continue to gain in importance. So, how should the food industry react to these seemingly disturbing suggested realities? Here a few thoughts, which attempt to answer to both of the assumptions.

- The R&D-driven and -run food company outsources manufacturing to brick-and-mortar locations of retailers, thereby reducing lengthy supply chains. The products will be manufactured under license and cater to more local tastes. I assume that due to increasing online purchases even in the area of food, retailers will have more empty space on their hands; they can either close those locations or alternatively keep them for manufacturing. Food companies will sell or close down their manufacturing facilities.
- As a consequence, the food company and their R&D community aggressively out-license their intellectual property to those who can better valorize it. This is, with very few exceptions, not in the genes of the food industry but is

a normal reflex for instance in the automotive industry. The food industry still believes that they have to continue to sit on their intellectual property and not let anyone come close. No pun intended, but in the future this will be a thing of the past.

- Manufacturing will become democratized. What I mean by this is that the consumers will increasingly turn again to preparing their own food. It is suggested that Millennials have more time on their hands and they will also use some of this extra time to cook, preferably together with friends or family.
- Kitchen robots will become "leasable" for the individual and thereby shoulder a large portion of formerly manufactured food products right at the consumer's kitchen.
- The 1,000-year-old approach of demase will make room for a new approach that I want to call "inapse" (innovate-apply-sell).
- Real innovation in the area of food is rather difficult to achieve; some would say impossible and therefore companies are satisfied with continuous improvement and renovation of products. Hence, the rise of "medical food," many times attempted to establish in the marketplace, might ultimately succeed. Real innovation is not only required here but may be afforded, given the expected higher sales prices and margins of such products. *Note: medical food has gone through multiple iterations of unsuccessful launches and relaunches. In my opinion, the main reasons for past failures are twofold: not enough patience on the part of management to wait for the success and not enough resources (people and money) put behind the development and marketing activities. An entirely new type of marketing is required.*
- To guarantee a successful launch of and sustainable future for medical food products the medical profession has to be fully involved. This in turn would also mean that physicians would receive additional training in food, nutrition and food, and health-related areas, which go beyond counting calories. This is not an easy feat to achieve because it takes a long time until such changes would finally make an impact. In the meantime, food scientists and medical doctors have to team up and work with what is available.

Table 9.3 illustrates these more disruptive, some would say utopian, steps forward.

9.3.2 The Return of Medical Food?

I have put a question mark behind the section title, simply because I am not sure what will happen in this segment of food products. Let me illustrate why I say this using the example of two companies, such as Novartis and Nestlé. Novartis had a line of medical nutrition and had created brands such as Boost®.

At the same time Nestlé was struggling to develop some meaningful medical and clinical products that consumers would not only understand and accept as such but more importantly also pay the price for. My former colleagues who were involved in this development will hopefully forgive me for writing all this, especially for my critical view on a development that was enthusiastically pursued by those who carried it out, and only half-heartedly supported by management, especially the middle management. These guys wanted to cover their backs in case the results were doubtful and wouldn't come fast enough. Timelines of development in this area of products, which ultimately would have to involve clinical studies were so out of sync with the typical food-related timelines that pressure grew and the food company decided to purchase the entire branch of clinical nutrition from the pharmaceutical company to get a head start in this area. This happened in July 2007, and the company created was "Nestlé Health Care Nutrition."

Without wanting to appear too critical, already the name reminded more of a health insurance company than functional, medical food. Much was said, written, and discussed in those years without much avail, and the old Boost is still in the marketplace and can be found in many flavor variations in C-stores for a price of around $7 for a six pack, the real six pack that is. $1.15 or so for a bottle of 8 ounces or 237 mL; not much of a premium. I know for a fact that in those years many attempts were made to push for developing medical foods worth not only the development but also a premium price. Again, the way forward was to create a new unit, the "Nestlé Health Science" (NHS), a few years later with its own research arm, the Nestlé Institute of Health Sciences (NIHS). Top management gave the latter an initial time frame of 10 years (sounds generous, doesn't it? Watch out for the catch!) to develop a series of medical food products and devices and service products that combine food and the medical world. While the NIHS was actively pursuing its mission to discover, the parent company NHS acquired relevant companies "left and right." I must assume that

Table 9.3 Possible ways out for agriculture and food industry: The more radical steps.

Assumptions	
1. Globalization will slow down	**2. Consumption of industrial food products will decrease**
Outsource manufacturing	
Aggressively outsource intellectual property and increase revenue	
Become part of the growing home-cooked trend	
Kitchen robots will become affordable and leasable	
From demase to inapse	
Innovation to develop efficient medical foods with claimable benefits	
Close collaboration with, and involvement of, medical profession	

there was a strategy behind all this, namely to enter into the world of medical and clinical products and services and use food and beverage products with clear-cut functionalities as the important link and connector, and all this at a faster pace than having to wait for 10 years; makes sense actually.

On their Web site, NHS (n.d.) writes the following:

> With advances in genomics and biotechnology, breakthroughs are being made in our understanding of the fundamental causes of disease and the potentiality of nutrition to better manage health. Through our external focus and research capabilities and network, we fuel our innovation pipeline of nutritional therapies.
>
> The role of nutrition is entering a new stage of discovery and innovation – not just in addressing nutritional gaps but with potentially direct therapeutic impacts – that is changing the management of health and forging an increasingly integral role in the management of health across the human life cycle.

When checking their brand portfolio, one can find 16 brands, one of which is a family of medical devices; many are drinks with functionalities, some are powders, again with specific health-related functionalities, and others are mixed formats such as powders, drinks, soups, and bars. This format mix is to be totally expected because the parent company has mastership in all techniques leading to these formats.

By the way, Boost is one of the products, still selling for approximately $7 for 6 bottles. I really do hope that all or most of the other products are so efficient and important to potential consumers, rather users or patients that a much higher price can be achieved for these.

Let me come back to the title line ending on a question mark. I still don't know whether medical foods and drinks will ever really exist as an important category in a food company's portfolio of products, unless the know-how selling part becomes an important accompanying element to the brand and thereby creates additional, non-negligible income for the company, any food and beverage company, that enters this field.

If not, those who always knew and always warned would unfortunately win, and the wheels would just continue to turn in the old demase mode, maybe for another 1,000 years or so.

9.4 Reality or Fiction? Reality and Fiction!

Let me end this chapter with some odd stuff, odd in the sense of really far out and maybe never going to happen. But then who would have thought that smartphones would ever exist back in 1980 or self-driving vehicles would eventually hit our streets even back in 1990? And there are many examples of unexpected step changes even in the area of agriculture and food. Most or rather all of them happened for two main reasons.

- First, societal changes took place caused by wars, pandemic events, or economic reasons.
- Second, out of need or serendipitously, important discoveries led to technological step changes.

There are again many examples in both, agriculture and food, such as new plan- breeding techniques discovered by Johann Gregor Mendel or the invention of modern cooling techniques, for instance the fridge, mostly attributable to Carl von Linde. There are dozens more examples, and I am certain that you have come across many of these yourself. So I won't go into this in more detail, just to say that surprises caused and stimulated by events described are most likely just around the corner.

Increased automation and robotization were already mentioned, and I could imagine that the large food companies of this world create their own technology and leasing finance branch, much like the big automakers have done for long time, to lease out the latest generation of smart kitchen robots,.

In a first phase, the food company adds the technology and the financial services to their portfolio and, like automotive companies lease expensive hardware, first to the early adopters, later to wider and wider circles of consumers. These robots will be branded, same as in the car industry, and are differentiated by functionalities, service, reliability, energy consumption, price and amount of monthly lease rates, and last but not least, by design, capturing the consumers emotionally. Given the actual costs of smart robots and even taking a very high multiplication factor of numbers of robot units built into account, I strongly believe that it might take another 10 or even 20 years until they would get to a price range that is affordable for the average person.

In the meantime, there is ample opportunity to create and successfully run a leasing business, which will always have the advantage of being able to lease the newest kitchen robot with enhanced functionality and modern design every 3 years or so, while the lease returns will be sold at affordable prices to consumers.

In my opinion, smart kitchen robots, multifunctional cooking arms, and similar devices will be a kitchen reality much sooner that we may believe and the large food companies who have industrially prepared (cooked) food products in their portfolio would be well advised to at least consider to get a foot into this business to offer a whole series of branded kitchen robots.

9.4.1 A New Manufacturing Reality

Previously I mentioned and discussed a possible democratization of manufacturing food that could catch the industry on the wrong foot if they are careless enough to only listen to their own, internal manufacturing gurus, who, for continued job security would never let go this large portion of the value chain. This territoriality is even more pronounced in the so-called branded food companies, who, to this day, try to hang on to exclusively produce their own brands in their own factories as much as they can. Yes, for the launch of new products they use co-manufacturing almost every time because they are uncertain whether the predicted volumes will be sold in the first or second year, and moreover they do not want to interfere too much in their own factories and disturb the flow of things, even if there would be free capacity. The latter may sound like a contradiction, although it's not; it's just a matter of not wanting to invest in necessary equipment yet, equipment that the selected co-manufacturer has at their disposal.

Once a certain volume is assured and looks sustainable, the big crunch and run for financing and installing and trialing equipment starts and cannot go fast enough. If too slow, fingers are always pointed to the engineers who were not expedient enough, and marketing always seems to get a free ride. Sounds like I am holding a grudge against marketing, which I don't, I just believe that most of them are counterproductive to the success of most large food companies, at least the ones I know.

But wait until you read what I suggest should happen to the manufacturing group inside a food company. As suggested before, I believe that manufacturing should take place outside the food company; in other words, it should be outsourced regionally, covering a large number of products and ideally at the future empty spaces of the large brick-and-mortar stores of retailers, because consumers will buy more and more online.

So, the sequence of events is most likely: more online shopping for food, retailers have large surfaces on their hands; food company reverts to other core areas seeing manufacturing as "outsourceable" and fills the empty space of retailers; manufacturing will more and more happen on demand; consumers will need more information about healthy food and nutrition; and the food company will ultimately sell such information and will increasingly go into the robot technology and financing field of activities.

You may want to either challenge this entirely or you might believe that I have not even gone far enough. Either way, the section of this chapter carrying "odd stuff" in the title, is the right place to speculate about future scenarios that may appear frightening, unrealistic, too far-fetched, or maybe just the right elements to take up, discuss, and embrace some or all of them. Because the food industry is traditionally slow, which also has its positive sides by the way, it might take a long while until any of what was discussed in this chapter will actually be thought of, and more so discussed and potentially acted on. The industry has a long history of sitting out disruptive events, and they may think that they can do it this time again. However, the cycles of societal and technological change have become so short that it might just not work this time around. Let's all watch this space!

9.5 Summary and Major Learning

This chapter was all about the writings on the wall, the desire to read them, and more importantly attempt to understand them and act accordingly and potential transformations of the industry that might or have to be a consequence of these writings.

The following topics were suggested, discussed, and analyzed:

- A new business model 2.0 for the food industry is based on these three elements: R&D represents the company's driver, open innovation and partnerships are the new normal, and retailers become real partners.
- The growing role of automation and increasing introduction of robots in the food industry as well as in the agriculture industry was suggested to be one of these writings on the wall that the industry at large has partly fully embraced for instance in logistics, supply chain, and manufacturing. However, automation will more and more take over our homes, and it is questioned, whether the industry will still remain this gigantic manufacturing and distributing Moloch.
- Automation in agriculture has entered the area of "orphan crops" in important ways, and today coffee as well as grapes, just to name two, are preferably harvested by fully adapted machines, except in cases where the topography of the respective fields would not allow this.
- The movement of buying local is an increasing element of distinction for today's consumers and is being politically exploited, not always in the most favorable ways. Local jobs in California, to quote just one example can pay as little as $3 per hour in the garment industry. In Germany you can find the "infamous" €1 jobs.
- It was discussed and analyzed that regulatory authorities related to agriculture and food have become increasingly powerful, which is good and bad at the same time; it can enhance sustainability in the industry but it can also hamper or at least slow down real innovation.
- The trend to all natural products and so-called "clean labels" on industrial food products will continue and may even increase.
- Shopping local is another important trend ongoing in the industry for quite some time already. It looks like consumers turn increasingly to food and food products that don't come from too afar.
- From product to know-how seller was again emphasized. The so-called business model 2.1 was again discussed and analyzed in some detail.
- It was suggested and discussed that consumers might be turned off by complex ingredients mixes listed in fine print on labels of packages. One important way out toward shorter and "cleaner" labels can be achieved by turning to creating flavors and tastes rather in the manufacturing and cooking process than by adding flavor concentrates. This might be of significant consequences for the flavor industry.

- Possible scenarios to anticipate certain inevitable challenges likely to happen in the industry were suggested and discussed in some detail. The following scenarios were discussed: population growth is likely to continue, arable land is likely to remain at the same size (not necessarily in the same regions as today), food security is going to be increasingly unbalanced, consumers' hostility toward processed industrial food will increase, the trend to home-cooked food will grow, diet mania will continue, consumers will become increasingly confused with multiple and sometimes conflicting nutrition and health messages, large corporations remain risk and change averse, and lastly smaller companies will lead the way to change.
- Possible ways to respond to these challenges were suggested and discussed: the agriculture industry rides the wave of change with consumers' benefits in mind, agriculture become more efficient in sustainable ways and responds to the real needs of all involved (people working in the industry as well as consumers), the food industry comes away from demase only and teams up with retailers, the value chain needs to be reevaluated, R&D personnel goes out of its own walls and meets consumers and sells know-how, and ultimately food companies are run by people who understand food!
- Possible future more disruptive responses were suggested and discussed, based on two assumptions: globalization will slow down and consumption of industrial food products will decrease in the more affluent parts of the world. The following was mentioned: manufacturing to be outsourced, intellectual property as much as possible to be outlicensed, home cooking will increase in importance, kitchen robots will become affordable at first through lease programs and later at decreasing prices through ownership, the old demase model will gradually be replaced by the inapse model, real innovation will be required to develop and successfully sell premium medical food with proven and claimable properties by increasingly collaborating with members of the medical profession.
- If medical food claims to be successful it must be sold at a premium so that higher development costs can be paid for.
- A few outlandish topics were finally discussed and analyzed in this chapter. Dramatic events in our past such as societal changes through wars, pandemic events, or pure serendipity have led to many of unanticipated changes and important innovations. The industries of agriculture and food are ill prepared for such changes because they are simply too slow to react. The rapid digitalization and especially the aggressive introductions of robots of all kinds in the industry itself as well as at people's homes will have dramatic impacts for which we all have to be prepared. Food industry can and must embrace these changes.
- Lastly, it was suggested that manufacturing will become increasingly democratized and will not remain an important element in the value chain of future food companies, which will increasingly become knowledge companies.

References

Adams, A. (2012). Making the switch. Available from: http://www.winesandvines.com/template.cfm?section=features&content=97355 [Accessed December 12, 2016].

Bulcke, P. (2016). *Nestlé Gazette*.148, 5.

Goldfarb, A. (2008). Moving toward mechanical. Available from: http://www.winesandvines.com/template.cfm?section=features&content=53452 [Accessed December 12, 2016].

"Industry 4.0." (n.d.). Available from: https://en.wikipedia.org/wiki/Industry_4.0 [Accessed January 4, 2017].

Kornberg, J. (2016). In L.A.'s garment industry, 'Made in the USA' can mean being paid $3 an hour. Available from: http://www.latimes.com/opinion/op-ed/la-oe-kornberg-garment-industry-wage-theft-20161214-story.html [Accessed December 14, 2016].

Nestlé Health Science. (n.d.). Available from: https://www.nestlehealthscience.com [Accessed January 5, 2017].

10

Food for the Future: A Future for Food

Part of the secret of success in life is to eat what you like and let the food fight it out inside.

—Mark Twain

10.1 Proactive Agriculture

Most of this chapter deals with the plant side of the agriculture industry and will suggest, discuss, and analyze topics and issues in this area. By definition, agriculture is proactive. In other words, it seeds far in advance, looks after the growth of whatever was planted, and harvests some 4 to 9 months later—some plants even only every other year. So, in principle, everything looks good and there should be no need to change any of this. However, this is a shallow look and misses the heart of the issue. The agriculture industry—all of its branches from the chemistry, to the seeds to the farmers and ultimately to the traders in between some of these steps—do not seem to have an open ear to the real needs of the people. Milk farmers are put under pressure by retailers, who claim that raw milk may not be charged higher than at a certain price. The seed industry is under pressure that it may be in conflict and competition with the packaging industry in the case of corn. The chemical industry is under pressure from the public for contributing to health problems stemming from plants not appropriately treated, and traders are accused, probably rightly so, that they deal with cereal contracts 40 times or even more often, thereby making the final product unaffordable for too many.

Well, this doesn't look like everything is in order; agriculture has not done its homework and does not appropriately anticipate people's needs in the most efficient and proactive ways. The situation is far from being OK, and it is probably extremely difficult to fix any of this any time soon or even in the long term. Let me, however, try to raise a few topics and possible actions and remedies for

Megatrends in Food and Agriculture: Technology, Water Use and Nutrition, First Edition.
Helmut Traitler, Michel Dubois, Keith Heikes, Vincent Pétiard and David Zilberman.

certain situations that are responding to societal and demographic changes, especially present and most likely future food trends.

The big issue seems to be that agriculture has gone the route of the well-liked and acclaimed economy of scale, and make crop fields profitable for every player in the value chain, which has grown larger and more automated.

Food trends on the other hand have gone, and will increasingly go, the other direction: more segmented, more personalized, and often claiming total individuality. So, in the not-so-distant past it looked almost impossible to harmonize these two opposing trends: toward larger for agriculture and toward more individual for food. How to bring this together is the big question. And it appears that people are finding answers for themselves; it's a process that is driven by individuals and small and medium-sized organizations and some smaller industry partners. It is not yet fully understood, let alone embraced by the large industry partners, both in agriculture as well as food. This is the real dilemma: industry representatives are too cozy in their present situation, after all they are making money and do not grasp or do not want to acknowledge some of the changes that will come, whether they like it or not. This is not a conspiracy theory; it looks more like refusing reality and closing one's eyes.

It would of course be easy to announce that change is imminent without offering some ideas as to how to ride the wave of change and suggest possible answers that lead to a new agriculture industry, which, together with a new kind of food industry can and will embrace such answers. It will still have a comfortable revenue that allows the industry to invest in continued innovation in all areas: agriculture, food security, food safety, food affordability, balanced food for many, and ideally for all, and a deep understanding—also understood by consumers—of the intimate relationship between food intake and health in its most natural way. All this is quite a handful, but if we don't link all these topics and work toward resolving this apparent split between industry and people, we will do a disservice to the industry itself and ultimately the people who expect to be fed properly and responsibly.

Let me again repeat the most important topics here and follow up with a more in-depth discussion as to how the agriculture industry might become really proactive and able to act in the best possible interest of everyone involved in the value chain and the ultimate outcome. So, here we go, these are the crucial topics when it comes to agriculture (both plants as well as animals) and food:

- Flexible agriculture: the industry responds to the real needs of people
- Food security: enough food for everyone, getting to all and not just a few
- Food safety: consumers can rely, always and everywhere that the food they eat is safe
- Food affordability: food raw materials (plant and animal) can be afforded by all who are hungry

- Nutrient balance for all: everyone should have access to a balanced diet
- Increased understanding of relationship between food and health: nutrition understood and lived

10.1.1 What If Agriculture Anticipated Real Food Requirements and Trends?

By asking this question in the title I suggest that the agriculture industry does not really think and operate in ways that allow it to respond to food trends, present and future, efficiently and timely. To me, it appears the agriculture always is running behind trends, although it would only require one growth cycle—for plants typically one year—to catch up. But it doesn't seem to happen. Yes, it happens on a small scale, always and everywhere but not with the big corporations, they just don't seem to show any kind of flexibility toward smart adaptation. If this was the case, we would have new staple crops, new types of animal proteins from animals or insects, and we would have a new approach to the dairy industry, which seems to be in perpetual disarray for many years.

So, here we have them again, the two big influencing factors of relevance for the entire industry:

- Society at large with changing needs and new food and dietary trends springing up every so often.
- Environmental changes, irrespective of the side you are on, happening before our very eyes.

It appears that the entire industry of agriculture is ill prepared to respond to both, or even worse, does not want to respond in creative, innovative, and responsive ways. Let me discuss and analyze a few examples here, beginning with the possible answers to the environmental challenges.

One of the great challenges (discussed at length in Chapter 2) is water, the right amount in the right arable areas that is. When I say "right," I mean the amount that is just needed to grow crops sustainably without using more water than can be naturally replenished. We should, however keep the option of technological "helpers" open. Many water aqueducts were built by the Romans more than 2,000 years ago with the goal to bring water from areas of abundance to areas of needs, mostly for drinking water purposes but also for agricultural ones. Here's a short historic overview of Roman aqueduct (n.d.).

> 312 BC Aqua Appia, Rome's first aqueduct is built by Appius Claudius Caecus, the aqueduct is nearly all underground.
>
> 144 BC Aqua Marcia, 90 km (56 miles) in length, construction starts.
>
> 33 BC Aqua Julia is built by Octavian (Emperor Augustus)
>
> 19 BC Aqua Virgo is built to supply the thermal baths in the Campus Martius.

38–52 AD Aqua Claudia built

109 AD Aqua Traiana brings water from Lake Bracciano to supply Rome's suburbs, now called Trastevere.

There are many more examples during times closer to our era. One of them is the "Wiener Hochquellwasserleitung" (Viennese aqueduct for mountain water from approximately 100 km away), built for parts of the city during the era of Emperor Ferdinand of Austria from 1835 to 1841. The present-day version of this aqueduct was built between 1870 and 1873 and was inaugurated at the occasion of the world exposition of that year by Emperor Franz Joseph I.

Another example of an aqueduct that was built closer to my home is the Los Angeles aqueduct built by a team around chief engineer William Mulholland between 1905 and 1907, mainly bringing in water from the Owens Valley just east of the mighty Sierra Nevada mountain range to the greater Los Angeles area. The project was, and is still, debated for its environmental impact, and it was accused of literally drying out Owens Valley, and later on, through an extension in the beginning of 1940 also Mono Lake on the northern part of the valley. Fact is that the "new" abundance of water in the early 20th century co-enabled the development of an entire region, both from a societal and an agricultural point of view.

I am mentioning these examples because water can be brought in from regions of abundance to places that are in dire need through many technological feats, aqueducts being one of them. Today we build pipelines mainly for oil but nobody holds us from channeling water from far-away places to crop fields that need such water. Today, we see not much of this yet happening, even though the agriculture industry of southern California and the San Joaquin Valley are in a state of drought. Even the very rainy winter 2016–2017 cannot fool us that we need more sustainable solutions to secure the water supply that any type of agriculture—both plants and animals—needs. To some degree, this is realized in the so-called "Central Valley Project" (n.d.), yet with big flaws such as transporting water in open canals in an area that gets extremely hot in summer and one-third of the flowing water can easily evaporate.

> The Central Valley Project is a federal water management project in the state of California. It was devised in 1933 in order to provide irrigation and municipal water to much of California's Central Valley by regulating and storing water in reservoirs in the water-rich northern half of the state, and transporting it to the water-poor San Joaquin Valley and its surroundings by means of a series of canals, aqueducts and pump plants, some shared with the California State Water Project.

It can be seen that many of these water supply projects were already undertaken many years ago, responding to a clear need of survival and growing food.

The city of Santa Barbara built a water desalination plant back in the 1980s after several years of drought in Southern California. The plant operated for a few months between March and June 1992 only to discontinue operation because significant rainfalls followed in the winters afterward. These years of more abundant winter rains forced the installation into an extended hibernation, or rather, standby until 2016.

Pernett and Mitra (2016) wrote about it in the *Daily Nexus*, a publication of the University of California. Here's a short excerpt:

> On May 5, 2015, the Santa Barbara City Council declared a stage three drought condition. This required a 35 percent citywide reduction in overall water use. To account for this and prior restrictions, the city decided to reopen the Charles E. Meyer Desalination Plant, set to begin operation by October 2016.

So finally, if nothing else worked any more, after 25 years (!) after its first and last operation, the plant was poised to put the pressure on again, literally that is. It is also clear that some overhaul was necessary after so many years of inactivity and for a technology that was initially installed in the 1980s. It sounds almost like restoring a classic car. I mention this example only to demonstrate the enormous inertia that the industry; in this particular case, the water industry, and it may take years, if not decades, to achieve substantial changes in any type of industry, especially agriculture and food. The only factor that can bring changes fast is of catastrophic nature, something that none of us wishes.

Let me get back to the ideal agriculture industry that makes everything, or at least almost everything, right and responds to what is happening around us and in many parts of the world in more inclusive ways and more in the interest of the people. The water example, more precisely the increasing scarcity and value of water, clearly shows that the industry must play a more active role in this entire equation. Farmers cannot sit on the sidelines and be the victims of drought without actively contributing to the solutions, such as growing plant varieties that are more drought resistant or use less water to begin with. If we understand that it takes around 1,000 liters of water to produce 1 kilogram of corn or several thousands of liters for 1 kilograms of meat we can begin to imagine the difficulties that we are facing. If we accept the fact that at least one-third of the open canals in the California Water Project are lost to evaporation (it's probably even worse for its brother, the "Arizona Canal," because it's even hotter there), and if we accept that many irrigation pipes are leaking, then a substantial amount of water is lost in this fashion and we have not really understood the problem. Agriculture has to contribute its part and plant varieties have to be developed that can be grown with less water and in increasingly dry environments. Yes, a lot seems to happen already, but it is neither fast enough nor solving all these problems.

One can argue that with the increasing observed frequency of catastrophic weather events such as flooding in some areas of our world, we should have enough water to easily grow all the food we need. Unfortunately we observe this discrepancy between where all the water comes down and where it is really needed. Moreover, too much water is not a good thing either because, for example, many farmers in rural Switzerland experienced it during spring 2016 when fields were flooded and crops simply didn't grow during the most critical growth period. And there are many such "Switzerlands" around the world where too much water hampers, critically reducing growth, and ultimately, agricultural output. Some of them are doubly punished by flooding from rain from above and flooding of low-lying land with water from the close by seas. Nothing grows on overly salty land with the exception of saltwater-resistant crops, which do not seem to be a focus point of the industry at all, and yet they probably should be!

So, in short, the proactive agriculture needs to respond to challenges of:

- Not enough water
- Too much water
- Develop drought-resistant plants that grow with less water
- Develop saltwater-resistant plants with good yield and growth frequency
- Plant and grow much closer to where food is needed to assure food security for all
- Develop and grow plants and breed animals that more efficiently reply to and anticipate trends

This looks like quite a handful, and it probably is. But it simply requires a repositioning of all players in the agriculture industry and an even closer collaboration with the food industry. *Agri-transformation* has almost become a un-word in the food industry because making specialty food products and medical foods has become so much more fashionable, at least on the management level of many food companies. However, this is what this is all about: Assuring food for all. It's as simple as this and does not take anything away from reaching larger and searching for better, more nutritious, and healthier foods from better agricultural sources. It is most and foremost about food security! And the agriculture industry, every player, from seeds and chemistry, to traders and farmers and retailers, has a role to play in this.

10.2 Democratized Agriculture

After some discourse on proactive and anticipating agriculture industry I shall discuss and analyze this increasingly important topic of agriculture that becomes more democratized again. This is not a call for back to nature; quite the contrary, it is a call forward to new ways as to how agriculture could develop itself into a bright and inclusive future. In a previous chapter I briefly discussed and analyzed the role of urban gardening and more importantly urban agriculture and the impact that it already has on agriculture and food security at large. And this role is only going to get bigger when we think of all the increasing needs, diversity in populations and diets, and climate changes, which will take an ever more important toll on all matters in agriculture and future food security.

10.2.1 Agrihood

The concept of *agrihood* has become more popular again in recent years. It is based on the concept of suburbia, first created in the middle of the 19th century with the idea to make urban homes look more rural by amply planting trees and shrubs; however no agricultural exploitation was planned then. Roth (2014) published a story on the topic of agrihood.

> Suburban development and open farmland have never exactly been on friendly terms. But earlier this year *The New York Times* reported on a new trend rocking the suburbs: agrihoods, or communities built around shared farms instead of golf courses. Urban planners estimate that there are at least 200 of these new 'burbs spread across the country, from 160-acre Agritopia in greater Phoenix to Serenbe, Georgia's idyllic 1,000 acre "hamlet."

Early in January I received an invitation from Polanui Gardens on the island of Maui, Hawaii, to visit their project. Let me quote the letter:

> Aloha e Neighbors:
>
> You are invited to stop by and meet the proposed "Polanui Gardens" affordable housing and agricultural project team at an informational open house on January 30th, 2017. Polanui Gardens is a proposed "agrihood" for West Maui's working families. An "agrihood" is a project that puts farming and housing side by side, creating an agricultural focal point and rural feel to a residential neighborhood. This small, affordable housing project strives to provide much needed workforce housing and diversified agricultural opportunities to West Maui and is located in the beautiful Pola Nui Ahupua'a. The project proposes between 46 and 50 workforce homes to be sold to qualified families and individuals making between 80% and 140% of Maui's mean annual income as well as 16 agricultural lots to be sold at fair market value. These homes will be clustered among a "Food Park" which will consist of edible landscaping and space or community gardens, walking and hiking trails, approximately 9-10 acres of farmland, open space corridors, and a community

> recreational park. This project will be located mauka (above) of the future Lahaina Bypass Phase 1-B2 and makai (below) the existing Launiupoko, Makila and Pu'unoa subdivisions. Efforts to utilize low impact design concepts, native and drought resistant landscaping, and solar water heating will help to reduce the project's footprint on the environment and the cost of maintenance and utilities for future workforce homeowners. (M. Gill, Project Manager, Polanui Gardens, Maui, Hawaii, January 11, 2017, personal communication)

This is a really comprehensive description of what these new forms of agrihood can contribute to agriculture in particular and to society at large. If properly done, it brings a whole new meaning to suburban living and can be an ideal and complementary contributor to agricultural output. It opens the same debate fault lines of small versus large, and decentralized versus centralized, which you can find in today's energy supply debate with opposing elements such as (many) individual solar panels versus large, centralized power generation units. In my opinion there is no simple answer like: this is better than that. However this is clearly complementary, and every effort to sustainably generate either electricity, or in the case of agrihood, agricultural output must be acclaimed and supported by individuals, society, and policy makers, who ensure the necessary frameworks—not more and not less!

The appearance of such agrihood projects is a real sign of democratized agriculture because all of a sudden growing food, even animals, is at the reach of everyone, and not only farmers, farm workers, and large farming conglomerates. Agrihood still has to demonstrate that it can really contribute to substantial agricultural output, in continued and sustainable ways and not just remain the romantic dream of a few "gentlemen/women" farmers who search for an alternative lifestyle and eventually get tired of it because, whatever you call it—agrihood or just farming—is a lot of work and not an easy undertaking, full of hopes and frustrations, and as already discussed, dependent on so many external factors not necessarily under our control.

10.2.2 Permaculture

Permaculture is yet another, fairly recently developed agricultural system, although based on principles that we discovered and individually applied for centuries, if not millennia. It is described as a "system of agricultural and social design principles centered on stimulating or directly utilizing the patterns and features observed in ecosystems. Permaculture was developed, and the term coined by Bill Mollison and David Holmgren in 1978. Permaculture is a philosophy of working with, rather than against nature."

The underlying arguments are rather multiple and complex, amply described by Holmgren in his book *Permaculture: Principles and Pathways Beyond Sustainability*, by 12 design principles:

> **Observe and interact**: By taking time to engage with nature we can design solutions that suit our particular situation.
> **Catch and store energy**: By developing systems that collect resources at peak abundance, we can use them in times of need.
> **Obtain a yield**: Ensure that you are getting truly useful rewards as part of the work that you are doing.
> **Apply self-regulation and accept feedback**: We need to discourage inappropriate activity to ensure that systems can continue to function well.
> **Use and value renewable resources and services**: Make the best use of nature's abundance to reduce our consumptive behavior and dependence on non-renewable resources.
> **Produce no waste**: By valuing and making use of all the resources that are available to us, nothing goes to waste.
> **Design from patterns to details**: By stepping back, we can observe patterns in nature and society. These can form the backbone of our designs, with the details filled in as we go.
> **Integrate rather than segregate**: By putting the right things in the right place, relationships develop between those things and they work together to support each other.
> **Use small and slow solutions**: Small and slow systems are easier to maintain than big ones, making better use of local resources and producing more sustainable outcomes.
> **Use and value diversity**: Diversity reduces vulnerability to a variety of threats and takes advantage of the unique nature of the environment in which it resides.
> **Use edges and value the marginal**: The interface between things is where the most interesting events take place. These are often the most valuable, diverse and productive elements in the system.
> **Creatively use and respond to change**: We can have a positive impact on inevitable change by carefully observing, and then intervening at the right time. ("Permaculture, n.d.)

I do not want to even attempt to enter into a full-fledged discussion of the topic of permaculture because it is far too complex to give it ample space in this book. Modern agriculture departments at universities offer the topic of permaculture as one of the study branches and students spend years in following courses both practical as well as theory. Let me, however, discuss some of the design principles and how I believe they fit into the overall topic of this book, especially on the food for the future and future for food.

Observe and interact seems to be a pretty straightforward principle, almost trivial and I would assume that much if not all learning of humankind is based on this. As I see it, it is rather a reminder to trust our instincts and observations of nature when making decisions as to how to design our lives around these and especially how we grow our food in ways that we can get enough balanced food and get it for everyone from an agriculture that does not exploit the soil in unsustainable ways. Although it is trivial it's important to always remind us!

Catch and store energy is another reminder to do the obvious. It's like an electrical car that stores recovered energy when braking back into the batteries. One of the downfalls of the sugarcane industry, especially in the state of Hawaii, was burning the sugarcanes in the fields before harvesting. This was done to reduce the volume of plants to be harvested, but at the same time, it took an incredible unnecessary toll on the environment. The air was smoke-filled, ash particles were flying to all places downwind, and caramel smells could be felt all around. Mice and spiders fled from the fires and were closing in on people's homes. Not only was this entire burning really bad for the environment and one of the main reasons, apart from decaying sugar prices for the downfall and ultimate shutdown of the industry in that part of the world, but it was simply stupid. Stupidity was mainly based on greed and the unwillingness to invest in modern harvesting and downstream techniques that would make use of the leafy parts of the sugarcane plant and use it to ferment it to valuable methane. The industry had the oldest machines you could imagine, and used techniques, trying to milk the factories as much as possible, until every last drop of profitability was pressed out of them.

For more than a decade, the town of Güssing in the eastern part of Austria uses green biomass (grass, etc.) to produce gas to be used in combination with natural gas to be fed into the households of their approximately 3,700 citizens. The technology is based on "fast internally circulating fluidized bed" and is described as follows.

> GRE DFB multi-fuel gasification
>
> The GRE DFB gasification system is the world's first functioning FICFB (Fast Internally Circulating Fluidized Bed) plant. The core of the plant is formed by the two inter-connected fluidized bed systems of the fluidized bed steam gasifier (reactor).
>
> In the gasification zone the cut-down biomass is whirled up and gasified in anaerobic conditions at approx. 850 °C in the shortest time possible by introducing steam. The bed material (olivine sand) has the function of a heat transfer medium and provides a stable temperature in the reactor.

> In the next step the resultant product gas is purified and cooled. The heat emitted during cooling is used for the generation of district heat. Subsequently, the gas is filtered, and tar scrubbed with bio-diesel. In this specialized process all resulting residual materials are recycled.
>
> In the gas purification neither solid waste nor waste water arises, and the product gas is completely free from nitrogen. Fluidized bed steam gasifiers work extraordinarily reliably and regularly (7,000-8,000 operating hours per annum) and are considered state-of-the-art (Best Available Technology). (Gussing Renewable, n.d.)

This example is one of many and shows that such complete use and transformation of all available biomass from every type of plant follows this design principle of catch and store perfectly well. It is actually so obvious to do this that one wonders what holds many communities and small towns up in not following a similar approach. In my opinion, it all falls back to the fact that we as a society have gotten so used to receiving energy, any type of it, from large providers, doubting in turn the capabilities of small and close by units to provide us with required energy 24/7, month after month, year in, year out. I predict that a new mind-set will get hold of our communities, however not without substantial, ongoing in-fighting, thereby delaying the transformation process.

Obtain a yield is, in my eyes, a two-sided principle because it would insinuate that more is always better. As I do understand it though, it means to make sure that your personal, your community's, input in terms of labor and raw materials into the agricultural value chain is rightly and correctly rewarded. This may not always be the case; hence, this call to vigilance.

Apply self-regulation and accept feedback is another rather self-explanatory principle; however, it is endangered by those who try to exploit the value chain, especially on the growth side. An interesting example here comes from the wine industry, demonstrating that less is actually more. Quite a few years ago, the industry in Europe was regulated in such a way that all those wineries who wanted to obtain a label status of "controlled origin" had to abide to a restriction of yield per vine, more simply measured in kilograms of grapes per square meter or a similar measure. There were two main ideas behind this: first, it ensured that the piece of land that the vine grew on was not overly exploited by chasing for the highest yield, and second, such self-inflicted yield restriction ensured a better nutrient feeding for the fewer grapes on the vine.

While I am writing this in January 2017, I am visiting with friends in Santa Maria in the San Luis Obispo county of California, and we are visiting wineries and the increasingly popular olive oil places in the vicinity. My observation of at least the wineries we visited was to see rows of vines being grown closer and closer, and at least on the surface, suggesting that the land may be exceedingly exploited, in ways that go totally against the above design principle.

Use and value renewable resources and services is the next design principle on Holmgren's list and it pretty much speaks for itself. It is interesting to mention that the food industry has come a long way in this area, probably much farther than most of the larger farms, let alone the smaller ones. The name of the game is "co-creation." Co-creation describes the process of burning leftovers from the process of creating food, for instance spent coffee grounds after extraction, to be burned and transformed into steam, which in turn is either directly used for generation of heat or drives turbines to produce electricity. Some coffee factories, especially the ones I do know better personally, do create between 50% and 100% of their energy needs through co-creation in more recent years.

Produce no waste is a logical consequence of the principle just discussed but encourages all players in the value chain from "field to fork" to increase their efforts and try to reduce waste as much as possible. On the one hand, no industrial operation has any intention in creating any amount of waste in their processes; on the other hand it is sometimes easier and more comfortable to just let go and throw away. From personal observation, I would say that smaller units tend to be following this design principle more closely and more successfully.

Design from patterns to details is the next on the list, and it almost is identical, at least in spirit to some of Dieter Rams' 10 principles of good design, especially the element of design being thorough down to the last detail and at the same time as minimalistic, honest, long-lasting, and innovative as possible. Good design according to Rams also has to be environmentally friendly, aesthetic, make any product understandable, and ultimately demonstrate the usefulness of the product to which the design is applied. Personally, I can see so many parallels between the 10 principles of good design and this as well as some other of the permaculture design principles described and discussed in this section. For more reading on the topic of food and design, please consult my book *Food Industry Design, Technology and Innovation* (Traitler, Hofmann & Coleman, 2015).

Integrate rather than segregate is not a complex message and rather self-explanatory, at least from an integrated agriculture and food domain point of view. The purist permaculture view relates to symbiotic situations where plants, animals, and insects (yes, I know, these are animals too) that can positively influence each other are grown and bred in proximity rather than incompatible ones.

Use small and slow solutions is a call to reducing complexity and working toward simple, straightforward, and easily manageable solutions. I often heard, after many years in the industry, "we have to reduce complexity" and how often have I seen the exact opposite happen. This principle is almost a call to arms and should help us to finally achieve more simplicity. Critics would say that the ultimate goal is not to reduce complexity but to manage complexity, thereby beating your competitors who cannot do the same thing. I am not sure whether this is not just another nice slogan for management textbooks.

Use and value diversity almost sounds like a societal punch line with hollow meaning; however in the context of agriculture, it gets a whole new (actually rather old) meaning. Diversity of fauna and flora in close proximity is probably the best way to avoid agricultural chemistry and to assure that bees and other useful and necessary insects can survive and strive. This principle, like so many others before makes a lot of sense, and in larger agricultural operations, especially, is not sufficiently followed.

Use edges and value the marginal simply translates into my long-standing and steadfast belief that real innovation can only happen at the interface of disciplines and at the fault lines of stubborn debate. Translated into agriculture and design principles it simply means to be open to the small things that can not only surprise us but also make all the difference in the world and can lead to successful new developments, especially in the world of agriculture and food and at its interface.

Creatively use and respond to change is the last principle in this long list and speaks for itself: just be prepared to change and see the opportunity that a new type of breed, a new type of agricultural practice, a new innovation in supply chain and distribution, a new type or class of food products, and ultimately, consumer benefits and demands can bring to this entire complex world of agriculture and food.

10.2.3 From Large to Small

In light of the topics of agrihood and permaculture, there is one more aspect that in my eyes merits some more in-depth discussion and analysis, namely the direction from large to small in agricultural and food operations. This is in no way a call toward back to nature or to the recreation of the utopian colonies of the late 19th century or anything like that. I really want to discuss the possibility of acting in "small ways," while being embedded in fairly or very large organizations. A former CEO of the Nestlé Company once described this as mentally, and why not operationally, "turning away from the all determining supertanker mentality to becoming an agile fleet of small racing yachts." Easier said than done!

> The Nestlé supertanker couldn't become faster and bigger. So the only way (forward) was to break it up into a very agile fleet of independent boats, with a common supply chain afterwards. The challenge is, how you manage that without losing coherence and strategic direction. (P. Brabeck Letmathe, several public communications between 2005 and 2008)

It was a nice suggestion and, as far as I could see, it was partially followed by some management but certainly not all. Management's first and foremost goal was, and still is to maximize, or if you want, optimize, profit and if that meant to apply the old ways, so be it. In the years following Brabeck's first mention of the "agile fleet," I was still active in various functions for the Nestlé Company, more specifically in the areas of packaging and innovation partners. I had the impression that these roles were members of an agile fleet; however, at the same time I could always feel the attraction of gravity of the large mother ship, the supertanker as Brabeck named it.

So, how could this work in ways that it was intended to work? An entire flock of small entities—all agile and super-fast—reacting and all "flying" for and toward the common goal of commercial success as well as consumer satisfaction in the case of a food company and food security in agriculture, just to name the two probably most important features in these areas. In my opinion, this can only work if large corporations trust their people in ways that they normally should trust them. Let me explain. What happens today probably looks similar to this: the company hires its employees based on functional and personnel needs; employees are hired after several rounds of interviews; more recently this hiring happens through committees. Once the employee is formally hired, he or she typically has a specific function to fulfill, a function that was part of the initial job description. I specify this here because this is not necessarily always the case: things change en route, and there could always be surprise changes.

The newly hired person starts his or her job and not only receives a salary of a certain size but is also put in charge of people and projects, resources in general and is expected to deliver as measured against a certain set of previously and jointly discussed objectives. So far so good, but then something strange starts to happen. The company typically invests in the order of $300,000 per year into one person overall (workspace, salary, social charges, insurances, etc.) and one would believe that because of this relatively hefty investment, the company should have full trust in its employee. During many years of personal observation and experience of such situations, I can say with some conviction that on a scale from 1 to 10 as to how well the company trusts its employees (newly hired or longer serving), I would give this a rather low 3, if at all. It always surprised me that employees had to justify any single step of work during any given project in review meetings, visits of high-ranking management representatives, or on similar occasions.

Not only was there no trust, but often also one could sense a demeaning and pejorative attitude on the part of the higher ranks, giving a kind of message that the employee was not really in charge, not really trustworthy. Do I paint a dark picture? Maybe, but I have seen this time and again and have seen many highly motivated employees literally deflate and lose their initial motivation. Why do I emphasize so much on this here? Simply because not trusting the people who were specifically hired to do specific jobs to further the well-being of a company and its customers and consumers is the safest way to block this highly desirable and repeatedly talked about transition away from the "supertanker" to an "agile fleet of independent boats or even small racing yachts," the proverbial racing yachts being even faster and even more agile than simply boats.

Trust in people, in my eyes, is *the* key element in this desirable breaking up of large entities to highly efficient, responsible, and self-organized units within the organization of a larger food or agriculture company. Trust in people is the key ingredient for this call "from large to small," because typically small can better understand the ultimate client because it can be so much closer to the client. This is not the proverbial rocket science; it is just simple common sense, and every manager in every large organization, agriculture or food based, should have a deep understanding of what it means to trust his or her people. By the way, this trust goes both ways: from top to bottom and equally from bottom to top. The latter, once disappointed can easily turn to frustration, loss of motivation, and ultimately, even hostility toward the organization. It is the closeness to the ultimate clients of the industries that makes this concept of "from large to small" or from "supertanker to agile fleet of boats" not only so attractive, but a real must-do for the entire industry if it's real intention is to ultimately serve their clients in the best possible and most profitable and sustainable ways.

10.2.4 The Growing Role of Urban Agriculture: Self-Centeredness or Community Driven?

The important role that urban agriculture plays in the overall agriculture industry today was previously discussed. What I intend to briefly discuss and analyze here is a potential and perceived hostility that one might sense between representatives of the traditional farming industry and urban agriculture. I recently had a discussion with an old friend, whose entire family, including herself, works in agriculture, more precisely in crop farming. They are active in some of the orphan crops such as citrus and avocado, as you find so many in rural California. We spoke about the topic of urban gardening and its big sister, urban agriculture. I was surprised to learn from her that apparently many in the farming community not only see urban agriculture as an unwelcome competition but also almost accused those who are actively pursuing any form of urban agriculture as egotistical and self-centered, only thinking about their own personal needs and not really catering to the needs and requirements of the larger public.

I did not want to discard this critical voice at all, although I had some difficulty listening at first. Although her argument might, at first sight have some merit, it is in my opinion far too simplistic and clearly reflects the views and opinions of someone having lived her entire life within the confines of the traditional farming industry. From where she comes from, it appears logical to see those smaller entities of urban agriculture as something rejectable, and outright dangerous, to the survival and sustainable success of many in the farming industry who struggle with high raw material, labor and water costs, and on the other end, “everyday low prices” on the part of the retailers.

The reality is that both parties have a point here: the traditionalists who fight for long-term survival of their industry and the up-and-coming urban agriculturalists who see their type of agriculture as being closer to the consumer with advantages such as known origin of what is produced, growing for the consumers’ needs—catering to their good nutrition, good food, and ultimately good health—and mostly or almost exclusively selling in neighborhood farmers’ markets. In all fairness, it has to be said that traditional agriculture and farming and its urban counterpart are totally compatible with each other and complement each other in perfect ways. No need to be afraid! What I could actually see is learning from each other, especially for the traditional farmers as far as growing more diverse crops as well as irrigation and water-conservation techniques are concerned. And for the urban gardeners, they could look at large-crop farming techniques and extract best practices and the love for the land.

Ultimately, one of the most important mega trends that may evolve from all this is simply a convergence of traditional and urban agriculture, every person becomes a farmer of sorts and looks after the land, even if it is a 3,000-square-foot roof garden of a high-rise building in an inner city. Another convergence one might expect to see is the one between farming and production of energy, especially electricity but not exclusively as the Güssing example given previously in this chapter. I have described and discussed the possibility of installing high-rise solar panels in crop fields and windmills of any size could be installed in other large crop fields, much like the olive trees in fields of wheat described by Homer some 2,500 years ago. The name of the game, or rather mega trend, is convergence of symbiotic entities, in this case agriculture of any size and type and manufacturing of energy, electricity and gas (methane) and all this in increasingly close proximity to the locations of where people live, even in inner cities.

The very creative inventor and color physiologist Erich Chiavi from Davos, Switzerland, suggested and created a model, which started with the idea of “Vision Hill” and what he called “Hügelstadt” (town on the hill) back in 2000. The fact that cities, especially medieval towns, were built on hills for easier protection and defense against marauders or pirates when close to the sea is nothing really new. What is new in Chiavi’s idea is his concept of creating an artificial and hollow hill. On the surface of this ideally semi-circular shaped hill he imagines housing for people as well as small terraces and gardens, ideally used for growing food, while the hollow interior is used for work, schools, shops, and parking and storage as well as all necessary utilities. Figure 10.1 shows a first concept drawing by Chiavi, and Figure 10.2 depicts a simplified model of such a town on a hill.

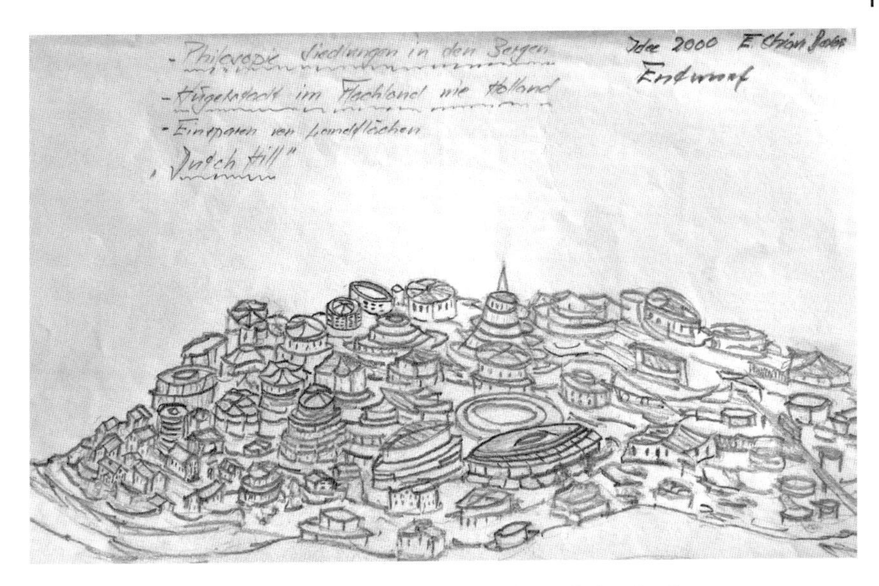

Figure 10.1 Town on the hill drawing. © Erich Chiavi, Davos Switzerland.

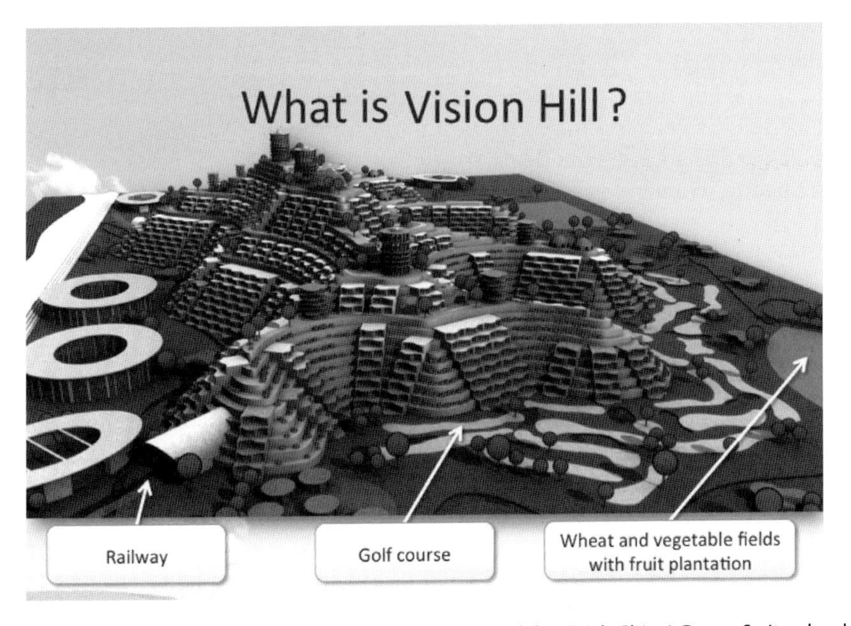

Figure 10.2 Vision Hill: Town on the hill simplified model. © Erich Chiavi, Davos Switzerland.

This concept is capable of integrating so many human activities and would especially be adapted to countries with flat land, which at the same time could help to deviate or even disrupt catastrophic weather events such as hurricanes. "Mutual encroachment" of diverse activities is the name of this new mega trend that could become an important element to a more sustainable society at large. One important answer to this mutual encroachment has to be given by architects. It was not so long ago that architects were building structures of any kind and discovered the "real south direction," so that solar panels could be installed harmoniously and logically. The next big thing is this integration of places for human habitat and agriculture, ideally including transformation of agricultural output to foods, both industrially and individually at home. The depicted hill town could be part of the solution.

10.3 Agriculture and Food Manufacture in Exotic Places

In the years between 1994 and 1996 I was flying back and forth between Switzerland and Florida—Orlando to be more precise. Orlando was and still is the location of Disneyworld, especially the EPCOT Center. EPCOT, short for "experimental prototype community of tomorrow," opened back in 1982. The opening date is slightly over 35 years ago and in those days all the different pavilions appeared to be futuristic and really prototype. Today, much of what was dreamed up then has become trivial reality today. One of the pavilions, the "Land Pavilion" was sponsored by Nestlé USA in the 1990s and, by some networking and chance, I was designated to be one of a small team of Nestlé and Disney representatives that had the task to look into the various attractions of that pavilion and possibly come up with some new, refreshing, and up-to-date prototype worthy ideas as to how these attractions could become even more attractive.

Apart from the usual food- and eating-related entertainment in the shape of several large food courts, there were mainly three attractions: a 15-minute film on the theme of the circle of life with music and song from the *Lion King*, composed and sung by Elton John. The second attraction was called "Food Rocks" and was performed, if one can apply this term, by animatronics figures, and lastly "the boat ride through the land," which was not only entertaining and interactive but highly educational. The circle of life short movie was just perfect, emotional and really showed this interconnectedness of everything living on our planet, from plants to animals to humans and food and stable and healthy environment for all. The food rocks animatronics show was kind of old-fashioned, already only 12 years after its inception and the message was "thou shall not" to every nutritional sin that one could possibly commit. Instead of giving out positive messages, it was very much biased toward the negative ones, almost like the German Struwwelpeter stories of 1845, which frightened the holy daylight out of every child, while the parents were hoping that their children would obey.

It didn't work then and we felt that it wouldn't work now, but changing this show was seen as too complicated and too costly. Our focus then turned to the boat ride through the land, and I was part of an effort to modernize and carefully rewrite the story that was told while gliding in a boat through a narrow canal inside and partly outside the building of the Land Pavilion. This was much fun as the story was all about humans and their relationship to the nature and especially also the onset of agriculture in the times when people, tribes, and communities began to settle down and needed to grow and breed their food in sustainable and respectful ways. A large portion toward the end of the ride was dedicated to vertical growing of plants to dig for new potential arable land to grow enough food for an increasing world population.

At the very end of this boat ride through the land was a small showcase laboratory from NASA, more like a shop window though, which had as its central theme growing food in space under low- or no-light conditions, and it covered the theme of growing food in space. I cannot recall all the details of what was showcased but it almost looked like much of what was one of the central themes of the 2016 movie *The Martian*: how to grow food in difficult and unusual environments, and basically recycling everything—really everything—to maximize the agricultural output. Space is such an exotic place and growing food in space could potentially teach us a lot how to optimize every single step in the agricultural value chain. Additionally the zero- or low-gravity environment can teach us additional elements in this entire chain of events of growing and possibly also manufacturing food. The target here is not so much, if at all, the small and number-wise restricted community of astronauts and people traveling into space but humanity at large, on our very earth, simply through learning through restrictions and the need for totally recycling of waste and nutrients of any kind. We definitely don't do this enough in today's agriculture and food manufacture. This topic of agricultural waste joins the one of food waste quite seamlessly and has to be tackled urgently if we want to feed all people on our planet.

10.3.1 An Ice Cream Factory in Greenland?

An ice cream factory in Greenland may sound a bit strange and may not really be doable. On the other hand, it is also a fact that one of the largest ice cream factories in the world is located in Bakersfield, California.

For those who may not know, and that's probably the majority of the readers, Bakersfield is probably one of the hottest places in the state of California, more specifically the Central Valley. There is of course a reason for its location, which has got to do with short supply chain of ingredients, especially milk and dairy products. Dairy cattle is close by, hence it seemed to be logical to put the factory there. It's similar to the locations of pet food manufacturing sites in the states of Kansas or Missouri, all sites in close proximity of slaughterhouses.

The question here is what makes more sense or what is easier to do: either use naturally present cold temperatures in the proximity of working temperatures to manufacture ice cream and accept long distribution distances both for incoming and outgoing materials or alternatively, what is done thus far, put a factory in a hot or very hot place because it is closer to the raw material streams as well as the ultimate consumer of the product. It's almost a thermodynamic decision as to how much of a temperature differential between exterior temperature (e.g. 40 °C) versus ice cream manufacturing temperatures (e.g., down to −40 °C) we are ready to accept, especially from financial and environmental aspects. Most or all industrial cold is produced via ammonia evaporation cycles and requires rather large amounts of ammonia in such factories. Ammonia inhaled in small quantities clears up your sinus, larger amounts however are rather toxic. You don't want to be around too much ammonia in the air. Its toxicity restricts the use of ammonia mostly to industrial applications, and ammonia is neither found in household cooling applications nor air-conditioning units. This is by the way one of the reasons why ice cream factories are always really big; it's the economy of scale, not so much of the product itself but the generation of cold and its preservation.

In either case, outbound logistics, or in simpler words, the distribution of final products requires an uninterrupted cold chain, so the real differences are limited to incoming supply chain (arrival of raw materials in the factory) as well as the creation of cold for manufacturing and initial, short-term storage in the factory. Given the rather low-world market prices and free capacities for shipping goods over long distances, it might be worthwhile to calculate these and compare these additional costs (mostly inbound supply chain) to the potential savings of a much lower temperature differential of maybe 20 ° or 30 °C instead of as high as 80 °C as would be the case for a factory such as Bakersfield. I do realize that proper insulation of building can do miracles these days; however I have seen ice cream factories in other places in southeast Asia, which did not seem to care much about heat loss (or rather cold loss). It could be an exciting topic for a doctoral thesis in chemical engineering to study such a case of a hypothetical ice cream factory somewhere in Greenland or another cold place, taking all these parameters into account and possibly be surprised by the outcome.

10.3.2 A Chocolate Factory in Ghana?

You think that making ice cream in Greenland was an outlandish example of manufacturing food in nontraditional, more exotic places? Let's try this one here: make chocolate in Ghana. Well, I say Ghana because I am from the old school, and Ghana used to have the highest quality cacao beans, especially cocoa butter. I could use Côte d'Ivoire or any other cocoa manufacturing country around the globe in this $+5^{o}/-5^{o}$ latitude around the equator. Let me stick to Ghana though, just to illustrate my example and let me quote a few cocoa-related facts from the *Ghana Business News* (2015):

- Most important export crop, accounted for 8.2 percent of the country's GDP and 30 percent of total export earnings in 2010.
- Total production from 450,000 tons in 2000 to 900,000 tons in 2010. 90% of total production grown by smallholder farms.
- All cocoa beans are sold to Licensed Buying Companies. Main export destinations: European Union, Japan and the United States.
- Ghana is the world's third largest producer.
- Second largest exporter of cocoa beans after Côte d'Ivoire. It has been estimated that in 2010/2011 Ghana's exports of cocoa reached 1,004,000 MT (GAIN, 2012). Ghana has maintained its position as the 2nd largest exporter (by quantity) of cocoa beans for the period of 2005–2011.
- The country ranked 8th, 9th and 7th in cocoa butter export in 2005, 2006 and 2009, respectively. In 2010 exports of cocoa butter and paste to the USA increased from $32 million to $86 million, most likely because of the higher quality of cocoa products produced in Ghana (GAIN, 2012).
- Cocoa is considered to be the highest export crop earner for Ghana accounting for 8.2 percent of the country's GDP and 30 percent of total export earnings in 2010 (Ghana Statistical Service, 2010; GAIN, 2012).

The point I am trying to make here is the absence of finished cocoa products, chocolate for one in this entire set of statistics. Why is that so? Can Ghana not make its own chocolate and export it to countries where chocolate is consumed? I do realize that consumers across the globe like to argue about which country makes the best chocolate, like Switzerland, Belgium, or the United Kingdom. And consumers seem to be proud that they only eat the best chocolate, coming from a place well known for its high-quality chocolate, simply because they believe so, and they have acquired the taste for it.

Why wouldn't Ghana make chocolate that goes out to the world—Chocolate that is made from the freshest cocoa beans and has not been filled and shipped in jute or other material big bags and are manufactured only months after they had been harvested?

I still recall a visit in the "Speicherstadt" in Hamburg, Germany, where a buffer stock dozens of years old was proudly pointed out to me. Robin Dand (2010) mentioned that for price stabilization purposes "during 1981 and 1998 up to 250,000 MT of cocoa beans were stored in northern European ports alone, at the time a significant component in overall cocoa stocks" (p. 209).

The idea behind such buffers is to store excess cocoa beans during years of good harvest and take them out of the demand stream and put them back into the stream when harvests were underperforming. On paper this looks like a good idea because it can, at least to some degree, keep prices more stable and less fluctuating, but on the other hand, it's pretty ridiculous when one looks at the potential degradation of raw material quality, even under the best storage conditions. From a pure quality point of view this approach of buffer stock is rather pathetic and just shows that the industry today uses fresher beans complemented with buffer stock of really old, definitely less fresh quality. I have worked in the business of chocolate and confectionery on the development side and have quite some experience and knowledge in this area. I can say with authority that there is a dramatic taste difference between cocoa beans not only based on their origin but especially also based on their degree of freshness.

Here a short version from the International Cocoa Organization (ICCO) Web site (2013) as to how cocoa beans are processed, and what is required to make good chocolate. Nowhere does it say that the beans have to be stored for a long time before they undergo the first step of roasting.

Step 1. The cocoa beans are cleaned.
Step 2. To bring out the chocolate flavor and color, the beans are roasted.
Step 3. A machine is used to remove the shells from the beans to leave just the cocoa nibs.
Step 4. The cocoa nibs (can) undergo alkalization, to develop flavor and color.
Step 5. The nibs are milled to create cocoa liquor.
Step 6. Manufacturers generally use more than one type of bean in their products.
Step 7. The cocoa liquor is pressed to extract the cocoa butter, leaving a solid mass (presscake).
Step 8. The cocoa butter is used in the manufacture of chocolate. The presscake is broken into small pieces, which is pulverized to cocoa powder.
Step 9. Cocoa liquor is used to produce chocolate through the addition of cocoa butter. Other ingredients such as sugar, milk, emulsifying agents and cocoa butter equivalents are also added and mixed.
Step 10. The mixture then undergoes a refining process through a series of rollers until a smooth paste is formed.
Step 11. The next process, conching, further develops flavor and texture.
Step 12. The mixture is then tempered.

Step 13. The mixture is then put into moulds or used for enrobing, and cooled in a cooling chamber.
Step 14. The finished product is then packaged for distribution to retail outlets.

So, let me go back to the simple and easily understandable argument: the fresher the raw material, the better the quality of the finished product. It's valid for my vegetable soup that I (or rather my wife) cook(s) at home, and it is equally valid for making chocolate with the freshest beans in reasonable distance from the harvesting sites. Ghana, or Côte d'Ivoire for that matter, could and should make their own chocolate and not send most of their cocoa beans to the "finishing country." When you look at the 14 steps from ICCO, you quickly see, or rather do not see, any substantial step of this value chain other than the very first one (cleaning) taking place in the origin country.

By making chocolate "on the spot," the quality of the finished product that you as a consumer will experience can be *so* much better, I can promise you this. Ask your preferred chocolate manufacture to move their production to origin countries and optimize their supply chains to get the chocolate to you and into your mouth in the fastest possible time frame. Here some short information on time frames. First, cocoa beans are shipped and stored in warehouses and silos for weeks, if not months. Second a chocolate product is typically consumed 4 to 5 months after its manufacture. The beans that were used to make your chocolate may easily be 9 to 12 months away from the day they were harvested by then!

10.4 A Future for Food

Food will always exist because people will always have to eat, unless of course we replace ourselves by robots and become ethereal creatures. Because there is no time frame on the latter that I am aware of, we can safely stick to the former: people will always have to eat. What does that mean for the future of food? How will food be in the future, will it be the same as today, will the food industry still exist like today, will agriculture still be the main supplier of food, and will agriculture and food industry merge in ways that we cannot even imagine today?

10.4.1 What about the Role of Restaurants?

As an attentive reader of this book you may have realized that I did not discuss nor mention one player of the food industry, namely restaurants. I didn't do this because I had forgotten but felt no real inclination until now. The question is simply: how do restaurant trends determine present and future food trends in sustainable and anticipatory ways? I do believe that restaurants all over the world play a role but, in my opinion more of a solidifying one rather than torchbearer of new things to come. And if they lead a movement, such as a molecular kitchen, then it can be seen that such a trend, or rather a fashion, is not embraced by many, and is rather elitist and possibly short-lived. To be frank, most restaurants and restaurant-like places do follow emerging or already established trends, modulate them, make variations of them, and ultimately solidly establish them in the mainstream.

Even if some chefs try to make a name for themselves because they may try out the "unspeakable," such as molecular cuisine or old embalming techniques to preserve meat and modify it in possibly surprising ways, very little of this has had, thus far, a real bearing on the food industry and has not really influenced food trends in a leading role and might not do this in the future. But who am I to see the future? I can only guess it. However, I guess it on certain indicators I can see in the industry at large, in the present "diet-mania," in consumers' desires toward a more vegetarian lifestyle, and certainly in the kind of fatigue that many consumers feel with industrial food that they see as processed. Again, there is a future for food and knowing the industry and also the ways as to how many consumers tick—they say and believe one thing, but they do not always react accordingly—the future of food might just be similar to the past and the present. It may be more refined, more personalized, more back to the roots, more individual input to the preparation of one's food, more respect to the environment and sustainability at large (whatever that means in the minds of different consumers) and maybe—hopefully—less of the approach of lower and lower spending on food, while the very same people might spend more and more on cars, vacations, and consumer electronics.

10.4.2 Pet Food Is Food, Too

Let me repeat the truism: people will always eat and food of one kind or another will always exist. I would, however not restrict this need for food to people, to human beings, because there is a growing trend in all societies to feed the pets we have at home. Hence, there is the growing segment of pet food—pet care in general with an increasing number of pet-related retailers such as Petco of California or PetSmart of Arizona. Both retailers operate basically countrywide in the United States and have grown substantially in the last couple of years. Petco, which was founded more than 50 years ago, also has subsidiaries in Mexico and Puerto Rico. PetSmart was founded some 30 years ago and has outlets in Canada and Puerto Rico. During my years with the Nestlé Company, I could experience firsthand the growing importance of pet food, which culminated in the acquisition of Purina to become Nestlé Purina around 2001.

Pet food is food, too, and has become so regulated that every pet food product has to also be safe for human consumption. The rumors, founded or unfounded, that I have heard time and again were that many elderly people with little income to spend on food would eat some type of pet food. I have never seen it personally, and maybe it never happens anywhere in this world but I would not be astonished if some of these rumors are true. Remember, pet food is food, too! I am in no way advocating that you should begin eating pet food. But if you would consider eating insect protein in one format or another (don't forget that insect proteins are always processed!), then maybe good nutritious pet food is not such a far cry anymore.

I emphasize this because of the amount of nutritional research findings that go into pet food and the strict food regulations around pet food. I realize that we do not necessarily require diets that avoid hair balls in our stomach so much; however not all of the nutritional findings that are translated into pet food products are so far-fetched that they wouldn't be good for us humans, too. The main reasons that should and would really hold us back from consuming pet food at all are taste and texture; they may not be to our liking. I had been around enough in pet food manufacturing sites to say this with conviction. You have to be a dog or cat to really like it, and that's a good thing.

10.4.3 Will We Eat Food in Pill Format?

Already in the 1950s and 1960s when the profession of astronauts came into being and was discussed by the public at large, much was speculated as to how astronauts would eat their food in zero- or very-low gravity. A lot was speculated and joked about differently colored pills for different vegetables and other foods: highly concentrated nutrients from various sources, in dry form to be taken like aspirin or any other pill. It so happened that pills became not really the preferred format, however, and concentrated nutrients dried or as paste came into existence and some of this has carried over to a more modern version of so called *performance foods*, the gels or bars of today's sports nutrition of the likes of Powerbar®. To answer the question posed in the title of this section in the simplest way, my answer would be no. No, we will, in all likelihood, not revert to pills instead of eating "normal" foods, and this despite all the talk about medical food, functional food, and health food, and so on.

Yes, these more fancy and to some degree functional foods will exist. On the other hand, what is more functional than a nice fish filet with some vegetables or whatever you prefer it with that satisfies your hunger and at the same time all or almost all your nutritional needs? So, one prime functionality is satiation, another one is covering nutritional requirements, and a third, maybe most important one is good taste, the pleasure of eating, the hedonistic element of eating, beyond simple refueling of energy. Pills cannot give you this pleasure, and neither can sports bars (other than the ones you can walk in) or medical food and nutrients dripping out of a sterile bag. I am not suggesting that the latter is not required for specific cases. but it's very much like food for astronauts: the market size is pretty limited and, with robots taking over many tasks in society, repairing robots may become more of an issue than medical and functional foods for the vast majority of people on our planet.

So, to come to a conclusion, let me repeat and summarize some of the megatrends as far as the area of food is concerned:

- Yes, food will always exist, and people will always have to eat and drink.
- The type of food that people will eat in the future will largely resemble what we eat today.
- The food industry will still exist but will most likely transform itself more toward a knowledge-selling industry, gradually going away from the simple demase model that is over 1,000 years old.
- Consumers today, and probably tomorrow, will gradually turn away from industrial food perceived as processed and will increasingly cook themselves.
- This will most likely accelerate the food industry's transformation to a knowledge-selling industry.
- Automation in the kitchen will be revolutionized and smart cooking robots will increasingly replace traditional kitchen machines that have not much evolved since they first came into existence in the middle of the 20th century.
- Consumers will be able to afford smart kitchen robots similar to how they lease a car today.
- Nutrition research will continue to come up with conflicting and confusing messages, not really understood by consumers, at best cherry picked: what I believe might be good for me.
- Retailers will only survive if they adapt to the new realities of online shopping and potentially transforming their brick-and-mortar sites to manufacturing and distribution sites for food companies. A type of new relationship between retailers and food companies is emerging, which is based on total cooperation and balanced sharing of the value chain.
- Functional foods and beverages of any type and functionality will have their space in the food industry, however, I believe that this space will be small and produce margins that might be too small to sustain this field of business.
- Given that the food industry is and always has been conservative, risk-averse, and slow to adapt and change, it will attempt to maintain its structure and business model, only giving in hesitantly to the required changes.

And don't forget the old saying: When the food comes after you and tries to eat you, serve it with red wine, when it runs away from you, serve it with white!

10.5 Summary and Major Learning

This chapter described, discussed, and analyzed the general topic of food for the future and more specifically included the role of agriculture, present and future in this discussion. The following topics were treated in this discussion in more detail.

- Proactive agriculture was the first element, which was looked into. As a general rule, agriculture can be considered to be proactive as it plans ahead and harvests later. However, often crops are harvested, which have been planted, ignoring changing food trends.
- One major food trend indicates increasing segmentation; traditional agriculture still relies on few mega crops, largely neglecting so-called orphan crops.
- Food security (i.e. the availability of food for all in just and balanced ways) is of highest importance.
- Factors influencing food trends are twofold, societal and environmental.
- Increasing scarcity or overabundance (flooding) of water in traditional food-growing areas is a growing problem, potentially leading to future shifts, both of arable land and populations.
- This will lead to a more careful selection and modern breeding based optimization of crops that will use less water than today.
- In summary, proactive agriculture needs to respond to challenges such as not enough or too much water, development of drought- and saltwater-resistant plants, crops growing closer to where they are consumed, and adaptive crop selection.
- Agriculture will become increasingly democratized; examples of this are the appearance of agrihoods and the increasing interest in permaculture.
- The trend "from large to small" is becoming increasingly popular. Large food companies such as Nestlé ask their employees to operate in ways to get away from the large supertanker mentality toward acting as a group of many small and agile boats. The same trend can be observed in the steady growth of interest in urban agriculture.
- Agrihoods and urban agriculture require new living and working solutions for people who want to become part of this trend. The example of a "hill city" was described, responding to these new requirements for proximity and partaking in growing food on more individual levels.
- As a seeming contradiction to the trend toward proximity, two examples of remoteness of food manufacture were discussed: an ice cream factory in Greenland and a chocolate factory in Ghana. The former would be a choice for energy savings and environmental reasons, whereas the latter would suggest to accept that raw materials should be not only harvested at the site of origin but also be manufactured to the finished product. Today, only one

step of a multistep process of chocolate manufacture takes place in the origin country, namely harvesting of the raw material.

- The future for food was discussed, including the role of restaurants, suggesting that, with few exceptions, they do not really play a leading role in creating new trends.
- The increasing importance of pet food in most developed societies was briefly discussed, and it was mentioned that pet food has to be safe for human consumption.
- Several times in this book the possible advent of medical food was mentioned and briefly discussed. It was suggested that there will be a role and space for functional and medical food, but it is expected to be rather small, given the limited premium that can be charged for anything that is called food.
- In conclusion the major food facts and trends were summarized as follows: Not so much a trend but a truism: food will always exist because people will always have to eat; food of the future will largely resemble today's food; the food industry will still exist but will have to transform itself to respond to new realities; these new realities will be an increasing trend away from industrial food and more home cooking; as a consequence, the food industry will become more of a food-related knowledge-selling service industry; smart automation will finally conquer the kitchen and cooking robots will become standard, not a must but a great to have; nutrition research will remain controversial; retailers will increasingly go online and potentially use their brick-and-mortar locations to manufacture food products in collaboration with food companies; functional foods and medical foods will, at a low level, coexist with regular food products; and every change in the food industry will take a long time, unless it is a forced one.

References

"Central valley project." (n.d.). Available from: https://en.wikipedia.org/wiki/Central_Valley_Project [Accessed January 13, 2017].

Dand, R. (2010). *The International Cocoa Trade.* London: Elsevier.

Ghana Business News. (2015). The sad story of Ghana's cocoa industry and the way forward. Available from: https://www.ghanabusinessnews.com/2015/06/22/the-sad-story-of-ghanas-cocoa-industry-and-the-way-forward/ [Accessed January 27, 2017].

Gussing Renewable. (n.d.). Available from: http://www.gussingrenewable.com/htcms/en/wer-was-wie-wo-wann/wie/thermische-vergasungficfb-reaktor.html [Accessed January 16, 2017].

Holmgren, D. (2002). *Permaculture: Principles and Pathways Beyond Sustainability.* Holmgren Design Services.

International Cocoa Organization (ICCO). (2013). Processing Cocoa. Available from: https://www.icco.org/about-cocoa/processing-cocoa.html [Accessed January 29, 2017].

"Permaculture." (n.d.). Available from: https://en.wikipedia.org/wiki/Permaculture [Accessed January 14, 2017].

Pernett, S., & Maitra, K. (2016). Desalination plant to reopen next fall. *Daily Nexus*, University of California Santa Barbara. Available from: http://dailynexus.com/2016-06-23/desalination-plant-to-reopen-next-fall/ [Accessed January 13, 2107].

"Roman aqueduct." (n.d.). Available from: https://en.wikipedia.org/wiki/Roman_aqueduct#Timeline [Accessed January 13, 2017].

Roth, A. (2014). Before "Agrihoods": America's odd history of planned communities. Available from: http://modernfarmer.com/2014/11/agrihoods/ [Accessed January 14, 2017].

Traitler, H., Hofmann, K., & Coleman, B. (2015). *Food Industry Design, Technology and Innovation.* Chichester: Wiley Blackwell.

11

Summary and Outlook

Saving our planet, lifting people out of poverty, advancing economic growth…these are one and the same fight. We must connect the dots between climate change, water scarcity, energy shortages, global health, food security and women's empowerment. Solutions to one problem must be solutions for all.

—Ban Ki-moon

11.1 Introduction

This is the first book of its kind that connects the roles of food and the food industry with agriculture and water resources and management in detailed and thorough ways. Most publications and discussions among policy makers, technical experts, and community representatives do not, at least most of the time, consider all three elements as highly interconnected as they really are: agriculture as the source of all food—basic or sophisticated, homemade, chef made, or industrial—and the very existential role in this equation that water plays. This book brought together a small community of expert authors on the topics of food industry, agriculture, both for plants as well as animals, and water and its role in a world of diminishing resources that we face today.

Let me end with a summary of the most important topics that were discussed in this book. As controversial as some of the proposals may have been, they all serve one common goal: thinking about the future of agriculture and food industries. This chapter will end with a short outlook based on many years of work in and for the food industry and many personal observations based on experience gained throughout these many years. This outlook should ideally lead to much debate and even better, controversy. This would be the ideal ground for moving forward in fast and fitting strides toward food industry and agriculture of the future.

Megatrends in Food and Agriculture: Technology, Water Use and Nutrition, First Edition.
Helmut Traitler, Michel Dubois, Keith Heikes, Vincent Pétiard and David Zilberman.

11.1.1 The Role of Agriculture in Today's Food Industry

Agriculture was the first topic of discussion and analysis. What it actually means and how it affects food security and society at large. Agriculture is based on several pillars such as

- Arable farming
- Cultivating the soil
- Growing many different crops for food, horticulture, fibers, forest products, and plants
- Breeding and raising livestock and animal husbandry
- Preparation and marketing of resulting products, especially food

And for all this, today's societies request agriculture to use fewer resources and be more careful with regard to the environment. This is not an easy task but luckily often goes in the same direction as the economic parameters surrounding the industry of agriculture. Another part of the equation for successful and more importantly sustainable agriculture is expressed by these four critical building blocks for growth and sustenance of life in general.

- Water; it appears trivial, and it is; however, there is an increasing disconnect between traditional areas of agriculture and availability of water for the purpose of growing food securely.
- The appropriate mix of the major "growth molecules and atoms," such as oxygen, hydrogen, nitrogen, carbon, sulfur, and phosphorus, all present in nature in different formats.
- Light, especially visible light, but not exclusively.
- Temperature, the most comfortable growth range that is, ideally between above 273 K (−0.15 °C) and below 325 K (51.85 °C).

Living beings and especially plants can adapt to many different, sometimes even extreme conditions, but a certain minimum or optimal range of all these four building blocks is required for growth.

Some history of agriculture was discussed. The very onset of agriculture in one form or another is based on the need of every living being to eat and drink. Humans in the beginning hunted and foraged for food of any kind. When they became settlers, first seasonal, later permanently, the need for growing crops and domesticating and raising animals became apparent and can most likely be linked to the appearance of agriculture as we know it today.

There is a first glimpse at nutrition in the old days. When people still had to hunt for food, it was not guaranteed that they would be successful at every attempt. So those who could eat most would be the ones who were more likely to survive and outlived other members of their families or tribes.

This does not really look like anything present-day nutritionists would recommend, nevertheless it's part of our food history: eat as much as you can while it's here and while it lasts. This might have been one of the reasons why people eventually became sedentary, and they turned to growing food and assumed one of the major principles of agriculture and food: food security, namely the permanent availability of food of any kind possible, depending on area and season.

11.1.2 Food-Preservation Techniques

Once growing crops and raising animals for consumption the next question of relative importance was how to make any type of food keep longer, become storable, and still be good for consumption after some time. So, preservation and appropriate techniques were needed. One of the oldest techniques is drying and is first documented as early as some 14,000 years ago. It might be much older; however there are no known documents to prove this. Other techniques that followed and were later introduced as an entire set of possible means of preservation, not necessarily in exact chronological order, were:

- Cooling
- Freezing
- Boiling
- Heating
- Salting
- Sugaring
- Smoking
- Pickling
- Alkalinization
- Canning
- Jellying
- Curing
- Fermentation
- Pasteurization
- Vacuum packing
- Food additives
- Irradiation
- Pulsed fields (electric, magnetic)
- Modified atmosphere
- Ultra-high pressure
- Bio-preservation
- Hurdle technology (hurdles to microbial growth)

Not quite as old as some of the food companies that are using some of these techniques to make their products, and one company even for 1,000 years. The main techniques used there are fermentation (beer), drying, smoking, and curing (salami and prosciutto).

11.1.3 Agriculture Is the Main Raw Material Supplier to Be Transformed to Food

Agriculture, more than ever is still the main supplier of raw material that is transformed to food, either individually or by the food industry, and this despite the rise of the industry of food ingredients. Some of their products are critically discussed by consumers as well as health professionals. A typical example for this is artificial color, which was, and still is, suspected to be responsible for certain child disorders such as attention deficit hyperactivity disorder. The food industry has made quite some strides to use natural colors for all products that require colors such as Smarties® or M&Ms®.

However, the vast majority of food raw materials stem from just a few crops, and especially dairy products. Here are some European Union numbers mentioned previously

- Cereals: 324 million tons
- Sugar beet: 128 million tons
- Oilseeds: 24 million tons
- Vegetables: 29 million tons (tomatoes, carrots, onions)
- Fruits: 14 million tons
- Grapes: 23 million tons
- Olives: 8 million tons (represents 75% of world production)

The 2014 milk production worldwide was suggested to amount to more than 735 billion liters.

11.1.4 Nonfood Uses of Agricultural Raw Materials

Although most of the agricultural output goes to food for human and animal consumption, there are some “exotic” outliers. Milk can, for instance, be used for the manufacture of plastics, textile fibers, glues, ethanol, and methanol. Corn is used for the production of plastic materials such as polylactic acid, and sugar cane is used for the production of ethanol for automotive usage. Both examples have some detrimental effect to the economy of agricultural raw materials.

Some alternatives were discussed and analyzed, most prominently the development of halophytic plants (saltwater resistant) for the purpose of fermentation to methane or ethanol.

11.1.5 Agriculture in a World of Rules and Regulations

Today, all food and beverage products are increasingly regulated for their safety and also for compliance with agreed-on manufacturing standards as well as allowed claims. Agriculture is at the beginning of this chain of regulatory steps and traceability of potentially dangerous and unhealthy components is a must. On the one hand, these regulations are really helpful and good; on the other hand they can have a tendency of exaggeration and may have a stifling effect on some innovative developments. On the whole, they are good and necessary and as the saying goes: better safe than sorry.

Almost every decade starting mid-20th century had its own set of outbreaks of food-related safety issues such as salmonella in eggs, *Escherichia coli* in undercooked meat, *clostridium botulinum* in improperly canned foods, *listeria monocytogenes* and so on. It is safe to say that there was always an outbreak first and reaction to it afterward.

11.1.6 Food Raw Materials and Process Became More Sophisticated and Complex

The entire industry of agriculture and food have become more complex and more controlled in the latter part of the 20th century; the industry finally became adult and had to comply not only with regulations as described but also had to respond to an increasing number of questions and pressure from consumer organizations and governments as well as nongovernmental organization other than consumer groups. Everything in the entire chain of growth, harvest/slaughter, transformation, and distribution and sales had to be controlled and, more importantly, had to become traceable, which was a fairly new requirement to the entire industry. This made all operations not only more complex and potentially costly but also more transparent.

Yes, food became safer, however, it also became too expensive for many, even in the more affluent societies. In the United States alone, approximately 44 million people receive food stamps because they cannot afford to pay for their daily food needs themselves. This will have to change, and both agriculture and the food industry can play a part in this change. Plants have to be bred that can withstand droughts and higher soil salinity, animals have to be bred and raised that have a better raw material-to-protein transformation rate, and new protein sources such as insects will have to be considered.

11.2 Water Management in Modern Agriculture

Water is essential for survival as well as economic activities as a whole. With population and income growth along with concern of climate change, there is a growing perception that water will be a binding factor on economic and agricultural growth. The chapter argued that water problems can be solved by better management and taking advantage of technological development. Reform of water management practices can increase both economic and environmental benefits of water. The following was discussed and analyzed in some detail.

Water resources are rich and diverse. It is important in analyzing water-resource management to identify sources of heterogeneity to develop focused strategies that address specific problems and integrated solutions that span multiple dimensions. The first distinction of water resources is by source, which can be distinguished by precipitation (rain-fed), surface water, and groundwater. Precipitation is a flow and groundwater is stock. Surface water can be divided to oceans and seas (salt), freshwater seas, and lakes, all of which are stocks, and then rivers, which are flows. This overview suggests that water-resource management has to recognize dimensions of space and time as well as randomness.

11.2.1 The "Water Reform"

A key to water reform is to introduce policies that will enhance incentives to increase water productivity and decrease pollution. Irrigation is crucial for production of food, but it is mismanaged. Allowing trading that allows water allocation between regions and pricing, and other mechanisms, that will allow water users to pay for the consequences of their action, both in terms of cost of water delivery system and impact on the environment and future resource availability, is likely to lead to increased adoption of efficient water technologies and use of irrigation when it provides value.

Economists have recognized the importance of institutions in shaping resource allocation and the important role that political economy considerations play in development of institutions, policies, and incentives. Institutions emerge in response to economic, technological, and political conditions and changes in conditions require institutional changes. However, institutional transitions are difficult to implement and they are the major challenge of water policy reform. Three major factors are likely to affect water institutions and organizations:

- Relative scarcity of water,
- Government ability to tax and finance public projects, and
- Policy objectives of government and society. The history of US water policies can illustrate these points.

11.2.2 Water Productivity

Water productivity can be enhanced by improved use of complementary inputs, including seed varieties, fertilizers, and other inputs. So investment in R&D and extension services, improved infrastructure, and regulatory systems that increase overall agricultural activity will improve productivity of water use.

Irrigation allows for the expansion of agriculture to arid regions that may be productive with irrigation. Adoption of technologies, like pumps, has allowed for increased productivity on farms as well as increasing the area of irrigated agriculture. Agriculture is the output of multiple inputs, often complementing one another (e.g., soil, water, fertilizer, seeds). Thus, increased productivity of fertilizer may increase the use of water as long as prices are the same. The adoption of irrigation leads to indirect gain in output when it leads to the increased use of other inputs. In some cases, though, the yield effect of improved irrigation may lead to increased supply and reduced prices, and that may curtail overall agricultural land.

Increase in water productivity, in turn, may reduce water use over time. Improved information technologies should be taken advantage of in design of water and agricultural technologies, in general. Expanding the integration of global agriculture within the knowledge economy can lead to increases in precision and efficiency.

11.2.3 Water-Related Government Policies

Government policies and water projects will continue to play a major role in water systems. Water project design should be a multidisciplinary exercise that incorporates structural as well as nonstructural measures.

Projects should be evaluated using benefit-cost analysis that takes into account environmental considerations, emphasizes adaptive learning, and is opportunistic with respect to timing of investment. Policy design should emphasize equity considerations and develop mechanisms, like tiered-pricing, as well as subsidies for adoption of technologies that allow lower income households and farmers to access and benefit from water.

Efficiency of agricultural water use is dependent on the pricing of variable inputs as well as the availability and timing of water. These factors are the outcomes of regional water institutions and policies. For example, the literature suggests that a critical element that affects water-use efficiency is the ability to trade water, and the allocation of water among farms. This is especially true when it comes to surface water. As we mentioned, in many regions of the world, water is allocated by water-rights systems (e.g., the prior appropriation system). A strict interpretation of this system disallows water trading because once farmers sell their water to others and don't use it, they lose their rights to the water. Thus, under water-rights mechanisms, the gains from trading are lost. Transition from water rights to water trading would allow higher efficiency.

11.2.4 Getting It Right: Policies and Price

Measurement of agricultural systems is essential for effective policies, and water policy should emphasize improving water quality and reducing the environmental side effect of water. Getting the price right is important for managing water quality as much as quantity. When possible it is useful to adhere to the polluter-pays principle or to use mechanisms like payment for ecosystem services to reduce pollution.

There is significant social cost for underinvestment or conveyance or for the use of uniform pricing for water over space. These suboptimal systems result in lower production levels, higher output price for consumers, greater conveyance loss, less adoption of modern irrigation technologies, and shorter canals. Uniform pricing results in underproduction close to the water source and overproduction further away.

Appropriate water pricing is important in the case of groundwater. In many regions of the world groundwater aquifers are shared by many producers; thus, their use patterns may suffer because of the "tragedy of the commons" problem, namely, each individual user doesn't take into consideration how their action will affect the overall dynamics of the source in the future. Thus, early in the use of an aquifer, cost of pumping may be low, farmers may overuse water, use cheap and inefficient pumping equipment, and that will result in the slow depletion of aquifers. For example, levels of aquifers have declined significantly in India as a result of overpumping.

One solution is that the government charge a price per unit of water pumped, reflecting the social cost of pumping. An alternative policy is creation of tradable groundwater quotas that limit the amount of pumping in a given region per time period. To introduce and enforce such policies, governments need to monitor groundwater use. Indeed, in some countries, for example Israel and Australia, a key element of water reform has been monitoring and control of groundwater use. These policies have led to improved productivity of water.

11.2.5 Controlling Water Quality

Control of water quality should recognize market structure considerations as well as information imperfections and develop mechanisms and technologies to overcome these constraints. Water policies must be modified periodically to take advantage of new technological capabilities that increase precision and attributional capability.

Maintaining water quality is a major challenge globally. Water-borne diseases account for roughly 4% of all deaths in developing countries. Further, 41% of people lack access to basic sanitation, and unsafe environmental conditions cause 25% of all children deaths. Water is contaminated by human, animal, and industrial by-products and design of policies to reduce environmental health risks has to control the risk-generation process that includes several processes: disposal, transfer, exposure, and vulnerability to waste. Each of these processes can be affected by policy. The disposal of toxic residues can be reduced both by reducing their use (e.g., by using more precise irrigation technologies, switching away from more toxic chemicals), which can be affected by regulations and incentives (e.g., pollution tax). The transfer of waste can be controlled by establishing effective sewerage and drainage systems, and subsequently either recycling the waste product or depositing them in areas where they cause minimal damage. The exposure to low-quality water can be reduced through filtering and control of water supplies.

11.3 Innovation in Plant Breeding: High-Quality Plant Raw Materials for the Food Industry

11.3.1 Agricultural Plant Output: The Essential Raw Material Source for the Food Industry

Plant products are an essential part of the raw materials used by the food industry either directly or after extraction of flavors and ingredients. They not only constitute the major part of the source of calories but also the basis for end product differentiation and indulgent products. Many different plant products must therefore be available to the industry in required quantity, quality, and safety but also when processing fresh products on a timely basis at the gate of the factory.

Plant products for the food industry are diverse from basic ingredients such as sugar or starch, potatoes or wheat, to spices herbs and flavors that can differentiate the final products. Some can be stored and are often traded as commodities, but others must be transformed as fresh products, and their supply has to be organized in some kind of virtual vertical integration to secure the seasonal supply.

The price of plant products might be a major part of the ex-factory cost or, on the contrary, a minor part of it, thereby opening new opportunities even at a higher initial price because it will not really impact the production costs of the final product.

11.3.2 Demand Forecast Based on Food Requirements

Chapter 3 was divided in three sections. The first one analyzes the forecast of the demand for plant products with regard to production of food. As it has often been shown, it appears that the world agriculture is facing a serious challenge for feeding the population in the next 20 or 30 years. The increasing demand for nonfood products from the agriculture (e.g., energy, industry) is making the situation even potentially more difficult.

11.3.3 Genetic Improvement of Cultivated Crops

The second section highlights the importance of genetic improvement of the cultivated crops used by the food industry. It will be the main approach for increasing the productivity and the quality of the products but also adapting the varieties to climate warming and to the natural and constant evolution of pest and diseases. It is not to say that agricultural practices will not have to evolve and take benefits from most modern technologies (e.g., precision agriculture, drop irrigation), but without genetic innovation and plant breeding, agriculture could probably not meet the needs of the population. This section is therefore focused on a simple description of plant breeding, its limitations, and all the technologies that have been developed for overcoming them. It shows that plant breeders and geneticists have made huge technical progress during the last 50 years and today, and new innovations will still be welcome, but breeders already have an amazing toolbox to tap into according to the crop, the objective, and the stage in the breeding process. Plant breeding is not much limited by science and technology.

11.3.4 The Major Crops versus "Orphan Crops"

The third section of Chapter 3 makes a simple observation: there are a lot of ongoing efforts in plant breeding, but they are focused on a limited number of crops that are profitable for the seed industry. Corn concentrates 30% to 35% of R&D on main crops and tomato 50% of R&D on vegetables simply because the seed market of these two species is competitive and profitable. Many other species do not receive any significant attention from the seed industry, and therefore they have been called "orphan crops," as in pharmacy and "orphan diseases." The future of their supply is clearly at risk for environmental reasons but also economic ones if not any more attractive for the growers. We analyze different options for giving more security to their future, and it appears quite clear that the food industry will have to get directly or indirectly involved in their genetic improvement. Without looking as vertically integrated because of image and public relations' risks, it will have to invent new business models by putting value on seeds or plants of these species and making them attractive for innovation by new seed companies that will focus on these crops and take the benefits of all plant breeding technologies already validated on few main field crops or vegetables.

11.4 The Agriculture of Animals: Valuable and Sustainable Sources for the Food Industry

Livestock and animal agriculture have a major impact on the global economy. Animals are the largest users of land in the world, and animal agriculture employs 1.3 billion people. It is easy to overlook the fact that in developing countries, ownership and raising animals can be the main livelihood for many people. Animals provide a substantial amount of dietary protein consumed globally, and it is a high-quality protein that provides essential micronutrients.

11.4.1 Growing Population: Growing Amount of Livestock

The world population is expected to reach 9.725 billion by 2050. Consumption of meat is expected to grow by 73 percent and 58 percent for dairy products during that time. Global livestock production is growing rapidly with faster growth in developing countries than developed countries. This trend will continue because of the driving forces in developing countries of population growth, income growth, and urbanization.

The way livestock is raised around the world is quite diverse, from large-scale intense operations to many small holders. This will continue, although a larger and larger share of production will take place on large-scale farms. Production in developing countries will increase, but there is opportunity because productivity and efficiency is low compared to the developed world. The competition for scarce resources will demand that productivity in developing countries must increase.

11.4.2 Animal Health and Intensive Farming

Everyone wants animals that are healthy, treated humanely, and do not harm the environment while at the same time contribute to the economic well-being of the farm. Farmers want to and will treat their animals well, but sometimes production methods are not understood. This in turn causes friction between the farmer and the consumer.

Intensive farming is efficient, not only from an economic point of view but also when viewed from the standpoint of how resources are used. Less total resources per unit of output are used in an intensive-management system versus an extensive-management system. Technology will emerge that can be used on farms to reduce the amount of methane gas released into the atmosphere. If economically viable, they will see widespread use.

Although there will be more production in intensive-management systems, production will occur in a wide range of methods because local conditions affected by economics, climate, politics, tradition, and a variety of reasons. Efficiency of animal production must increase if the projected increase in production is to happen. There is not enough land to support two-thirds more animals. Humane treatment of animals is important and will become increasingly more important in the consumers' view.

11.4.3 Animal Breeding

Animal breeding will focus more on total animal well-being and economics rather than so much focused on individually measured production traits.

The ability to do a better job of selecting for health and wellness traits will emerge and become a major part of breeding strategies. Breeding for efficiency of production in the use of resources will also occur. This will reduce the environmental impact of animals, but there will also be the possibility to breed and manage for lower methane gas emissions. Smaller dairy animals will be bred which will reduce their environmental footprint.

Genomics will play a major role in breeding with the emergence of the ability to have genetic evaluations for health and fitness traits early in an animal's life. This will speed up genetic progress and also reduce treatments that are required for sick animals. Other areas of technology such as gene editing and the study of epigenetics are emerging and hold the potential for making a large impact on the production and well-being of animals.

Data is important in animal agriculture and is needed for both managing a herd and to be used in the calculation of genetic evaluations. Phenotypic data is the basis for all genetic evaluations and has been collected for many years in developed countries. Previous focus has been on production focused traits, while in the future more traits will be measured that are related to animal health and well-being.

11.4.4 Good Farm Management: Good Data Management

Good data is important for a farm manager and must be available in a way that is easily accessible and can be used for making fast management decisions. Emerging technologies like sensors will be commonly used on farms to collect much more data than is now being gathered. Sensors will be integrated across the farm and will provide the farm manager with accurate details regarding the animals on the farm so they can maximize production and minimize negative health events. In developing countries, lack of a data-collection system is a large problem and decreases the opportunity for the use of local breeds in modern breeding programs. Although local breeds offer genetic diversity as well as disease resistance and heat tolerance, there will be little genetic progress as long as a method to collect phenotypic data does not exist.

11.5 The Food Trends—the New Food—Enough Food?

While the first chapters of this book dealt with specific aspects surrounding agriculture and agricultural practices, including the role of water and technologies of importance in plant breeding and animal husbandry, Chapter 5 focused much on food and especially historical food trends. This look back served for a better understanding of present and possibly predicting of likely future trends. Predicting the future is at the same time easy and difficult, especially when it comes to trends in the area of food. Easy because forecasts for a period of say 20 years that turn out to be false are largely forgotten until then and difficult because correctly predicting trends is of enormous importance, both on public health and economic bases.

11.5.1 Food and Beverage Fashions and Trends of the Past

Chapter 5 looked back, discussed, and analyzed food and beverage habits as far back as the times of Hellenistic Greece during the 5th century BC and later. This comes as no surprise: people ate and drank in these days, and some of the habits such as dipping bread in wine can still be found today in some form in some rural and remote villages in modern-day Greece. Breakfast consisted of barley bread, often dipped in wine, complemented by figs, olives, or both. A kind of breakfast pancake was fashionable already then. Other meals were composed of snacks such as chestnut, beans, toasted wheat, and honey cakes. Wine was the big accompanier to every meal. In general, it can be said that nothing really stands out in terms of food components that would be totally surprising from today's eating perspective. Vegetables, grains, fruits, meat, and fish were largely consumed then; however many of the recipes were different from present-day ones.

Food and meals during the times of the Roman Empire were similar in composition, although habits of eating times and recipes again had their local touch. The evening meal, the "cena" grew larger and became more important and also more diverse under the Greek influence and was often eaten as early as 2 PM in the afternoon. Here again, no big surprises to be seen as far as composition of meals is concerned, if not for certain recipes and the timing of their meals.

During medieval times in central Europe, societal disparities seemed to have an increasing influence on who ate what and could afford what. This is not really a new phenomenon and has carried on to this very day. Typical food components of these days were bread, meat, pottages, fowl, fish, vegetables, sauces, salads, desserts, and cakes. It all sounds modern. Interestingly enough, in those days every potable liquid seemed to be fermented because water was generally considered not pure, probably rightly so then.

The following centuries saw a kind of agricultural revolution with the advent of crops that were unknown in Europe. One important element to improved food security and food safety was the invention and introduction of refrigeration into the agriculture and food value chain, thereby making safe storage of most food products a simple reality.

11.5.2 The "Real" Food Revolution of the 20th Century

Despite being critical when it comes to nutrition and its findings, I strongly believe that some of these findings have really revolutionized how we grow, make, and consume food and beverages today. Taking the risk that I am seen being biased, my favorite nutrition-related finding is the first discovery and description of the existence of and need for essential fatty acids. This example is just one of many; some of the others are the discovery of vitamins, essential amino acids, the role of micronutrients, and a few more.

The critical question I would ask however: are the more essential food elements that provide health and wellness, such as proteins, essential fats, micronutrients, purposefully expensive and thereby kept out of reach of those who cannot afford them? I know it might sound a bit like a conspiracy, but it's an honest question to which I have no honest answer yet.

11.5.3 Present-Day Food and Nutrition Trends

This is one of my favorite topics: diets. When looking up on the Internet one finds more than 100 diets with all different names, yet some similar in recommendations and ingredients. People who follow such diets, other than the very famous one that in German is called "*friss die Hälfte*" (eat half), have to have money to do so. I have not seen any of these diets that was really cheap and affordable to the masses. The simple conclusions are: you have to have a relatively good income to afford following any of these diets and you have to have the stamina to hold on to the results achieved. Not really easy!

I suggested that there exist only two major food trends: Healthy food for those who can afford it, and enough food (any food) for those who have very little or nothing. This distinction is further aggravated by the vast amounts of perfectly edible food (raw materials as well as prepared dishes) that go to waste; the FAO estimates global food waste to be in the order of 1.3 billion tons or on average 35% of all available food. So, one-third gets lost or is simply thrown away!

11.5.4 New Food Sources: New Protein Sources

It appears to be more difficult to fix the waste than to look for new and possibly more affordable sources for food, especially proteins. Farming and use of appropriate insects is seen as one possible solution. Today it is more of a curiosity than a real fix; however many see a big potential in this approach. Many cultures around the world have always eaten insects as part of their diets and all those who like chocolate have eaten some minor amounts of insect parts in their preferred goodie.

Today, it is estimated that approximately 2 billion people are consuming insects and insect proteins as part of their diet. So, we know that insects are and can be eaten; the big question is really: do regular consumers really want to eat insects? The other question is of course whether a life cycle analysis of growing and transforming insects is any much better than for instance farming fish.

There are other potential sources such as microalgae, seaweed, duckweed, and rapeseed, which are looked into and which might be a good complementary source to the existing set of protein sources. The most difficult steps in using any of these sources, especially insects, is to how to fit them in existing manufacturing and supply chains.

11.5.5 Vegetarian Food and Its Impact on Society

A brief discussion on some of the major societal impacts of vegetarianism was done. I made a clear distinction between vegetarianism and veganism. Risking that the discussion was not really liked by some readers, I want to summarize my main point of analysis. The majority of potential critiques are all health based, especially when it comes to vegans and children and adolescents, because, despite pointing to the availability of dietary supplements especially of the vitamin B family, there is a strong risk for malnourishment in these age groups and also "mature" adults. I would also add that pretending to eat all naturally and healthily does not really go together with swallowing pills that contain necessary micronutrients that are not to be found in a vegan diet.

The other point briefly discussed was rather in the format of a question: If veganism would prevail what would happen to all our domesticated animals that we have had for thousands of years if not longer? They would have no reason to exist any longer, unless every vegan has a cow or a pig as a pet at home. Just one additional point: today, livestock contributes to approximately 40% of the global vale of agriculture, and a large portion of 1.3 billion people depends on it.

11.5.6 The Role of Urban Agriculture and Bees

Chapter 4 ended with a short discussion and analysis of the role of urban gardening, which I prefer to call *urban agriculture*. Today urban agriculture is practiced by 800 million people worldwide. They grow vegetables or fruits and raise animals in cities or other urban environments. For reasons that need more research, urban garden plots can be up to 15 times more productive than rural holdings; as an example, an area of just 1 square meter can provide up to 20 kilograms of food per year.

Finally, the important role of bees and their careful and sustainable treatment was discussed. It was mentioned that some cities already build so-called bee highways across their urban environment to assure not only safe passage but also more importantly improved diversity by planting many different flowers and other plants. A 2003 Franco-German study estimated that the work that bees carry out pollinating, especially food plants, amounts to a total value of more than €150 billions per year!

11.6 New Business Models for the Food Industry

The topic of new business models for the food industry is a central one in this book. One of the central elements of every food company's business model is the one that I termed *demase*. Demase stands for "develop, manufacture, and sell" and is around one millennium old, at least that's when this model was first mentioned in written documentation. The oldest known company in the world of food and beverages is the German beer brewery Weihenstephan first mentioned in written documents back around the year 1,050, and it probably existed even before that date. Nothing has really changed in the business approach since, and demase is still the name of the game. This book suggested that a real good overhaul of that model would not only be required but is maybe long overdue.

One of the important reasons why such change would be needed is based on an increasing mistrust in the industry by a growing number of consumers. Industrial food is increasingly seen as possibly not totally safe, even if this was only a perception, and several food-related scandals and negative events since the 1970s have not helped to regain trust from the vast majority of consumers. Change is needed and several ways forward are suggested and discussed and analyzed throughout several chapters of this book. This is not just a call to arms but also includes suggestions as to the type of arms and the strategic directions and goals that such change should strive for.

11.6.1 From "Consumer Is King" to "Customer Is King": Retailers Become Real Partners

Up to now, the food industry prides itself that everything that is done, developed, improved, and optimized is done for the benefit of the consumer—and of course, the profit margin. This is not a cynical remark, just a realistic one. There is nothing fundamentally wrong with this approach if it wasn't for the fact that retailers, who literally stand between the food company and its consumers, still today are not fully integrated as equal partners. They are often seen as a nuisance, a casual partner at best. Today, apart from some branches such as food service operations, the business model of a food company, apart from being demase-based, is mainly business to consumer, largely neglecting the highly important overall business-to-business axis.

Retailers need to become real partners in all aspects of strategy, choice of innovation direction, resource allocations, manufacturing sites, logistics, and especially marketing and sales. Today's food industry managers are mainly efficient in administration and think more about restructuring, "taking costs out," moving headquarters, and similar nonsense than investing into the business by investing into its people, great new products, and healthy solutions for the ultimate target, the consumer. There is still a strong belief that one can "save one's way to success," which in my opinion and based on many years of observation, has never worked and has most often led to further erosion of the business.

In this line of argumentation, instead of "cutting costs," it is advisable to team up with the most important partner, the retailer, and jointly conquer the consumer. It is the retailer who has most direct access to the preferences and shopping behavior of consumers, which is data that the food company dearly needs. Sometimes such data can be made available, sketchy but better than nothing. Often, expensive consumer research data are used for the purpose of better understanding the consumer, an approach that often misleads into believing that one has "the best-ever feedback," only to find out that it was simply not true. The glory of the moment is quickly forgotten by the reality check.

Retailer partners, real partners that is, are the way out of this dilemma. A possible pathway of events was painted that includes a real, mutually beneficial approach to collaboration and is depicted in Figure 6.1 (refer chapter 6 - page 148).

Two important examples were described and discussed, examples in which I was personally involved. The first one is the story of the development and successful launch of KitKat Chunky®, which was largely done in close collaboration with a retailer, especially the commercial side of this launch. The second example was the possibility of a closer collaboration with a large German retailer; ultimately for several reasons there was no real follow up, largely because of mutual mistrust and because the time wasn't right yet. I strongly believe that such collaborations are overdue.

11.6.2 Good-Bye to Selling Products and Hello to Selling Know-How

It was suggested that food companies, to this day, work on a business model that is about 1,000 years old and that this might be a good time to mull over some new models and directions. The first suggestion was the recognition of retailers to become real business partners. Here it is now suggested to go one step further and respond to an increasingly critical, sometimes even hostile, consumer of industrial food products. Demase, at least as the only approach, will become a thing of the past and must be improved. One such important step is to add the dimension of earning money by selling know-how about health- and wellness-related topics and insights. This can only be done, if the old ways of thinking are at least complemented and the R&D arm of the company takes on a more important leadership role similar to what is the norm in the pharmaceutical industry. I called the combination of three key elements, the new food industry business model 2.0, which does not yet include the selling of know-part yet. Business model 2.0 lays the groundwork for model 2.1.

So, model 2.0 requires:

- R&D of the organization drives the company.

- Open innovation and partnerships are part of the new DNA of the company.
- Retailers become real partners for the food and beverage industry.

As a consequence of this, the company can go the next step and become really knowledge-centric and finally recognizes the latent value (and I mean dollars) of its human resources, especially its scientists and engineers. Now the company is ready to enter the next phase and adopt business model 2.1. This model embraces a new reality, namely that the new product will be know-how. The major elements of model 2.1 are:

- The new food company sells know-how.
- Experts are trained to become communicators.
- Experts are part-time offsite at retailers to talk to consumers.
- Experts are also online to communicate with consumers.

It was suggested that such experts require a new type of training, during which acquiring great communication skills plays an important role in the educational curriculum.

11.6.3 Consumers Become Involved

One might argue that consumers, by buying food products and caring about their safety, health, and nutritional impact, are already involved today. Yes, they are, but unfortunately in more antagonistic than constructive ways. To escape the food industry's simplistic approach to new-product development via "renovation," it would be about high time to involve consumers in new-product development in much deeper ways and let them actively contribute to every aspect of food product innovation. By having the experts out there, as suggested in business model 2.1, this should become a much easier task and the future of "crowd development" of food and beverage products should be a bright one. I do realize that this has taken some roots on a smaller scale already; however, the large food companies still operate pretty much within their own enclosed walls.

This would also give consumers a direct impact into developing products that follow or even anticipate the major food trends, a topic that was extensively discussed later in the book. The major aspects are:

- Personalization of food to respond to allergies and food intolerances in general.
- Demand from agriculture to better respect nature and become even more sustainable.
- Demand that growing food, making food and consuming food will get closer.

11.7 The Internet of Just about Everything and What This Means for Agriculture and Food

The topic of connectivity in the world of food in an era when everything seems to be connected is of course an important one and required quite some in depth discussion and analysis. While writing this, I was especially astonished when I watched a TV commercial advertising for more gigabytes from your cell phone provider. During said commercial, the protagonist, a young and apparently very cool professional goes through life by having his eyes constantly focused on his phone screen, totally oblivious to what is going on around him. He is permanently connected and constantly checking his screen, seemingly living a parallel life without even realizing a single event going on around him. It borders to a miracle that he does not constantly bump into other people who he does not see or gets run over by a car or a bus. If this were a true reflection of today's connected life, it would be rather frightening. We may as well transform ourselves into robots, or alternatively let robots do our work, giving us the time to constantly check our cell phones. We will always be somewhere, just not here. Luckily, there is food, and it still has to be eaten in rather traditional ways, from our hands—with or without tools—into our mouths.

11.7.1 Modern Cooking: A Brief Look to the Past

Enough bickering, let's cut to the chase and get back to the discussion on food and cooking. Cooking has three dimensions:

- The process of combining various food ingredients in smart and personal ways to achieve the preferred outcome (taste, texture, and pleasure in general).
- The use of certain practices and techniques to achieve this.
- The emotional and social dimensions of bringing people together, during preparation and especially during consumption.

This was discussed and analyzed in some detail. A maybe overly simplistic view on the social component of eating was suggested for cultures and societies in which Christianity plays the dominant role. It was discussed that in predominantly Puritan (Calvinistic) cultures food mainly serves to refuel our body, whereas in Catholic environments food was and is more of a feast and, at least in part, explains the difference between burgers and fries and bistecca Fiorentina. Some more elaboration and analysis was given to support this hypothesis.

11.7.2 Robotics and Connectivity

Kitchen robots have existed for quite a few years, especially since the middle of the 20th century. They are more of the type of kitchen helpers and rely pretty much on transforming electricity to kinetic energy of the stirring, mixing, kneading, juicing, slicing, and you-name-it type. They are helpers but do not possess any real smartness. These kitchen helpers are simply not smart unlike real functional robots, which can become tremendous helpers in the kitchen to the point that they might eventually take over the majority or totality of the cooking process. Today, this is certainly out of reach, not because such robots could not exist, but because they are simply far too expensive. Just look at the kind of robots in the automotive or electronic appliances or even medical industry: they can do some pretty outlandish stuff. Their drawback is their price, which is far beyond the reach of most consumers and cooking aficionados.

It was discussed that there was a time when cars were rather out of reach of many if not most, similar to the situation of smart personal robots today. There are possibilities to overcome this by mainly two trends: first, robots will eventually become much cheaper and therefore affordable, and secondly, financing models such as leasing can be used to "own" a smart kitchen robot. The latter has the great advantage that I might get the newest, improved model every 3 years or so.

11.7.3 Food and Agriculture: Big Data

Big data has become a buzzword, used by many, understood be fewer than many. Big data is a term for data sets that are so large or complex that traditional data-processing applications are inadequate to deal with them. What is big is simply the mountain of data. Their interpretation is yet another story and is not only not easy but often can become controversial. Just look at the amount of climate-related data that scientists have collected and still do, and yet there appears to still be room for maneuvering and twisting toward preferred outcomes.

Big supporters of collecting such big data for the agriculture industry are the large space agencies, such as for instance the Jet Propulsion Laboratory JPL/NASA, who dedicates expert resources to survey crops and many growth-related parameters from satellites, low-flying planes, or even lower-flying drones. NASA is not the only one in this field.

On the other hand, the industry of agriculture depends not only on the data but in their correct interpretation when it comes to predicting local or regional climate deviations. As was discussed, big data are not only needed, but they are here to stay and will become more important, especially as the food industry tries to better understand the health and wellness needs of their consumers.

11.7.4 Will There Still Be Agriculture and Food Industries?

Admittedly, this is a provocative question and should rather be ending on "as we know them today?" It was stated several times that the business model of the food industry is close to 1,000 years old, and agriculture in its simplest form is much, much older. Agriculture has made many strides toward improvements and efficiency enhancement but is still basically based on planting seeds, letting them grow, and harvesting them or breeding animals to eventually slaughter them. So, not much has changed there.

The simple answer is: yes, these industries will still exist, but both need to learn a lot. How come urban agriculture can apparently have yield of up to 15 times higher compared to regular, traditional growing on large fields? This book has no answer to this question, but even if it may be exaggerated by a factor of 3 or even 5, there is still a 3 to 5 times better yield of one versus the other. There is much room left for graduate students and agriculture departments in universities to find answers to this apparent enigma.

This leads to another topic, namely that bigger is not always better. Bigger gets maybe away with fewer people looking after it; however, bigger typically means slow and less efficient. This should be a wakeup call for both agriculture and food industries. It was suggested that the agriculture industry is still in the phase of consolidation and growth toward larger units, and the food industry seems to be stagnating as far as such agglomeration to larger units (companies that is) is concerned. The industry of producing electricity was given as an example of decline in size when it comes to the actual production units.

11.7.5 What Will Remain, and What Will Disappear?

It was said time and again: hindsight is the only exact science, so an answer to the question in the title has a kind of crystal ball quality to it.

Let's give it a first try for agriculture.

- The large agriculture conglomerates that control seeds, chemistry and biogenetics of plants, and feed side of animals for meat will again transform to smaller units.
- Industry and farmers will collaborate even more closely.
- R&D in agriculture will be intensified and will become a cornerstone of the industry.
- Necessary yield improvements of crops will be increasingly based on modern breeding, non-GMO based.
- Urban agriculture will play an increasingly important role.
- GM plants will find growing resistance in all parts of the world.
- The approach to pesticides will be revolutionized, and they will eventually become all natural.
- Bees will be bred and developed to fulfill specialized tasks in the optimization of crops both for yield and nutritional value of plants.
- As robots and drones will increasingly become part of the agriculture industry, they will take on many roles; so why not potentially replace bees?

- Animal farming will be under increasing public scrutiny, and consumers will become more demanding with regard to how cattle is bred, fed, grown, and slaughtered.

And here a suggested list for the food industry.

- Food companies will sell mainly know-how together with products.
- Manufacture of food will be closer to consumers and consumers will have a greater impact.
- Large factories might still exist but they will be filled by many different product lines producing regionally rather than continentally or even beyond. This trend can already be observed.
- Consumers will be more closely involved in the development of great new food products.
- R&D will play a pivotal role in the food industry being simultaneously involved with consumers, selling their know-how and gaining new knowledge, especially regarding tangible nutritional value of food and food products.
- Processed food, as we know it today, will increasingly disappear.
- All-natural and organic will become the new normal.
- Food companies will become more specialized with regard to the type of products they make.
- Food companies of the future will be smaller than they are today.
- Robots will replace more and more manufacturing personnel and why not the CEO of the company? After all, emotions are not required for that job.

This is quite a handful and the readers are encouraged to critically look at these points and come to their own conclusions as to how the future of these industries may look. But one thing is clear: the Internet of everything, the total connectivity—desirable or not—will, and must, have an influence on the outcome. And there is one more thought: whichever outcome there will be, it's never going to be final; it's always going to be a continuous transition and progress.

11.8 Nutrition: What Else?

You certainly have heard this line in rather famous commercials for the Nespresso® brand. There it is meant to be the great product easily recognized because it can't be anything else. This is not exactly what is meant in the title, rather in the sense: nutrition and what else is out there to help us choose good and healthy food, irrespective or even despite the often confusing and controversial nutritional messages. The reader may have guessed from what was written in specific chapters of this book that I am not particularly fond of nutritional research and even went as far as denying the field of nutrition status as a science. Entire armies of nutritionists may argue against this and I am certain that they do. However, I stubbornly stick to my opinion in this matter. The main argument here is that nutrition never can say anything, draw any conclusion with authority and absolute or even relative assurance. It's all about might, could, should, would, maybe, and most prominent of all: we need to do yet another study.

It is however of little importance whether we consider nutrition to be a science or not; fact is that nutritionists need to get their act together and can increasingly deliver messages and results that are clear, unambiguous, and less prone to easy misinterpretations and faulty or so-called "alternative" understanding.

Reporting on food trends is much loved by many newspapers and magazines; some of them literally make all their money on the topic of "what's hot in food." It was suggested that probably the most reasonable nutritional recommendation that can be given is the one of a "balanced diet," or in other words, an equilibrated and not overly rich diet, composed of many, most or all of the macronutrients and essential micronutrients. My former colleague always used to say: "There is no bad food, only bad diet." What he meant was simple and that any given food eaten at any occasion per se is not bad, even if it was burger and fries, as long as the more medium to long-term food intake, say over a period of a week or two reflects a more balanced overall food intake. In all fairness, it has to be mentioned that there are some critical voices with respect to this position, mainly based on the fact that obesity has rather grown or at best stagnated in many affluent societies.

11.8.1 Healthy and Happy Eating

So, in reality everything should be simple, and yet, it is not or at least is made believe to be rather complex. Let me reiterate a typical suggested energy-based recommendation for combination of macronutrients intake:

- Proteins 10%–35%
- Fat 20%–35%
- Carbohydrates 45%–65%

There is nothing complicated about it, and yet it is apparently not so easy to follow these recommendations.

As mentioned, talking about and discussing the latest food trends has become a preferred pastime for many, and it was mentioned that one can find more than 100 fashionable diets that are based on many of these trends. Again, a balanced diet with the right amount of energy—adapted to your lifestyle—is the best way to healthy and happy eating.

Here is a summary of simple to follow tips—nothing really new, nothing earth shattering yet most effective.

- Try to avoid eating the same food all the time.
- Do eat just enough and stop eating before you will feel full.
- Have a glass of wine from time to time; it's the alcohol, regardless of color, that is positively correlated with improved cardiovascular health.
- Ideally eat and drink in company because eating, after all, is also a social event bringing people together.
- Find your personal eating schedule or in other words listen to your body and to your stomach. Begin thinking with your stomach.
- Do not use pills to complement your food; it's not really necessary.
- Listening to your stomach and your cravings will tell you more about what food you should eat than any dietary recommendation, even the most advanced one.
- Eat when you feel hungry, and drink (water) before you become thirsty.
- Cook yourself more often or at least take a greater interest in cooking.

11.8.2 A Short History of Nutrition

The history of nutrition may be as old as the history of humankind but was not really recognized as such. As early as 400 BC foods were often used as medicines and cosmetics. In the 1500s Leonardo da Vinci compared metabolism in our body to burning a candle. In 1747 James Lind performed the first documented experiment in nutrition by studying scurvy in sailors. The year 1770 was when Antoine Lavoisier was the first to discover the mechanism by which food is metabolized. In 1840 Justus Liebig described the chemical composition of carbohydrates, fats and proteins. In 1912, Casimir Funk was the first to coin the term *vitamins*, and in 1930 William Rose discovered the essential amino acids. In the same year the Burrs first described the role of essential fatty acids, and ever since the 1950s, the roles of essential nutrients as part of metabolic processes have been brought to light. The potential role of probiotics is one such example.

More recently a kind of backlash by regulatory bodies especially in Europe could be observed, putting a lid on the claims of the functionality of many micronutrients. The dispute is still ongoing. This includes the trends of "low" or "reduced," which sometimes doesn't sit so well with consumers: why would I pay more for less? Typical candidates for low or reduced are fat, saturated fats, salt, and especially sugar. In my opinion, it's only reduced sugar that makes some sense, but in a way that it brings our palates back to a healthy and more normal perception of sweetness.

Figure 8.2 (refer chapter 8 - page 207), good eating habits prioritized, attempts to indicate good, if not best, eating and consumption habits.

11.8.3 Nutrition Controversies

It was mentioned a few times already that nutrition, not really being an exact science, often leaves space for many different interpretations of its findings. There are many lists and tables showing dietary recommendations to consumers, and there should be some good consensus on these. Controversies simply arise from using for instance a number of total recommended fat intake being in the order of 15% to 30% leaving a span from single to double open for interpretation. Or take carbohydrates: 55% to 75%, which is either approximately half or three quarters of your diet...which is quite a range. Many epidemiological studies were and still are undertaken, and *meta-analysis* has become quite the buzzword. There are many more controversies, for instance, around blood cholesterol levels, for which doctors' recommendations have become gradually lower over the years. My diabolic suspicion though is the more people comply with suggested healthy lifestyle, the more natural lipoprotein levels will become "normal" and the next campaign toward even lower levels will follow suit.

11.8.4 Claims and Benefits

For many years, the food industry attempted to make their products look healthier by trying to tell the public that a particular product has certain health benefits such as lowering blood pressure, reducing body weight, regulating digestion, preventing cardiovascular diseases, and probably a few more. Many clinical studies were carried out to support certain claims and the entire field became confusing. So much so that for instance the European Food Safety Authority struck down basically all claims that were put forward by the food industry with regard to probiotics in 2009. This was a big blow for the industry because millions of dollars went straight down the drain, all money spent on not only the studies behind but also product development, formulation, and marketing efforts behind probiotics. So you can imagine the deception on the part of companies' managements when this verdict was published, and I have personally seen, from afar, some "heads rolling"...not to the Gulag though.

11.9 The Company Transforms Itself

This title is a mix of hope and euphemism. Hope because positive transformation beyond head count and savings and toward investing in innovation might just happen, and euphemism because transformation sounds so benign and could actually resemble more a kind of revolution, driven by society, nongovernmental organizations, governments, regulators, and why not by the fact that food is simply not distributed in just and fair ways. It is not suggested that all of this will happen, but already one element out of this list can be rather disruptive and can seriously disturb a continued smooth ride of any food company. So change is needed and change that goes beyond the habitual alibi action that is casually called "renovation," From personal experience, I can say that most new food product development in the majority of food companies can be called renovation and is often just a new package design or a relaunch in a new format, new flavor variant, or simple marketing gimmicks such as rebates or shop events.

This book does not pretend to have a one-size-fits-all solution, yet has suggested a few possible remedies to escape stand still or worse decline. One such suggestion is reflected in the new food industry business model 2.0 that finally recognizes a few realities, which present management does not want to do for reasons that I dare not to speculate.

What are these new realities? Business model 2.0 is based on the recognition that food companies should be run by people who understand food, and not by accountants, lawyers, or business people who can only count, legislate, or talk—pretty much like in the successful pharmaceutical industry. The new drivers should be people from the technical, especially scientific, community but also practitioners who understand what food is all about and especially what a transformation of agricultural raw materials to healthy, well-tasting, and affordable food products means.

Business model 2.0 is also based on the idea that not all ideas and solutions to issues are necessarily to be found inside one's own company and the likelihood that the answers are found externally is much higher; hence, the model issues a call to adopt the idea of open innovation and innovation partnerships as part of the new DNA of the new food company. And finally, business model 2.0 recognizes that fact that a new type of partnership with retailers is not only necessary but will become highly beneficial and profitable. There are some timid signs of adoption of some of the principles to be seen, but they are by far not extensive enough but do need to be supported and encouraged.

11.9.1 The Role of Automation: Threat or Blessing?

There is quite some debate ongoing about the threats that overly ambitious automation in the format of robots is endangering the very core of humans' rights to work and support families and communities. There is, as this is written, an ongoing debate, whether robots should pay taxes because after all they replace humans who, for the same type of labor would have been remunerated and would have paid income taxes.

However, automation in the food industry is probably almost as old as most food companies in existence today and is nothing really new, neither in agriculture nor food industry. Even such traditional industries as coffee or wine increasingly rely on harvesting their crops with especially designed machines and have reduced the role of manual labor substantially. This book has no magic bullet to answer the question as to what workers who are replaced by machines, let alone robots should be doing, if not becoming valuable and crucial members of an industry that still needs to grow food, possibly in new ways or become an expert in all topics food. And there is no guarantee, but one thing is rather clear, at least to me: in the future there will be more people working in developing healthy and nutritious new foods and beverages, even with specific functionalities that may or may not be claimed but will become known to the consumers through other than officially claimable channels.

11.9.2 Regulatory Involvement in the Industry

Regulation is not only of influence and importance in finished food and beverage products but also when it comes to planting the right diversity of crops, based on widely accepted new and beneficial plant-breeding techniques. Regulation needs to help put an end to trading and speculation in the major food crops; thereby helping to create a better standing and balance with regard to the agriculture industry, which in the last 20 or so years has come under much fire from the public.

11.9.3 The New Business Model 2.1

While business model 2.0 calls for subtle yet important changes such as "let the company be run by those who understand food", model 2.1 goes one step further. It suggests that a typical food company should not only rely on making and selling products but their experts, typically their scientists and engineers, should become communicators and sell know-how on food, health, and nutrition. Some thought was given to the question how to make money with this model, and simplified, high-level financial considerations showed promising results. Nothing is carved in stone here, and many more financial scenarios can and should be tested; however, it is predicted that they all will show a positive outcome through two axes: direct financial gain from consumers and indirect improvement of standing of the company in the community that it serves.

Such a model would not only have a direct impact on the food company itself but also on its partners and suppliers, who will need to have to adapt to such a model too, by increasingly selling solutions based on their know-how and less ingredient mixes to be added to the recipe. An example of the new role of the flavor industry was discussed and analyzed in some detail.

11.9.4 Scenarios of Relevance for Food and Agriculture

It was discussed that there are some "writings on the wall" that would and should push agriculture and food industry to read them, understand them, and act on them as they seem best fit. These scenarios as well as possible pathways toward applicable solutions were summarized in three tables in Chapter 9.

Table 9.1 summarizes scenarios of relevance.

Table 9.2 proposed a few possible ways out of what was called the "impasse" in more nondisruptive ways.

Table 9.3 represented the more radical and disruptive steps. (Refer chapter 9 - pages 235, 237 & 240).

11.9.5 Medical Food: A Future?

For many years, food companies have attempted to establish a footprint in the area of medical food and they have never really succeeded in doing so. Most often, the margins for such products, foods or beverages, are simply not important enough to warrant expensive development and clinical support as well as surmounting regulatory hurdles to make this a profitable proposition. The example of Nestlé's product Boost® was given, and it was simply shown that such a product sells in the format of a six-pack for an average price $7 in C stores (i.e., approximately $1.15 per bottle of 8 ounces). Let's watch the space and see, whether there is a future for medical food as food and beverage companies envision it.

11.9.6 Reality or Fiction?

Technological discoveries as well as societal changes have always led to important advancements in our lives. The industries of agriculture and food are no exception here, and they do have their share in these step changes. However, and this is the prediction, such important changes might not even happen in the core business areas of these industries but might rather happen in the fields of automation and robots. Pretty much like robots have profoundly changed the automotive industry, something similar is expected, especially in the food industry. The large food companies could jump on this development and have their own, branded cooking and food preparation robots and own related Internet platforms that combine health and wellness information with suggested recipes, ideally using their semi-finished or finished ingredients and products and link it to automation and robotics for the individual consumer at home. The big companies would have the power and means to create their own leasing departments that help consumers to always afford the newest model of the trendy and branded kitchen robot. Without wanting to really suggest it: this could become another great way to create an additional loyal consumer base.

Finally, the manufacturing division of the food company will undergo further overhauls toward even more automation, and most likely, more outsourcing of product manufacture, possibly in retailers' brick-and-mortar sites that will become increasingly available because their online business will pick up even more speed in the near future. The food industry, however, has a long and not unsuccessful history of "sitting out" disruptive events, and they might think they can get away with it again, but it might just not work this time.

11.10 Agriculture Listens, Finally?

I look forward to a time when the industry of agriculture—from farmers to breeding and chemistry—actually starts to listen to the final consumer of its products. They simply don't listen, or at least not good enough when it comes to desires and wishes of the consumer. They do listen to their own people who are involved and to the farmers who are in despair of an insect infestation or crave water for more irrigation. And yes, they listen to the seed and chemicals industry players who have come up with pesticides that inadvertently might kill a crop or harm the farmer or new hybrid seeds with no real advantage to the consumer. And yes, they listen to shareholders and traders who want to make more money by increasing the trading cycles of major crops and add to speculation, and sometimes even artificial shortages.

And they do not listen enough to what's cooking in the area of new food trends and how they might best be able to support these and the consumers following or even anticipating such trends. This would not only mean that the industry listens better but actually is willing to make changes and become more flexible.

Here the major topics of importance towards a new and modern agriculture:

- Flexible agriculture: the industry responds to the real needs of people.
- Food security: enough food for everyone, getting to all and not just a few.
- Food safety: consumers can rely, always and everywhere that the food they eat is safe.
- Food affordability: food raw materials (plant and animal) can be afforded by all who are hungry.
- Nutrient balance for all: everyone should have access to a balanced diet.
- Increased understanding of relationship between food and health: nutrition understood and lived.

So, agriculture not only becomes more responsive but also will eventually become proactive by acknowledging and tackling the following challenges:

- Not enough water.
- Too much water.
- Need for development of drought-resistant plants that grow with less water.
- Need for development of saltwater-resistant plants with good yield and growth frequency.
- Plant and grow much closer to where food is needed to assure food security for all.
- Develop and grow plants and breed animals that more efficiently reply to and anticipate trends.

11.10.1 Agriculture and Farming at the Fingertips of Everyone

Much has been said about "democratization of...," and we should add "agriculture and farming" as an ideal candidate for democratization. Urban agriculture has to be mentioned here. One intriguing example for urban agriculture can be found in the so-called agrihoods. By definition, an agrihood is the extension of suburban living. Available land in between and around homes is used for agriculture instead of horticulture and golf courses. Both plants and animals can be part of such agrihoods.

Permaculture is another example of the democratization of agriculture. Its principles can be used by everyone both the urban agriculturist and the larger farms and cooperatives. Permaculture "is a system of agriculture and social design principles centered on stimulating or directly utilizing the patterns and features observed in ecosystems," and there are 12 design principles for the basis of permaculture: observe and interact, catch and store energy, obtain a yield, apply self-regulation and accept feedback, use and value renewable resources and services, produce no waste, design from patterns to details, integrate rather than segregate, use small and slow solutions, use and value diversity, use edges and value the marginal, creatively use and respond to change.

When looking at these principles, some similarities to other areas than agriculture come to mind. For instance, catch and store energy is an important part of hybrid or electric cars; creatively respond to change is one of the more important principles taught in management courses, produce no (or less) waste has become a theme throughout many societies, and obtaining a yield can be seen as strive for better profit margins in business. On the whole, all principles do seem to make a lot of sense and, more importantly all seem to be practicable.

11.10.2 Small Is Beautiful

Smaller is not always better, and many economists swear on the opposite, namely large and call it "economy of scale." However, large is almost always inflexible and slow and, based on many years of observation and experience, the economy of scale (large) is often counterbalanced by large inertia in organizations, especially in large food companies. Companies do realize this dilemma, and call for the action of their employees to act as if they were operating in small, agile, and flexible units. It sounds good but is mostly held up by upper management who wants to be part of every decision that costs more than $100. That said, trust is the one element that can help foster this transition from acting large, grand, and slow to acting small, agile, and fast. The trust issue is even more surprising when looking into the reality that a company is willing to pay a fairly large sum to one workplace—in the order of $300,000 per year—and then wants to know every little detail as to how each of their employees has spent their time and resources. This is what you get when a company is run by accountants and business experts, who have little to no knowledge of agriculture itself, or even worse, of food and beverage products.

11.10.3 Is Urban Agriculture a Sign of Self-Centeredness or Is It Community Driven?

This is an important question because some in the traditional agriculture industry believe that it is rather an act of self-centeredness and potentially takes work away from those who would really need it. It's a fair criticism and should not be dismissed lightly. However, it can be shown that especially as a result of some important advantages of urban agriculture over traditional, large-scale agriculture, for instance, far better yield and improved diversity, urban agriculture and its little sister, urban gardening, have an important role to play and can substantially add to the livelihood of many.

A slightly utopian example, Chiavi's "town on the hill" takes urban agriculture one step further. Because of the artificial nature of the suggested hollow hill on which the town is built, there is an even closer proximity between city and country, and living, working, studying, relaxing, and gardening/agriculturing all live and coexist right next to each other.

11.10.4 Manufacturing Food Where It Makes Sense

Surprisingly enough, one of the world's largest ice cream production plants is located in a very hot place such as Bakersfield California, which in the top 10 of hottest cities in the United States and is not far behind Phoenix, Arizona, Palm Springs, California, and Las Vegas, Nevada. It's similarly hot in Bakersfield and yet, ice cream is made there, requiring quite a lot of coolant (liquid ammonia) and extremely good insulation. The maybe outlandish suggestion was made to profit from natural cold and produce ice cream in Greenland. Yes, agreed, it's far-fetched but why on earth would the opposite situation (Bakersfield) be good and acceptable? You may say logistics and supply chain and you may have a point there. But just think for a moment of a factory that is entirely run by robots and that ships goods out in regular airfreight mode. What worked for Berlin during the blockade could work for an ice cream factory in cold and remote places.

Rather for reasons of proximity to raw materials, the other example that was suggested was to manufacture chocolate right where cocoa trees grow and pods are harvested and beans are made ready for export. Out of a multitude of more than 10 steps that lead to chocolate as we know and love it, today just the first one is carried out in the country of origin. This should change and the example of a chocolate factory in Ghana was briefly discussed.

11.10.5 What Role Do Restaurants Play?

Not much if anything at all was thus far discussed in this book about the role of restaurants when it comes to the future of food. Running the risk to get a lot of flak about this statement, I truly believe that restaurants mainly translate trends into something pleasant, often even highly pleasant; however, with very few exceptions, they are rarely at the forefront of innovation in food and how it relates to health and wellness. Maybe this would be a worthwhile topic to write another book about?

11.10.6 The Role of Pet Food in the Food Industry

Pet food is food, too, and regulatory authorities do insist on almost the same kind of safety and ingredient-related rules as for human food. After all, some types of pet food are "covertly" eaten by humans. Therefore, the product has to be safe for human consumption. There are no numbers to be found as to what percentage of all pet food sold goes on anybody's plate, and it is assumed to be low. The taste profile may not be exactly what we would expect, but then there are consumers who have neither enough money to buy regular food nor receive food stamps.

However, the pet food industry plays an increasingly important role in the entire food industry, simply based on the fact that, for instance, some 80 million households in the United States own a pet. And the tendency is rising, and pet food development has become more and more sophisticated and applying many important nutritional findings of relevance to cats or dogs rather expediently.

11.10.7 Food in the Format of Pills? Will Consuming Food Pills Be Part of Megatrends?

The last topic discussed was whether or not food will ever come in concentrated pill format, similar to the imaginary astronauts food in many science fiction novels. Although there might be specific reasons for having "food pills" and occasions that require such formats, it is unlikely that this will ever be mainstream, at least in any foreseeable future.

Let me end this book by reiterating some of the most prominent and already happening ***food-related megatrends*** laid out the following through the entire book, and especially in Chapter 10

- Yes, food will always exist and people will always have to eat and drink.
- The type of food that people will eat in the future will largely resemble what we eat today.
- The food industry will still exist but will most likely transform itself more toward a knowledge-selling industry, gradually going away from the simple demase model that is more than 1,000 years old

- Consumers today, and probably tomorrow, will gradually turn away from industrial food perceived as processed and will increasingly cook themselves.
- This will most likely accelerate the food industry's transformation to a knowledge-selling industry.
- Automation in the kitchen will be revolutionized and cooking smart robots will increasingly replace traditional kitchen machines that have not much evolved since they first came into existence in the middle of the 20th century.
- Consumers will be able to afford smart kitchen robots similar to how they lease a car today.
- Nutrition research will continue to come up with conflicting and confusing messages, not really understood by consumers, and at best cherry picked: what I believe might be good for me.
- Retailers will only survive if they adapt to the new realities of online shopping and potentially transforming their brick-and-mortar sites to manufacturing and distribution sites for food companies. A type of new relationship between retailers and food companies is emerging, which is based on total cooperation and balanced sharing of the value chain.
- Functional foods and beverages of any type and functionality will have their space in the food industry; however, I believe that this space will be small and produce margins that might be too small to sustain this field of business.
- Given that the food industry is and always has been conservative, risk averse, and slow to adapt and change, it will attempt to maintain its structure and business model, only giving in very hesitantly to the required changes.

So, there is really hope that agriculture will recognize its role to serve people and the food industry will recognize its role of not only helping to feed us but also give us great and valuable advice regarding our health and wellness, so intimately linked to food and beverages.

Until we eat and drink again!

Index

Locators in **bold** refer to tables; those in *italic* to figures

a

Academy of Nutrition and Dietetics (AND) 195
adaptations to climate change 45
additives, food **8**, 19
adult nutrition 154, 158
affordability of food 248
agricultural revolution 119
agriculture
  challenges 23–25, 247–252
  consumer involvement 236, 305
  definition 3
  democratization 252–264
  diversified 253, 255, 259
  economy of scale 248
  food safety 248, 280–281
  food security 176, 235, 248, 252
  future scenarios **235**, **237**, **240**, 303–304
  future trends **181**, 183, *184*, 188, 308–309
  history 5–9, 278
  integrated 258
  intensification 92
  mechanization 88, 223
  nonfood applications 14–16, 280
  precision 37–38, 47, 60, 80, 286
  productivity 23, 32, 80
  R & D 183, 184, 297
  raw material supplier 11–12, 280
  satellite technology 177–179
  smaller units 235, 259–261, 297
  sustainable 163, 236, 250, 253, 255, 265, 278
  transformation possibilities **237**, **240**, 244–245
  urban 183, 236, 253, 261–264, 291, 297, 306, 307
agrihoods 253–254, 306
agri-transformation 252
alcohol, Medieval Europe 118
algae 16, 130
allergies 129, 162, 294
Alliance for a Green Revolution in Africa (AGRA) 56
allotment gardening 134
alternative food sources 128–132, 137
amino acids, essential 200
anaerobic digesters 95

Megatrends in Food and Agriculture: Technology, Water Use and Nutrition, First Edition.
Helmut Traitler, Michel Dubois, Keith Heikes, Vincent Pétiard and David Zilberman.

animal husbandry 87–88, 95–96, 278. *see also* livestock
  big data 102–106, 288
  employment 93, 133
  history 88–89
  humaneness 93–94, 96, 287
  sustainability 93–94, 288
animal proteins 89–93, 106. *see also* meat consumption
animal waste management 43–44
animal welfare 93–96. *see also* livestock (health)
aqueducts 249–250
arable farming 3, 24, 133, 278
artificial insemination, cattle 96, 97, 100
artificial ingredients 10–11
artificial sweeteners 204–205
aspartame 204
Atkins diet 197, 208
automation 242, 302–303

b

baby milk 142, 143, 158
balanced diet 193–195, 215, 249, 299
Beechnut's fake apple juice 143
beef industry, genetic evaluations 103. *see also* livestock
bees 136, 138, 183, 291, 297
benefit-cost analysis, water projects 47
beriberi 200
beverages
  history 9, 289
  milk-containing 13
  reduced sugar 195, 202, 204–205
big data 174–182, 187, 296
  animal husbandry 102–106, 288
biochemical markers 65
biofuels 14–16
  impact on climate change 46
  risk to food security 77
biomass 14, 256–257
bovine spongiform encephalopathy. *see* BSE
bread, Medieval Europe 116
breeding
  dairy industry 98–99
  livestock 97–102, 287–288
  plant 53–54, 59–68, *61*, 80, 242, 285–286
breeds, development of 89
BSE (bovine spongiform encephalopathy) 19, 143
building blocks of life 4–5, 278
Burr, George and Mildred 201
business model, food industry 141–165, 221–222, 228–234, 292
business model, plant breeding 78–80
butanediol 202
butter 13

c

cacao 12, 67, 267
Cailler chocolate 233–234
California
  Central Valley 39, 250, 265
  walnut industry 212–213
  water management 24, 31–37, 250–251
calorie intake by crop **81–84**
calorie-reduction 161–163
carbohydrates in diet 123–124, 208–210
carbon dioxide 5
cardiovascular health problems 208–209
cattle, artificial insemination 96, 97, 100
cattle, high energy input 77
cell biology 61, 67–68, 80
Central Valley Project 39, 250
cereals 12, 55–56
cheese 12–13
Chiavi, Erich 262, *263*

Chilean wine industry 39
chocolate manufacture 233–234, 266–269
cholesterol 210–211
citrus fruits 122
climate change 44–46, 77, 128, 249, 251–252
Clostridium botulinum 19
cluster regularly interspaced short palindromic repeats (CRISPR) 64
cocoa 233, 266–269, 307
coconut water 193
coffee 12, 75, 223–224, 258
colorants, food 11
companies, smaller 235
complexity of food production 21–22
connectivity 172, 186, 237, 298
consumer/s
  benefit 144, 155–156
  confusion 235
  engagement 158–160, 164–165, 236, 292–295, 302–303
  mistrust 156–157
contaminants, food 19
contaminants in food 19
cooking, modern trends 167–169
corn cob 12
corn yield *60*
cost-benefit analysis, water projects 42
cream 13
crops 3, 278
  production statistics **81–84**
  sustainability 59, 118, 249
  yield improvement 59–60, 77–78
crowd development 160, 294
cultivated land, total 54
cultivation, soil 3, 278

d

dairy consumption 91, **92**
dairy industry 12–14
  breeding values 98–99
  efficiency 94
  electrification 89
  farm sizes 93
  fertility rate 98–99, *99*
  genetic evaluations 103, **104**, 105
  nonfood use 13–14
de-horning cattle 102
delivery of food 173–174
desalination plants 42–43, 251
develop-make-sell (demase) model 141–144, 152, 164, 179, 222, 236, 239
diet. *see also* nutrition
  20th century 121–124, 289–290
  ancient Greece 111–113, 289
  influence of wealth 112–113, 115, 118, 123
  Medieval Europe 115–118, 289
  post Renaissance 118–121
  present-day trends 124–126, 290
  proteins vs. carbohydrates 123–124
  Roman Empire 113–115, 289
  specialized 162, 235
  through the ages *121*
dieting 125
di-haploid (DH) plant lines 66–67
dioxins 19
diversified agriculture 253, 255, 259
DNA-based technologies 62–64
DNA markers 65
domestication 7
drought resistance 36
droughts 24, 31, 34, 39

e

eating habits, good 195–198, *198*, *207*, *209*, 299–300. *see also* nutrition
E. coli (Escherichia coli) 18

economics
  animal breeding 287
  animal husbandry 94
  irrigation 35–36
  seed industry 80
economy of scale 248
ecosystem 254
ecosystems 57
efficiency, dairy industry 94
egg consumption 211
Eijkman, Christiaan 200
electrification, dairy industry 89
embryo implantation 100
embryo rescue 63
emotional element of food 167, 174
employment 120–121
  animal husbandry 93, 133
  urban agriculture 135
energy production 56, 257, 262. *see also* biofuels
energy use, reduction of 56–57
environmental issues 57. *see also* climate change
EPCOT (experimental prototype community of tomorrow) center 264–265
epigenetics 102
erosion, soil 57
Escherichia coli 18
essential fatty acids 121–122, 201–202
ethanol production 14
European Food Safety Authority (EFSA) 212–213
exercise 162
extreme weather events 45

f

Fabaceae 58, 69, 77
farmers' markets 262
farmers, relationship to seed industry 72–73, *74*
farming. *see* agriculture
farm sizes 89, 93
fast food 167–169
fasting 168, 196–197, 207, 215
fat-burning 197
fat, reduced 202–203
fats, saturated and unsaturated 206, 207
feasting 168
fertility rate, dairy industry 98–99, 99
fibers 3, 13, 88, 278
field trials 67
fish consumption 211
fish in diet 206
flavor industry 231–234
flavorings 19, 231–234
flexible agriculture 248
flexitarian diet 58
flood protection 45
food. *see also* nutrition
  additives 19
  affordability 248
  chain 58, *71*, 71–73, 281
  contaminants 19
  delivery 173–174
  emotional element 167, 174
  expected demand *55*
  fast vs. slow 167–169
  flavorings 231–234
  functional 271–272
  habits. *see* diet
  healthy 11, 125, 137, 153–154, 249
  home-cooked 156–157, 169, 194, 235, 238–239
  insects 129–131, 290
  intolerances 157, 162–163
  marketing 79, 102, 147, 151–153, 238, 278
  medical 239–241, 304
  miles 163
  natural vs. processed 226–227
  novel sources 128–132, 137, 290–291
  oil-derived 10

organic 95, 162, 169, 184
performance 271–272
pills 271–272, 308
preservation 7–9, **8**, 279
products 3, 21–22, 226–227, 230–234, 278
safety 18–21, 225–226, 248, 280–281
security 128, 176, 235, 248, 252
stamps 22
trends 111–124, 247–274, 288–291
waste 126–127, *127*, 137, 290

food industry
business model 141–165, 221–222, 228–234, 292
consumer benefit 144, 155–156
consumer involvement 158–160, 164–165, 236, 292–295
consumer mistrust 156–157
develop-make-sell (demase) model 141–144, 152, 164, 179, 222, 236, 239
future scenarios **235**, **237**, **240**, 303–304
health claims 211–214, 301
innovate-apply-sell (inapse) model 239, **240**, 293–294
knowledge based business 152–155, 183, 227–230, 237, 293–294, 303
local produce 224–225, 227
locations 264–269, 307
mechanization 223
models 238, 242–243, 247–274
processed foods 156–157, 169, 184, 226–227, 235, 238
R & D 150–152, 221, 236, 298
raw material sources 11–12, 280
regulations 20–21, 225–226, 280–281, 303
retailer involvement 147–149, *148*, 152–154, 292–294
robotics 304
scandals 142–144
smaller companies 235
sustainability 141, 151, 161, 270
transformation possibilities **237**, **240**, 244–245, 301–302
trends 161–163, **181**, 184–186, *185*, 188, 308–309

forest products 135, 278
fridge, smart 187. *see also* refrigeration
fridges, smart 172
fruits 12
functional foods 271–272
Funk, Casimir 122, 200

g

gasification 256–257
gender sorted semen 100
gene editing, livestock 101–102
gene edition 64
genetically modified organisms (GMOs) 19, 63, 183, 187, 297
genetic diversity 62–64
genetic evaluations
beef industry 103
dairy industry 103, **104**, 105
genetic innovation 59–68, 80
genetic resources, plant breeding 53, 59, 80
genome wide selection 66
genomics 97–106, 288
genotyping by sequencing (GPS) 62
Ghana, cocoa industry 266–267, 269
globalization of seed industry 70
gluten 211
gluten intolerance 162
GMOs (genetically modified organisms). *see* genetically modified organisms (GMOs)
Greece, ancient, food habits 111–113, 289
greenhouse gases 46. *see also* climate change
livestock 95

Green Revolution 35–36
growth of plants 278
growth trends, industries **181**

h
halophytic plants 14–16
harvesting, mechanical 223–224
HDL (high-density lipoproteins) 211
health
  claims for foods 211–214, 301
  concerns 16–17
  food 11, 125, 137, 153–154, 249
  issues 57–58
  livestock 96, 287
healthy eating 195–198, *198*, *207*, *209*, 299–300, *see also* nutrition
Hepatitis A virus in shellfish 19
high-density lipoproteins (HDL) 211
Hippocrates 199
history
  agriculture 5–9, 278
  animal husbandry 88–89
  beverages 9
  nutrition 199–202
home-cooked food 156–157, 169, 194, 235, 238–239
horsemeat in beef products 143
horticultural crops 3, 278
Hügelstadt 262, *263*
humaneness in animal husbandry 93–94, 96, 287
hunter/gatherers 6–7
husbandry, animal. *see* animal husbandry
hybridization 62–63
hygiene rules. *see* food, safety
hyperactive disorders 11
hypoallergenic formula for infants 158

i
ice cream 181–182, 265–266
industrial food products 156–157
industrial revolution 119
infant formula 142, 143, 158
infrastructure, water resources 42
innovate-apply-sell (inapse) model 239, **240**, 293–294
insects as food 129–131, 290
integrated agriculture 258
intensification of agriculture 92
International Cocoa Organization (ICCO) 268–269
internet, impact on agriculture and food industry 167–188
internet of everything (IOE) 172, 186, 237, 298
internet sharing of recipes 171
intolerances, food 157, 162–163
irrigation 31–33, 35–38, 46. *see also* water, productivity
isopropyl thioxanthone (ITX) contamination 143

k
kale 193, 214
ketosis, caused by fasting 196–197
KitKat Chunky 148–149
knowledge based business 152–155, 183, 227–230, 237, 293–294, 303

l
lactose intolerance 162
land management practices 38
Lavoisier, Antoine 122, 200
LDL (low-density lipoproteins) 211
legumes (Fabaceae) 58, 69, 77
Liebig, Justus 122, 200
life maintenance 4, 278
light, essential to life 4, 278
Lind, Dr James 199
linoleic acid 131, 206
lipids 15, 131
lipoproteins 210–211, 301
Listeria monocytogenes 19
livestock 87–88, 278, 286–287

automated data collection 103–105, 108
breeding 97–102, 287–288
gene editing 101–102
global production 91
health 96, 287
impact on global economy 106
pedigree 97
performance 97–98
phenotype 97–98
source of micronutrients 90
source of proteins 89–90
sustainability 93–94, 286–287
local produce 224–225, 227
locations, food manufacture 264–269, 307
Los Angeles, water supply 34, 39, 250
low-density lipoproteins (LDL) 211
low fat 202–203
low salt 203–204
low sugar 204–206, 207–208, 216

m

macronutrient compositions **194**
mad cow disease. *see* BSE
maize 12, 56, 66
drought resistance 36
monoculture 76
R & D 70, 73, 80
marker assisted recurrent selection (MARS) 66
marker assisted selection (MAS) 66
Marker, Russell 200
marketing, food 79, 102, 147, 151–153, 238, 278
McCollum, E, V. 122, 200
meat consumption 54–55, 89–91, **92**. *see also* animal proteins
mechanization, agriculture 88, 92, 223
mechanization, food industry 223
medical food 239–241, 304
Medieval Europe, food habits 115–118, 289
melamine contamination in milk powders 143
methane 14–16, 95, 100, 107, 133, 256, 287–288
micronutrients from animals 90, 106
milk-containing products 13–14, 280
milk industry. *see* dairy industry
milk powders 12–13, 142–143, 158
molecular biology 60, 61, 67–68, 80
molecular cuisine 270
molecules, essential to life 4, 278
Moley's kitchen arms 170
monoculture 76
monounsaturated fats 206
mutagenesis 63–64
mutual encroachment 264
mycotoxins 58

n

natural food 226–227
Nestlé 222, 228, 231–232
baby milk scandal 142
medical food 239–240, 304
nitrogen fixing 58
nonfood products 13–16, 280
nutrition 191–216, 249
contradictory findings 192, 298–299
controversies 208–211, 300–301
definition 199
dietary recommendations **209**
health claims for foods 211–214
history 199–202, 300
modern trends 192–195
nutritional information 155
nutrition science 121–124

o

obesity 161–163, 195, 204, 215
oil-derived food 10
oilseeds 12, 70, 280
oleic acid 131–132
olestra 202
omega-3 fats 206

organic cereal yields 56
organic farming 176
organic food 95, 162, 169, 184
orphan crops 12, 68, 70, 76, 78–79, 286
orthomolecular nutrition 201
out of home (OOH) food consumption 173, 187
oxygen 5

p

Pauling, Linus 201
pedigree, livestock 97
performance foods 271–272
permaculture 254–259, 306
pesticide management 38, 183
pesticide residues 19, 58
pet food 270–271, 308
phenotype, livestock 97–98
phenotypic data 61, 64, 102–103, 105–107, 288
physical activity 162
pills, food 271–272, 308
plant breeding 53–54, 59–68, *61*, 80, 242, 285–286
Polanui Gardens 253–254
polled cattle 102
pollination 136
pollution
  air 199
  soil 57, 199
  water 43–44, 199, 282–284
poly lactic acid (PLA) 14
polyunsaturated fats 206
population growth 23–24, 235
  expected 54, *55*, 91
potatoes 12
pottages
  Medieval Europe 117
  Roman Empire 113–115
precision agriculture 37–38, 47, 60, 80, 286
preservation, food 7–9, **8**, 279
probiotics 212–213, 301
processed foods 156–157, 169, 184, 226–227, 235, 238
processed milk 12–13
production statistics, crops **81–84**
productivity, agriculture 23, 32, 80
proteins from animals 89–90, 106. *see also* meat consumption
proteins in diet 123–124
protoplast fusion 64
puls. *see* pottage
pumping costs, water 31

q

quality management 58
quantitative trait loci (QTL) 65–66

r

R & D 236
  agriculture 183, 184, 297
  food industry 150–152, 221, 236, 298
  seed industry 69–71, **70**, 75–76, 78
ranches 88
raw material sources, food industry 11–12, 280
real food revolution 121–124, 289–290
recipes, shared via internet 171
recycled waste 265, 285
recycled water 31, 43
reduced fat 202–203
reduced salt 203–204
reduced sugar 204–206, 207–208
refrigeration 120, 223, 242
refrigerator, smart 187
refrigerators, smart 172
regulations, food industry 20–21, 225–226, 280–281, 303
Renaissance, diet since 118–121
renewable resources 258
restaurants 269–270, 308
retailers, food industry involvement 147–149, *148*, 152–154, 292–294
rice 12

robotic milking systems 105
robotics 223, 242, 297, 304
robotics in cooking 167, 169–171, 174, 239, 295–296
Roman Empire
  food habits 113–115, 289
  water supply 249–250
Rose, William 122, 200

s

safety of food 18–21, 225–226, 248, 280–281
Salmonella 18
salt, reduced 203–204
saltwater resistance. *see* halophytic plants
Santa Barbara 24, 251
satellite technology 177–179
saturated fats 206, 207, 209
scandals, food industry 142–144
scurvy 122, 199, 200
sea level, rising 45
security, food 128, 176, 235, 248, 252
seed industry 53
  economics 80
  globalization 70
  R & D 69–71, **70**, 75–76, 78
  relationship to farmers 72–73, *74*
seed production 67
seed varieties, modern 35–36
seed varieties, transgenic 36
segmentation 248, 272
selection, plant breeding 64–65
self-sufficiency 59
semen, gender sorted 100
Serenbe, Georgia 253
sexed semen 100
slotting allowance 145–146
slow food 167–169
small-holdings 88
small is beautiful 306–307
small scale companies, food industry 235
smart factory 237
smart fridges 172
social changes 249
soil cultivation 3, 278
soil erosion 57
soil quality 57
soybeans 12, 69, 76
Spanish cooking oil disaster 142
specialized diets 162, 235
spirulina 193
Staphylococcus aureus 19
staple foods, production statistics 68–69
stevia 204
suburbia 253–254
sugar beet 12
sugar-cane 36, 46
sugar consumption 211
sugar, reduced 204–206, 207–208, 216
superfoods 192–193
supply chains 72, 126, 238, 259, 265–266, 269
sustainability
  agriculture 163, 236, 250, 253, 255, 265, 278
  crops 59, 118, 249
  dairy industry 13
  food industry 141, 151, 161, 270
  livestock 93–94, 286–287
Sustainable Agricultural Initiative (SAIN) 80
synthesis 58–59

t

tall oil 131–132
temperature increase 44–45
temperature range, essential to life 4–5, 278
Three Gorges Dam 33
tilling 63–64
tomatoes
  processed 60, 75
  R & D 70, 73, 80, 286
toxic residue 43

toxins, food 58
traceability, food chain 58
transformation possibilities **237**, **240**, 244–245, 301–302
trends, food. *see* food, trends
Trichinella spiralis 19
20th century diets 121–124, 289–290

u

United Nations Refugee Agency (UNHCR) 128
unsaturated fats 206, 207
urban agriculture 183, 236, 253, 261–264, 291, 297, 306, 307
urban and peri-urban agriculture (UPA) 134
urban gardening 133–136, 138
urban population 54
utopian ideals 123

v

vanilla flavor 232
variety in diet, Medieval Europe 118
vegan diet 132–133, 138, 162, 291
vegetable proteins 58
vegetarian diet 55, 58, 138, 157, 162, 270
  impact on society 132–133, 291
vertical gardening 136
vertical integration 79–80
vitamins 122, 200

w

Wall Street Journal 192–193
walnut industry 212–214
waste disposal 95
waste food 126–127, *127*, 290
waste management 43–44, 265, 285
water
  conveyance systems 40
  demand 29, 31
  essential to life 4, 278
  institutions 33–35, 38–41
  management 29–30, 46–47, 281–285
  markets. *see* water, trading
  policies 33–35, 38–41, 282–284
  pollution 43–44, 282–284
  pricing 40–41
  productivity 36, 38, 282–283
  projects 42, 283
  pumping costs 31
  quality management 31, 43–44, 47, 284–285
  recycled 31, 43
  rights systems 38–39
  Roman Empire 249–250
  shortages 24, 31, 34, 39
  supply management 31, 249
  trading 38–39, 46
water resources 30–33, 277
  infrastructure 42
  management 44
wealth, influence on diet 112–113, 115, 118, 123
Weihenstephan brewery 9, 26, 141, 292
wheat 12
wine industry 224, 257
working out 161–163
World Bank 56
World Coffee Research 80
World Health Organization (WHO) 91
World Wars 123

y

yield improvements 77–78, **83**, **84**, 119, 127, 137
  corn 59–60, *60*
  due to irrigation 32, 35, 36
  due to plant breeding 70–71, 78, 183, *184*, 297
  milk 94
  urban agriculture 307
yield statistics **81–84**